AUTOMOBILE ENGINEERING

AUTOMOBILE ENGINEERING

By

Sudhir Kumar Saxena
Professor
Department of Mechanical Engg.
National Institute of Technology,
Kurukshetra, Haryana

UNIVERSITY SCIENCE PRESS
(An Imprint of Laxmi Publications Pvt. Ltd.)

BANGALORE • CHENNAI • COCHIN • GUWAHATI • HYDERABAD
JALANDHAR • KOLKATA • LUCKNOW • MUMBAI • RANCHI
NEW DELHI

Published by :
UNIVERSITY SCIENCE PRESS
(*An Imprint of Laxmi Publications Pvt. Ltd.*)
113, Golden House, Daryaganj,
New Delhi-110002
Phone : 011-43 53 25 00
Fax : 011-43 53 25 28
www.laxmipublications.com
info@laxmipublications.com

First Edition : 2009

OFFICES

✆ **Bangalore**	080-26 61 15 61	✆ **Chennai**	044-24 34 47 26
✆ **Cochin**	0484-239 70 04	✆ **Guwahati**	0361-254 36 69, 251 38 81
✆ **Hyderabad**	040-24 65 23 33	✆ **Jalandhar**	0181-222 12 72
✆ **Kolkata**	033-22 27 43 84	✆ **Lucknow**	0522-220 95 78
✆ **Mumbai**	022-24 91 54 15, 24 92 78 69	✆ **Ranchi**	0651-221 47 64

UAE-9356-195-AUTOMOBILE ENGINEERING **C—17101/09/02**
Typeset at : Kalyani Computer Services, New Delhi.
Printed at : Repro Knowledgecast Limited, Thane

Dedicated
to

Late Dr. S.C. Srivastava,
Professor, Mechanical Engineering Department
National Institute of Technology
(formerly Regional Engineering College)
Kurukshetra

CONTENTS

PREFACE

The number of textbooks available on 'Automobile Engineering' is not very large. The field and scope of automobile engineering is enormous. Every manufacturer of automobile has his own resources and is trying to improve the product in his own manner. Number of automobile manufacturers being very large, the development in this field is very quick.

The present book attempts to explain different aspects of automobile engineering in a simple way. The diagrams that have been included in the text are simple and can be followed easily. Every attempt has been made to include the latest development in this field.

It is hoped that the book would prove useful to the undergraduate/post graduate engineering students particularly those studying mechanical engineering.

It is very much possible that some mistakes have been left in the text of the book though every effort has been made to remove them. Author would feel oblige if the mistakes are brought to notice. Author would also welcome the comments and suggestions by experts in this field, particularly those in the industrial organizations, to improve the book in forthcoming editions.

—Author

1

INTRODUCTION

au'tō–mō–bile'" (or a–tō–mō' bil), n. [*auto-*, and L. *mobils*, movable.] a car, usually four-wheeled, propelled by an engine or motor that is a part of it, and meant for travelling on streets or roads; a motor car.

— ***Webster's New Twentieth Century Dictionary–Unabridged, Second Edition.***

The meaning of an automobile is a four-wheeled vehicle, carries a small number of passengers, driven by an engine or motor and independent of rails or tracks. The subject of automobile engineering is the study of engineering aspects and technical details of cars. There are hundreds and thousands of different types of cars that are available today. During the past century, tremendous development in this field has taken place. Apart from cars, vehicles for different usage have been developed. These range from commercial vehicles of different types—to public utility transport systems such as buses and coaches—to the carriers of different types—to tractors and similar machines meant to be used for specific purposes which are propelled by an engine or motor that is part of these machines. Though this study is meant for automobiles only but in the present scenario it can not be strictly confined to cars, and therefore, wherever felt necessary and relevant the other forms of automobiles meant for allied purposes will also be included.

1.1 CHRONOLOGY OF DEVELOPMENT OF AUTOMOBILES

The history of development of automobiles is not very old. The first self-propelled vehicle was built in France in 1769. In those times, steam engines were used as source of power to move the vehicles. In 1895, **Stanley** built a steam car 'Stanley Steamer' which could gain a considerable speed. Prior to this vehicles used to move at a slow speed. In France, **Lenoir** drove the first automobile with gas engine.

In those days, means of communication were not fast and there were not many interactions among the engineers and scientists. The development took place simultaneously in different parts of the world. In Germany, in the year 1885, two engineers built gasoline-powered automobiles. These engineers were **Daimler** and **Benz,** very big names in automobiles even today. They used engine based on 4-strokes developed by **Otto. Louis Renault** worked in France to develop the automobiles, and some have opinion that he was the first to drive an automobile. Whatever the truth may be but the last decade of nineteenth century witnessed the development in the field of automobiles, which formed the basis of what we see today.

In America, **Henry Ford** built a car running on four bicycle wheels, powered by two-cylinder gasoline engine. (shown in Fig. 1.1). **Seldon** built the first automobile manufacturing plant in New York in 1895. Another plant, at Detroit, was started around the same time. In 1897, this plant produced 425 cars. In London, motor bus service appeared in 1898. It became popular immediately. The reason being that in those days everybody could not afford to buy a car. The bus service was cheap and provided a faster mode of transport as compared to horse carts. This popularity of bus service resulted in quick increase in number of buses and in a span of five or six years, the number of buses rose considerably in London. The number of manufacturers of car rose to more than hundred in 1905. Some prominent manufacturers included **Cadillac, Ford,** and **Buick**—the names well known even today and some other names such as **Chrysler, Nash** and **Hudson.** In 1908, Buick company was converted into General Motors and absorbed Oldsmobile, Cadillac and nine other companies. The first World War, which continued for four years from 1914 to 1918, proved to be boom for automobile industry because quick movement was very important for any army to win.

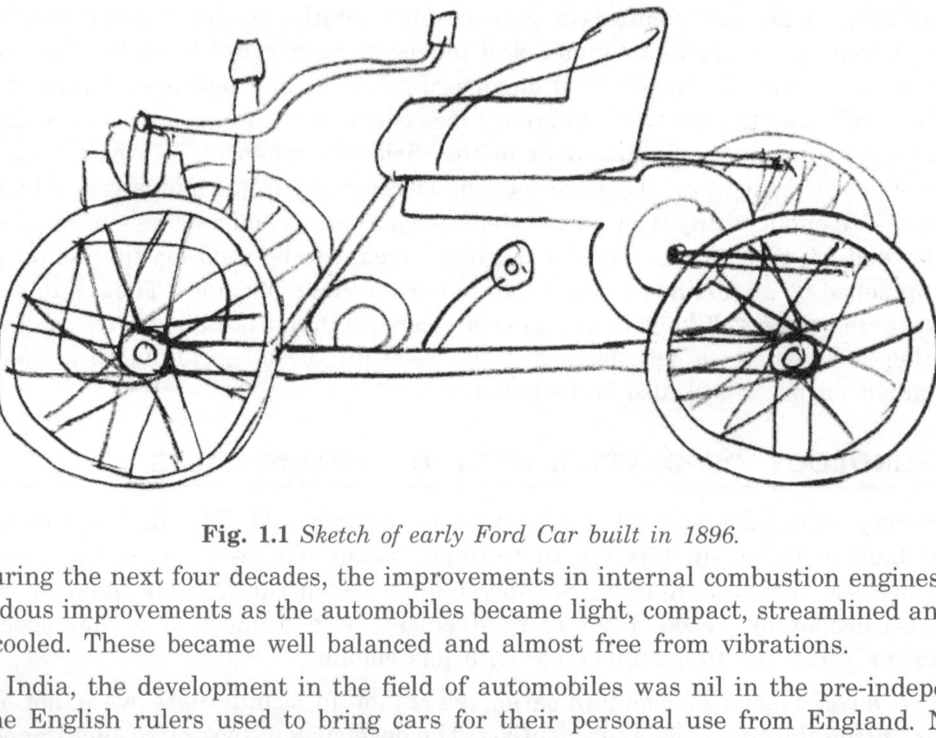

Fig. 1.1 *Sketch of early Ford Car built in 1896.*

During the next four decades, the improvements in internal combustion engines caused tremendous improvements as the automobiles became light, compact, streamlined and water or air-cooled. These became well balanced and almost free from vibrations.

In India, the development in the field of automobiles was nil in the pre-independence era. The English rulers used to bring cars for their personal use from England. No body attempted to make the cars here until in 1942 when Hindustan Motors was established. Another company Premier Automobiles Ltd. (PAL) came into existence in 1944. Initially both of them were engaged in the manufacturing of auto parts. They started manufacturing cars later. In 1945, **Mahindra brothers, Kailash** and **Jagdish Chandra** formed Mahindra and Mahindra. They indulged in the manufacture of utility vehicles. They tied up with Wiley's and launched jeep in India. In the fifties another company Standard Motor Products of India Ltd. came into being and launched their car. But this firm could not continue in the face of competition offered by Hindustan Motors and Premier Automobiles Ltd. Another company Ashok Motors, which was later converted to Ashok Leyland, started manufacturing heavy vehicles in India. It was in 1954, when Tata, in collaboration with Mercedes-Benz, launched their trucks. During the same time, Hindustan Motors started manufacturing cars in collaboration with Landmaster and Premier Automobiles Ltd. tied up with Fiat and started manufacturing Fiat cars. In 1957 Hindustan Motors launched Ambassador Mk I. Premier Automobiles Ltd. launched Fiat 1100D in 1964. In 1972, the collaboration with Fiat ended and Premier Automobiles Ltd. Launched Premier President which was later renamed as Premier Padmini. On the other side Tata continued to manufacture commercial vehicles, their only rivals being Ashok Leyland. After the collaboration with Mercedes-Benz ended they started selling vehicles with their brand name, Tata.

Eighties saw the sea change in Indian market as far as automobiles were concerned. Oil crisis in seventies caused the necessity of fuel conservation throughout the world. Big cars, which consumed more fuel, were becoming unpopular. The concept of small car took shape. The car, which was fuel-efficient, low priced, and served the purpose of a small family, was the need of the day. In India, this dream was realized in the form of Maruti cars introduced in 1983. Suzuki of Japan provided technical collaboration. Maruti became a car of common man. Even middle class people could think of having a car, which was only a dream in seventies. The number of cars on Indian roads increased to manifolds during the period from 1985 to 1995 courtesy Maruti Udyog Limited.

The middle of nineties saw the opening of economy and the process of globalization. This brought several foreign companies in India, which came with luxury cars. Mercedes, Honda, Ford, Hyundai, Daewoo are the few names who have launched their vehicles in recent past. In near future some more manufacturers are likely to enter with new models of cars.

1.2 CONCEPT OF AN AUTOMOBILE

The course of an automobile engineering intends to provide knowledge and information about automobiles from engineering point of view. It also involves the inclusion of technical details of all the components of an automobile. To explain the engineering and technical details of an automobile, there can be a general approach where explanation belongs not to any particular type of automobile. The fact is that there are numerous automobile manufacturer and every individual manufacturer is attempting to provide best product to his customer. In that context they are trying to improve each and every component of their product. Inclusion of details of each product, thus, is quite complicated and is not feasible. The approach in the book will lead to take up details of automobiles in general manner.

1.3 COMPONENT OF AN AUTOMOBILE

An automobile has several number of components. But there are four basic components. These are:

(*i*) The Chassis

(*ii*) The Engine

(*iii*) The Transmission System

(*iv*) The Body.

Apart from these four basic components, there are controls and auxiliaries. The controls are meant for controlling the movement of vehicle. The auxiliaries are additional components meant for providing comfort to the user of the automobile.

1.3.1 The Chassis

The chassis of an automobile incorporates all the major assemblies consisting of engine, components of transmission system such as clutch, gear box, propeller shaft, axles, control system such as brakes and steering and suspension system of the vehicle. In other words, it is the vehicle without its body.

The chassis of an automobile has the frame, suspension system, axles and wheel as the main components (Shown in Fig. 1.2 (*a*)). The frame could be in the form of conventional chassis or unit construction may be adopted. In conventional chassis frame, the frame forms the main skeleton of vehicle. It supports engine, power transmission and car body. The frame is supported on wheels and axles through springs. The frame carries the weight of the vehicle and passengers, withstands engine, transmission, accelerating and braking torques. It also withstands the centrifugal forces while cornering and takes up stresses due to rise and fall of axles. In unit construction type there is no frame (Shown in Fig. 1.2 (*b*)). The structure of body of the automobile is first formed and then different components such as engine, transmission system and other parts are placed at suitable places in the body structure. The transmission system itself consists of a number of parts such as clutch assembly, gear box, propeller shaft, differential and axles. The other parts include the interior details which are utilized by the passengers and driver of the vehicle. Through suitable designing, the parts are so arranged that they provide maximum comfort and make journeys in the automobile enjoyable.

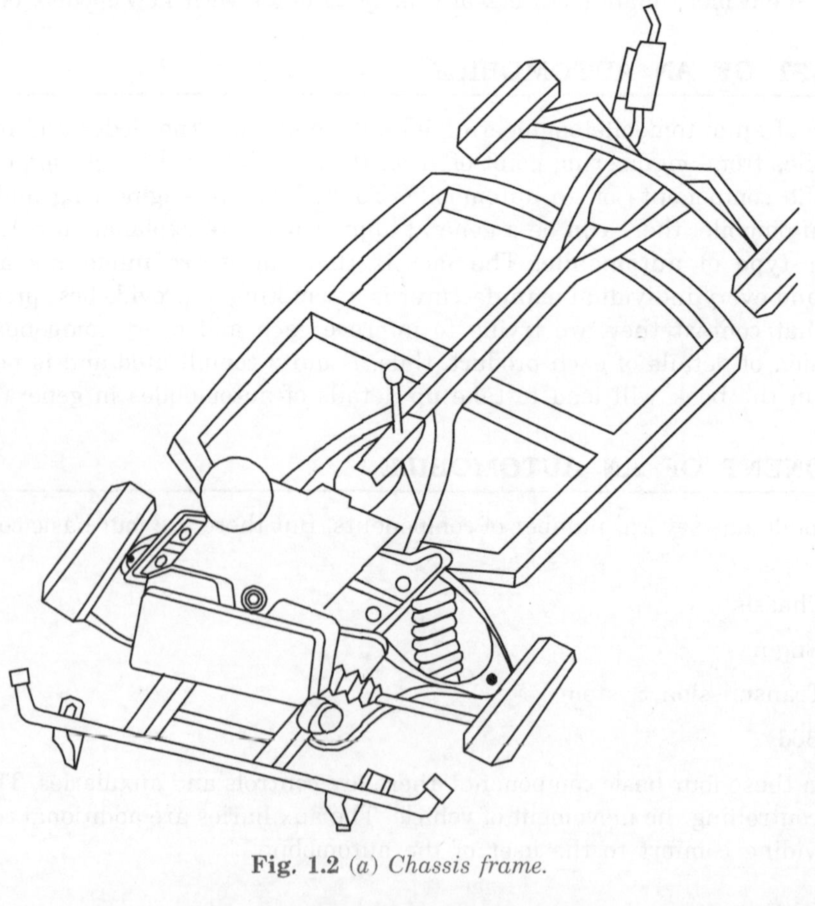

Fig. 1.2 (*a*) *Chassis frame.*

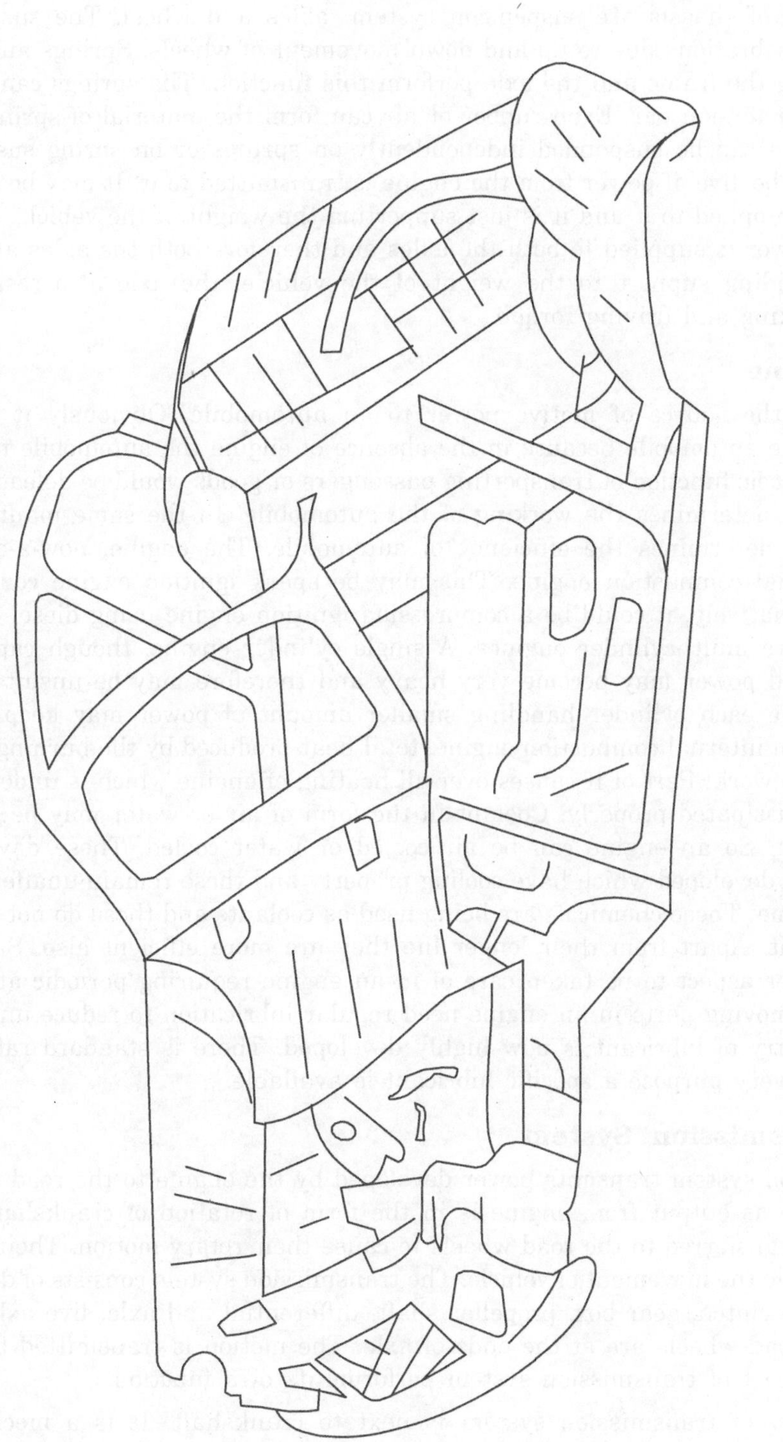

Fig. 1.2 (*b*) *Unit construction (structure of body of the automobile).*

The other part of chassis are suspension system, axles and wheel. The suspension system absorbs the vibrations due to up and down movement of wheels. Springs and shock absorbers connecting the frame and the axle perform this function. The springs can be leaf spring, coil spring or torsion bar. Even rubber or air can form the material of springs. The wheels of the vehicle can be suspended independently on springs or on spring suspended axles. The axle may be 'live' if power from the engine is transmitted to it. It may be a 'dead' axle if no power is supplied to it and it is just supporting the weight of the vehicle. In 'four wheel drive', the power is supplied to both the axles and therefore both the axles are 'live'. In addition to providing support to the weight of the vehicle, the axle also resists the stresses due to braking and driving torque.

1.3.2 The Engine

The engine is the source of motive power to an automobile. Obviously it is very important part of the automobile because in the absence of engine the automobile may not move at all and its basic function of transporting passengers or goods would be defeated. The power of the engine determines the working of the automobile. In the same manner, the efficiency of engine determines the efficiency of automobile. The engine, now-a-days, is invariably an internal combustion engine. This may be spark ignition engine consuming petrol as fuel. Alternatively, it could be a compression ignition engine using diesel as fuel. The engines used are multi-cylinder engines. A single cylinder engine, though capable of providing the desired power may become very heavy and therefore may be unsuitable. In multi-cylinder engine each cylinder handling smaller amount of power may keep engine light in weight. In an internal combustion engine, total heat produced by the burning of fuel is not converted into work. Part of it causes over-all heating of engine which is undesirable. This heat is to be dissipated properly. Coolant in the form of air or water may be used to take away this heat. So an engine can be air cooled or water cooled. These days some chemicals have been developed which have cooling property and these remain unaffected for a longer period of time. These chemicals are being used as coolants and these do not require frequent replacement. Apart from their longer life they are more efficient also. Similarly lubrication is another aspect to be taken care of in an engine requiring periodic attention from the user. The moving parts in an engine need regular lubrication to reduce unwanted friction. The chemistry of lubricant is now highly developed. There is standard rating for lubricants and for every purpose a specific lubricant is available.

1.3.3 The Transmission System

The transmission system transmits power developed by the engine to the road wheels. The power available as output from engine is in the form of rotation of crankshaft. This movement is to be transferred to the road wheels to cause their rotary motion. Their rotary motion makes possible the movement of vehicle. The transmission system consists of different parts. These include clutch, gear box, propeller shaft, differential and axle, live axle to be more precise. The road wheels are at the ends of axle. The motion is transmitted through these parts. Every part of transmission system performs its own function.

The clutch, part of transmission system is next to crankshaft. It is a mechanism enabling the rotary motion of one shaft transmitted to the second shaft 'at will'. When the engine starts it should not be connected to road wheels *i.e.,* these should not start moving as soon as the engine starts. Secondly, this motion should be transferred smoothly so that passengers in car do not feel discomfort and its mechanism is not spoiled. It case of vehicles used for the transportation of goods smooth transmission process is essential as otherwise it may cause damage to goods.

Gear box is the component of transmission system next to clutch. It has got gear train and it provides different gear ratios. These ratios determine the rotary speed of output shaft from gear box. The torque transmitted to the road wheels give rise to a propulsive force (or tractive effort) between these and the road. When starting from rest large tractive effort is required. This makes essential the introduction of considerable 'leverage' between engine and the wheels so that torque from engine, which is almost constant, produces large tractive effort. This 'leverage' is provided by the gear box. Different gear ratios available in the gear box can provide the required tractive effort to overcome the resistance faced by the automobile under different conditions.

Propeller shaft transmits the output from the gear box to the axle. This axle may be in the rear or in the front or in some cases both the rear and front axle may receive output from gear box (Shown in Fig. 1.3 (*a, b, c*). The output from the gear box is in the form of rotary motion of the shaft and this motion is transferred to the axle.

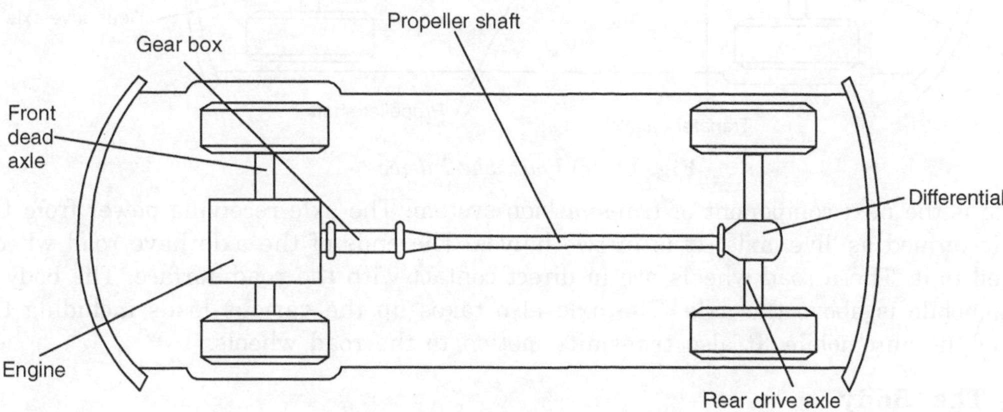

Fig. 1.3 (*a*) *Rear-wheel drive.*

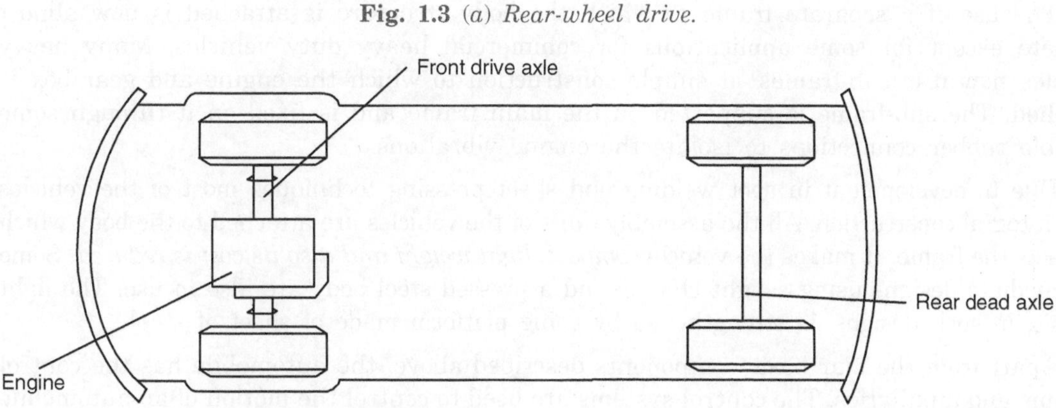

Fig. 1.3 (*b*) *Front wheel drive.*

Differential is the next component of the transmission system. The motion of propeller shaft is fed to the differential which turns it through 90 degrees. This is essential as the axle is at 90 degrees to the propeller shaft. The function is performed with the help of a pinion and a gear. Another important function of differential it to reduce the speed of inner wheels and at the same time enhance the speed of outer wheels by the same amount. This is required when the automobile is moving on a curved path. On a curved path, the outer wheels are required to traverse a circle of bigger radius than the inner wheels. This means that the outer wheels are required to traverse larger distance as compared to inner wheels. As the automobile is to move as a single unit, all the four wheels must travel together. Therefore the outer wheels should travel larger distance and inner wheels should travel smaller distance in the same time period. Hence the variation in speed of inner and outer wheels is needed. This is performed by the differential with the help of sun and planet gear system. Further details of differential to be taken up later.

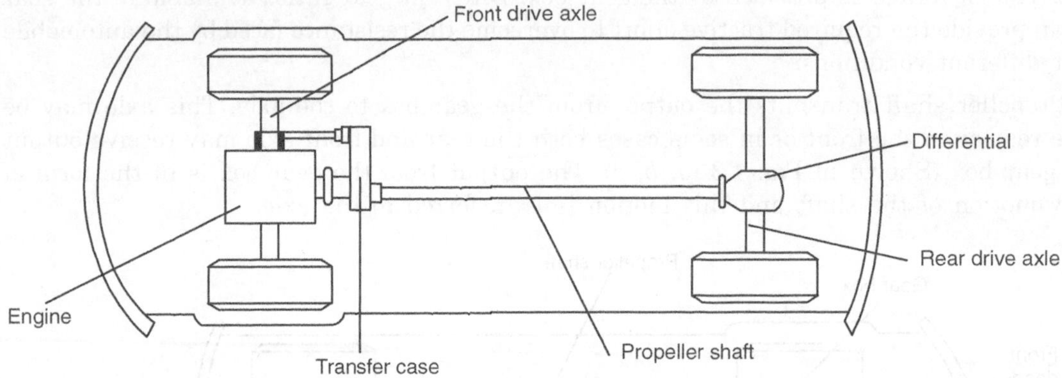

Fig. 1.3 (*c*) *Four wheel drive.*

Axle is the next component of transmission system. The axle receiving power from the engine is termed as 'live' axle. It is in two halves. The ends of the axle have road wheels connected to it. These road wheels are in direct contact with the road surface. The body of the automobile is above the axle. The axle also takes up the various loads including the weight of the automobile. It also transmits motion to the road wheels.

1.3.4 The Body

The use of a separate frame to which the body structure is attached is now almost obsolete except for some applications for commercial heavy duty vehicles. Many heavy vehicles now use 'sub-frames' of simple construction to which the engine and gear box is attached. The sub-frame is supported on the main frame and is fixed on it through some suitable rubber connections to isolate the engine vibrations.

Due to development in spot welding and sheet pressing techniques most of the vehicles have integral construction. All the assembly units of the vehicles are attached to the body which also acts the frame. It makes the vehicle *compact, light weight and also its cost is reduced*. Some intermediate designs using a light chassis and a pressed steel body are also in use. The light chassis, in such designs, is strengthened by using platform made of sheet of steel.

Apart from the four basic components described above, the automobile has the control systems and auxiliaries. The control systems are used to control the motion of an automobile and therefore are essential in an automobile. These include (*a*) Steering system and (*b*) Braking system or brakes.

Steering system

The automobile while moving may be required to traverse a circular path. It has to be turned through some angle, if the path is not straight. There may be other situations also when the road is turning towards left or right and an automobile is required to turn to left or right. This turning of automobile towards left or right or on the curved path is provided through steering mechanism. Steering system is required to be quite accurate as the automobile should turn accurately along with the path.

Braking system

This causes the reduction in speed of the vehicle and brings it to rest when necessary. Bringing an automobile to rest is as important as its movement. Obviously, when we have reached our destination, we would like to stop; and therefore, vehicle should come to rest. Also, there may be some kind of emergency and vehicle may be required to slow down or stop on the way. At that time also its motion is to be controlled. This control on motion is provided with the help of brakes.

1.3.5 The Auxiliaries

These are the components of an automobile which may not be essential but it can make the driving more comfortable. The fact is that with passage of time some auxiliaries become essential. Few years back, the indicators—to indicate the turning vehicle—were not used. But now these have been made mandatory by the government. Though air-conditioner is not essential and is just to provide comfort conditions. Now it is provided in every vehicle in developed countries and is being adopted by more and more people in India and in other developing countries also.

The study of automobile engineering involves going through the in-depth study of all the automobile components. These include engine, transmission system, control and auxiliaries system. The engine used in an automobile is an internal combustion engine. The transmission system consists of a number of parts, introduction of which has already been presented. Suspension systems, wheels and tyres are the components of automobile, details of which are to be elaborated. Study of steering mechanism and brakes is also important as these form the control system in an automobile.

QUESTIONS

1. Give the meaning of an automobile.
2. Explain the contribution of automobiles in the development of mankind.
3. What are different components of an automobile?
4. Explain the functions of chassis.
5. What is the purpose of engine in an automobile?
6. Explain briefly different components of Transmission System.
7. Explain front wheel and rear wheel drive.
8. Explain the difference between front wheel, rear wheel and four wheel drive.
9. What do you understand by "Auxiliaries"?

2

INTERNAL COMBUSTION ENGINES

2.1 INTRODUCTION

An internal combustion engine is the source of 'power' in an automobile. Internal combustion engines can be broadly classified on the basis of their working, number of cylinders and fuel used. On the basis of their working the engine can be two stroke or four stroke. In two stroke engines the working cycle is completed in one revolution (360°) of crankshaft. In four stroke engine the working cycle is completed in two revolutions (720°) of crankshaft. In automobiles four stroke engines are used. As far as the classification of the basis of number of cylinders is concerned the engines can be single cylinder or multi-cylinder. A multi-cylinder engine is more suitable for automobiles. Conventionally the engines used petrol or diesel as fuel and accordingly these are classified as petrol engine or diesel engine. Lately, some alternate fuels such as compressed natural gas, hydrogen etc. have also been used. Few hybrid cars with twin engines are also being developed. These cars are equipped with electric motor running with battery alongwith conventional engine.

2.2 ENGINE COMPONENTS

2.2.1 Cylinder Block

This is the main component of the engine. It forms the middle portion of the engine. On its lower end crank case is located and on its upper end the cylinder head is located. The block has cylinders in which pistons slide up and down. The number of cylinder may vary 4, 6, 8 or sometimes even higher depending upon the size of the engine and power output needed from it. These cylinders may be in a single row (line) (Fig. 2.1). or may be arranged in a manner that forms two banks (Fig. 2.2). These banks are at a certain angle to form V.

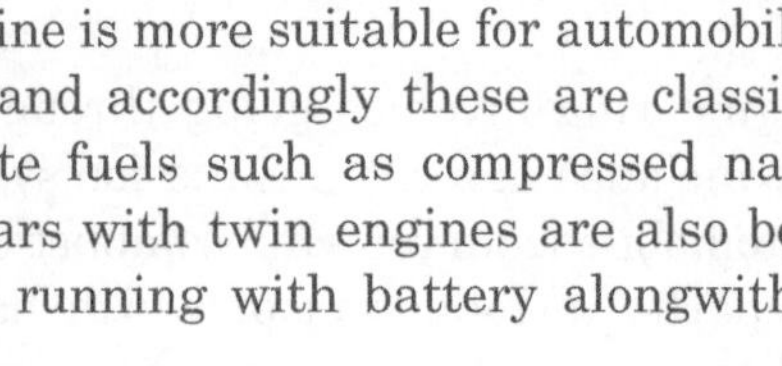

Fig. 2.1

The engines having all the cylinders in a row (line) are known as in line engine and in those where cylinders form a V are known as V-engines.

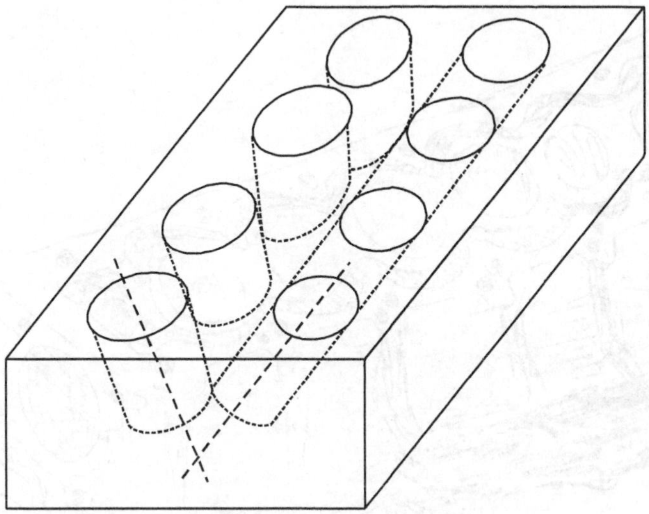

Fig. 2.2

The cylinder blocks are also provided with passage for coolant. The hot gases at high pressure push the piston down in the cylinder. This causes maximum heating of cylinder block due to unutilised thermal energy. The coolant flowing through the passage takes away this heat.

Most cylinder blocks are made through casting using cast iron. Sometimes iron mixed with nickel or chromium is also used. Aluminium alloy is also used to cast the cylinder block.

2.2.2 Crank Case

It forms the lower part of the engine. It accommodates crankshaft. *The crankshaft is a long straight piece of metal in a vehicle that connects the engine to the wheels and helps turn the engine's power into movement.* It is supported at its ends in the walls of crank case. The crank case has the provision to support the crankshaft. In some engines where the crankshaft is too long, it is supported in the middle portion also and crank case has provision to provide support. Apart from this, the crank case acts as sump for lubricating oil. Generally the upper part of crank case is an integral part of cylinder block, the lower part of crank case is bolted to it.

Aluminium alloy is the most suitable material for crank case. It is light weight and has good thermal conductivity. Earlier cast iron was also used as material for crank case (Fig. 2.3).

2.2.3 Cylinder Head

The cylinder head has provision for fixing the inlet and exhaust valves. This also forms the top of combustion chamber. The combustion chamber is given different shapes. Each shape produces effective combustion of fuel.

The material of cylinder head is cast iron or aluminium alloy. Machining is done so that various components can be installed smoothly.

To prevent noise and vibration reaching the body of the automobile the cylinder heads are provided with cover. The cover is made of three-layer sheet. The outer two layers are metallic and middle layer is that of plastic. This plastic layer does not allow the transmission of noise and vibration from the engine.

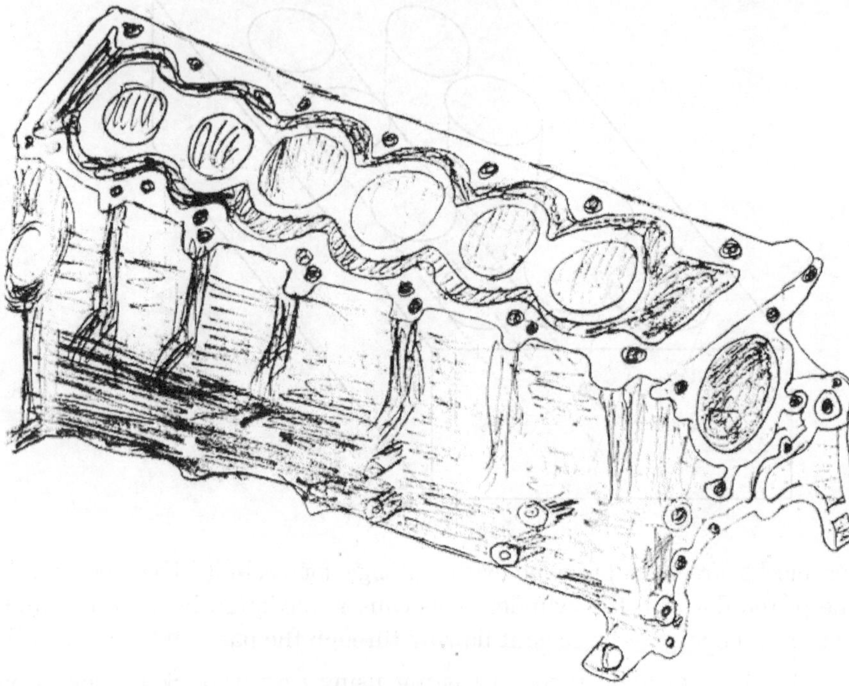

Fig. 2.3

In wedge shape combustion chamber the turbulance of burning mixture is enhanced. The cup shaped combustion chamber prones turbulance and is particularly suitable for diesel or turbo-charged engines. In hemispheric combustion chamber combustion of fuel is relatively slow. Figure 2.4 represents wedge shaped and hemispherical combustion chambers.

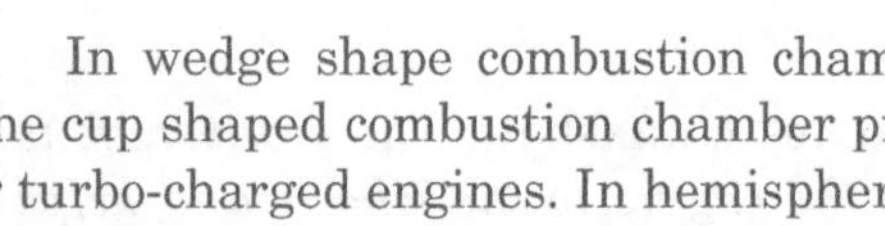

(a) Wedge shaped

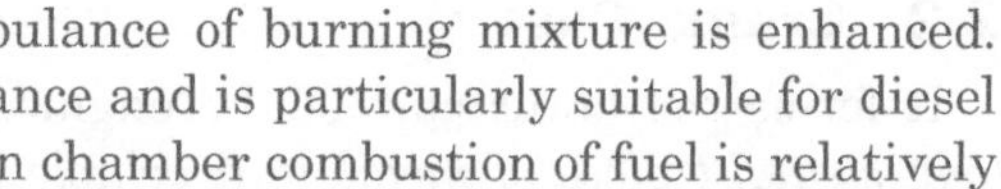

(b) Hemispherical

Fig. 2.4

2.2.4 Piston

The piston moves inside the cylinder and has reciprocatory motion. It is air tight and does not allow the leakage of charge and hot gases with the help of rings which form a part of piston ring assembly. It also transmits the impact produced by the gases at high temperature and pressure. It accommodates the small end of the connecting rod and takes the lateral thrust due to obliguity of connecting rod.

The top portion of the piston is known as crown. Piston rings and piston pin form integral parts of piston assembly. The lower most part of piston is known as skirt. A slot is provided in the piston. The slot accommodates the expansion of piston material when hot. This helps in reducing the clearance between skirt and cylinder bore when cold. When hot, the slot accommodates the expansion of piston material.

Aluminium alloy is the most commonly used material for piston. The alloy contains about 12% silicon which has less co-efficient of expansion as compared to aluminium and cast iron. Addition of phosphorus enhances the fatigue resistance. Addition of cobalt and chromium further reduces the co-efficient of expansion which makes piston suitable for use when proper cooling many not be possible.

Piston rings and piston pin are the components of piston sub-assembly. The outer surface of the piston ring is in close contact with the inner surface of the cylinder. The ring becomes perfect circle in the cylinder. In unassembled state, when left free, it is slightly oval. The rings are known as compression rings when these prevent leakage of hot flue gases during expansion. In addition to compression rings, there are oil rings meant for lubrication of piston sub-assembly, particularly in big engines.

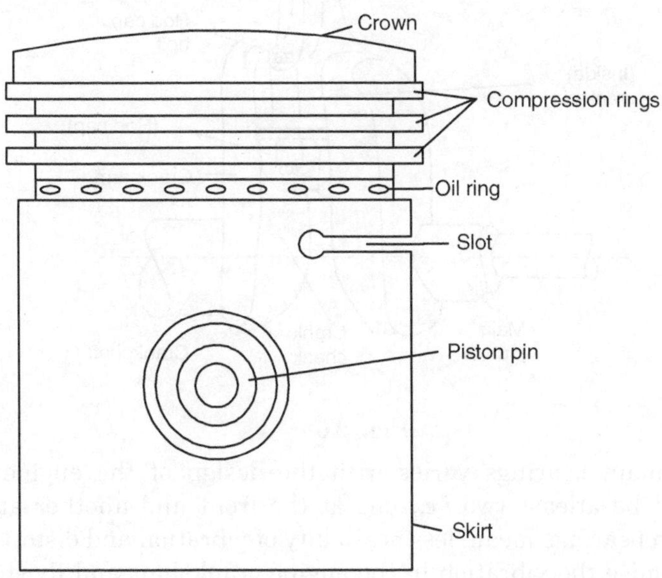

Fig. 2.5

Piston pin, also known as gudgeon pin is third component of piston sub-assemble. It is supported at its both ends in the circular wall of piston. Its middle portion holds the small end of the connecting rod. It is a simple cylindrical component with provision of lubrication inside the small end of connecting rod.

2.2.5 Connecting Rod

As the name indicates it connects the piston with crankshaft. Its one end holds the piston pin and is known as small end. Other end, known as big end, holds the crank pin. It may have circular, rectangular, I, T or even H section. It is a steel forging and is highly polished for enhanced endurance strength. It is provided with a passage for transferring lubricating oil from the big end bearing (crank pin) to small end bearing (piston pin).

2.2.6 Crankshaft

The crankshaft requires detailed study as it is important component of the engine used in an automobile. Multicylinder engines are used in automobiles. The crankshaft of a multicylinder engine due to its shape and loading requires special attention of designers.

Basically a crankshaft converts the reciprocating motion of the piston(s) to rotating one. It applies the principle of simple machine known as wheel and axle. The crankshaft is made from steel forging or casting and is machined to provide suitable journals for connecting rod and main bearings. The parts of the crankshaft from main bearing journal to the connecting rod bearing journals are called crank arms or cheeks. The length of the crank arm determines the stroke of the engine. From the centre of the main bearing journal to the centre of the connecting rod bearing journal is half the engine stroke. The part of the crank shaft inside the connecting rod bearings is called the crank pin and those inside the main bearings are called the main journals (Fig. 2.6).

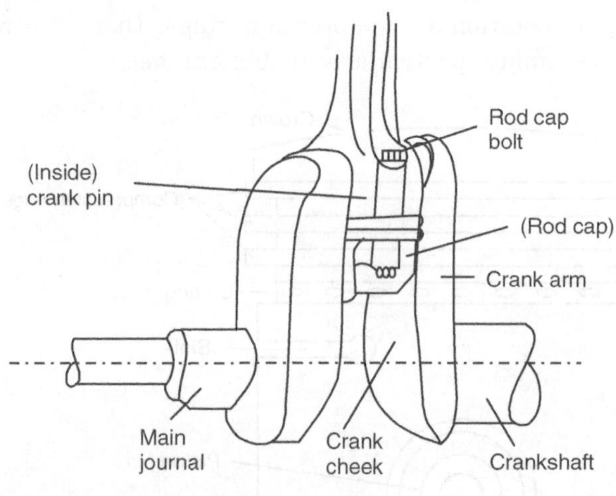

Fig. 2.6

The number of main bearings varies with the design of the engine and number of cylinders. There must be atleast two *i.e.,* one at the front and another at the rear of the crankshaft. More main bearings mean, less possibility of vibration and distortion of crankshaft of given size. To minimise the vibration in the engine crankshaft and flywheel are balanced separately and then often tested for balance when mounted together.

2.2.7 The Crankshaft of a Four Cylinder Engine

In four cylinder engines, with low rating, two main journal bearings may serve the purpose. But in highly rated engines there is usually one bearing between each pair of crank throws. In some four cylinder units, there may be three bearings. In this case the shaft has to be very stiff, otherwise at certain speed and load, it would whip, and causes heavy load on central bearing which may fail. The tendency to failure can be eliminated or reduced by balance weights on each side of that bearing. Inadequate stiffness of the shaft transfers high proportion of the load to the edges of the plain bearing at each end.

With the larger number of bearings, which support the shaft uniformly along its length, the crankshaft can be made slender and lighter without risk of whip due to bending. More bearings, however, need a stable crank case structure under variations of temps and loads. It makes the crank case more costly also.

Frictional drag of the bearings on the shaft increases as square of diameter but in proportion to length. Torsional stiffness is a major factor governing the dimensions of the shaft. Torsional stiffness increases its natural frequencies of vibration, together with their harmonics, above the rotational speed of the engine. For any given length of shaft both the bending and torsional stiffness depend on the diameters and overlap of the main and big end journal and the thicknesses and widths of the webs. The lengths of the journals are a function of their loadings and the strengths of the bearing materials. For compactness and stiffness, the lengths of the bearings and thicknesses of webs are kept as small as possible. If the bearing shells are very short, the lubricating oil may squeeze out before it can spread right around their working surface. The strength of the shaft depends on its material. To improve the fatigue strength fillet radii are kept more between webs and journals. Heat treatment and hardening of shaft also improves fatigue strength.

Forged crankshaft for a four cylinder compression ignition engine is shown in the Fig. 2.7. There are no balance weights in the stiffly designed five bearing shaft.

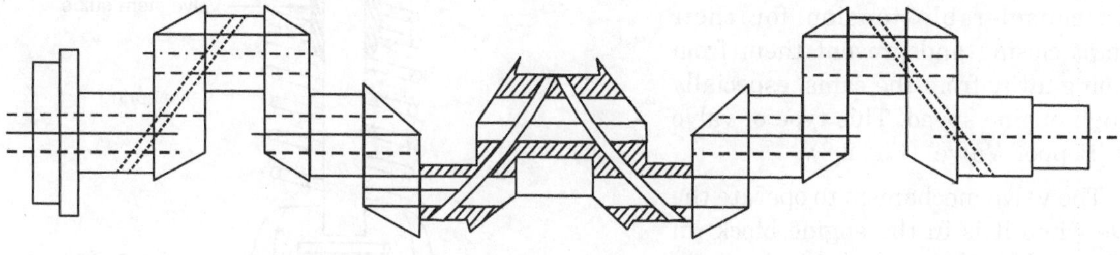

Fig. 2.7

The primary requirement being minimum weight as this is for compression ignition engine, the structure is stiff enough to take up the opposed revolving couples of each half of the shaft.

2.2.8 Crankshaft Material

Generally steel forgings are used. High carbon, high copper, chromium silicon iron have also been used. Best steels for crankshaft are the nitrogen-hardened type, though costly. Less costly and inferior also are the high carbon or alloy steels, surface hardened by the flame or induction method. Least durable are high carbon or alloy steels that have not been surface hardened.

Cast iron crankshafts have low cost. It also has a high hysterisis and damps out the vibrations. The complex shape can be given (through casting) to the shafts without the need of costly tooling or machining. Hollow journals needed for light weight can be achieved by casting.

Spheroidal Graphite (S.G.) irons have better tensile strength and fatigue resistance and better bearing qualities than cast irons containing graphite in flake form. They include the following grades: 600/3, 650/2, 700/2 and 800/2. In these numerator (600) represents the ultimate tensile strength (N/mm^2) and denominator (3) represents the percentage elongation. The graphite is converted to S.G. form by adding magnesium innoculants during smelting and by controlled cooling.

Salt bath nitriding increases the safe working stresses by 15%. Wear resistance increases by more than 200%.

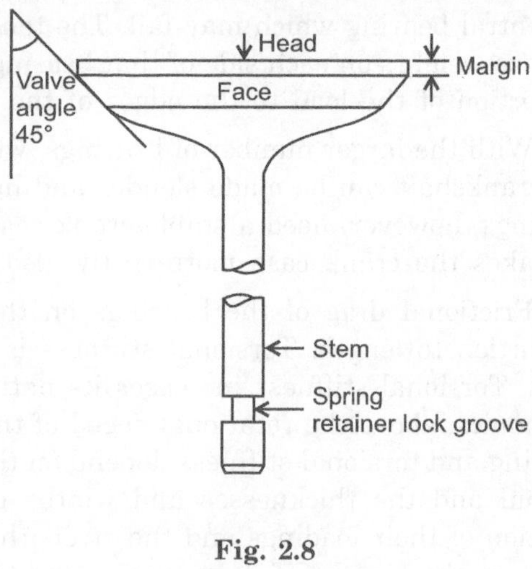

Fig. 2.8

2.2.9 Valve and Valve Mechanism

There are two commonly used valves one the inlet/intake and other is exhaust. The valve movement is actuated by an eccentric projection called cam which moves on a rotating shaft—the camshaft. As the camshaft turns, the cam lifts the valve but its closing depends upon the spring. These springs must have considerable tension for their prompt closing and prevent them from jumping away from the cams, especially at high engine speed. This type of valve is a 'Poppet Valve' (Fig. 2.8).

The valve mechanism to operate the valve when it is in the engine block (in L, T and F-head design) is as shown in Fig. 2.9.

Clearance is provided between valve lifter and valve stem so that expansion due to heat generated in the engine does not affect the working adversely. Clearance is more in exhaust valve as it is at a higher temperature. The clearance can be adjusted with the help of a nut that is provided.

Improved systems have also been designed such as hydraulic valve lifter and Eccentric Rocker arm. These automatically compensate the difference in clearance.

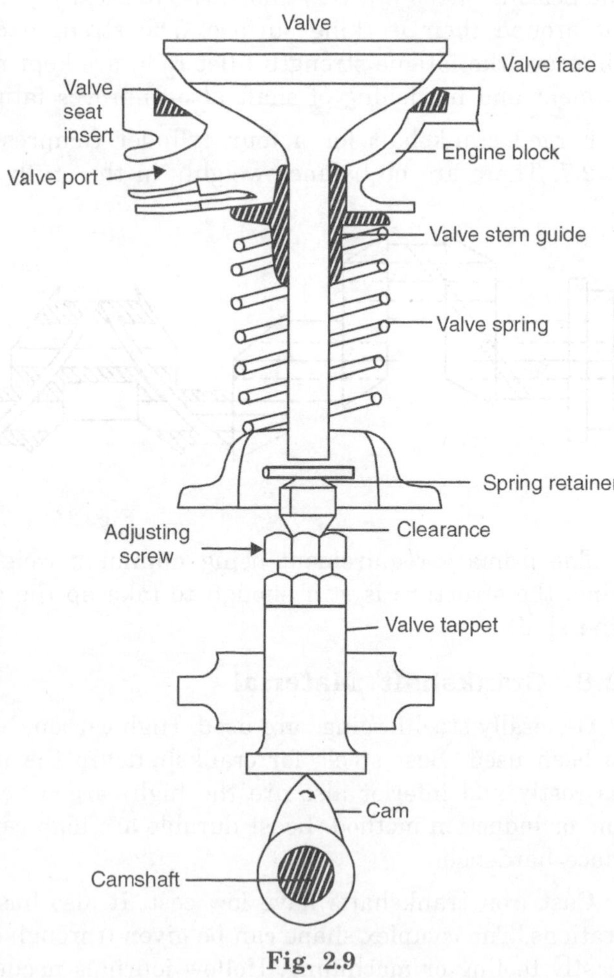

Fig. 2.9

2.2.9.1 Camshaft

It is a long straight piece of metal with a cam on it joining parts of machinery, especially in a vehicle. It is responsible for opening/closing of valve. It carries one cam for each valve to be operated and valve lift is determined by the distance that the toe of the cam projects above the rounded base of the cam. The camshaft is driven by the crankshaft either by a pair of meshing gears or by means of a pair of timing sprockets connected by a chain. The camshaft rotates at 1/2 the speed of crankshaft as the valves open only once in complete cycle that require two crankshaft rotations in 4-stroke engine.

2.2.9.2 Overhead camshaft

The overhead camshaft (OHC) is a lately introduced modification in engine design. Earlier valves were operated with the help of camshaft located in the lower portion of engine. The push due to cam lobe was transferred to valve through push rods and rocker arms. The push rods and rocker arms resist the changing speed and direction due to their inertia. Sufficient force is needed to move them. Due to this, the push rod bends or flexes slightly. Bending and flexing also occurs in rocker arm but as the push rods are thin and long it is significant in them. At lower speed of engine the flexing of push rod has little affect but at high engined speed it causes a lag in valve action. This causes a limit to the engine speed while designing an engine.

With the camshaft located in cylinder head, cam lobe directly pushes bucket tappet or rocker arm and necessity of push rod is eliminated (Fig. 2.10).

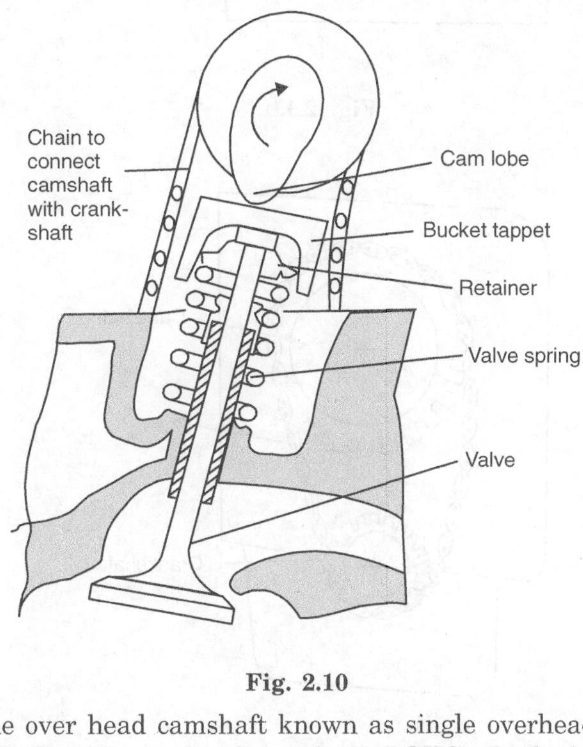

Fig. 2.10

There may be one over head camshaft known as single overhead camshaft (SOHC) or two overhead camshaft (Double overhead camshaft, DOHC). In double over head camshaft engines, one camshaft operates intake valves and other operates exhaust valves.

To drive the camshaft either timing gears (Fig. 2.11) or sprockets and timing chain are used (Fig. 2.12). The use of sprocket and chain is preferred as these provide silent operation. Here the camshaft and crankshaft rotate in the same direction where as with gears the direction of movement is opposite.

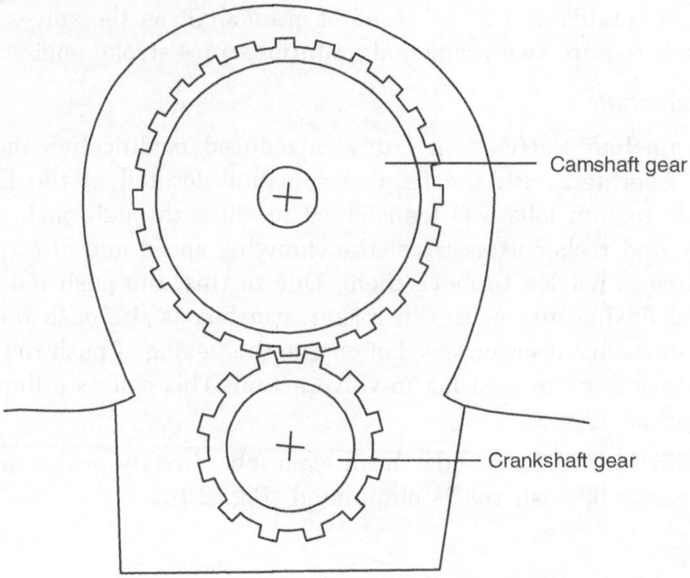

Fig. 2.11

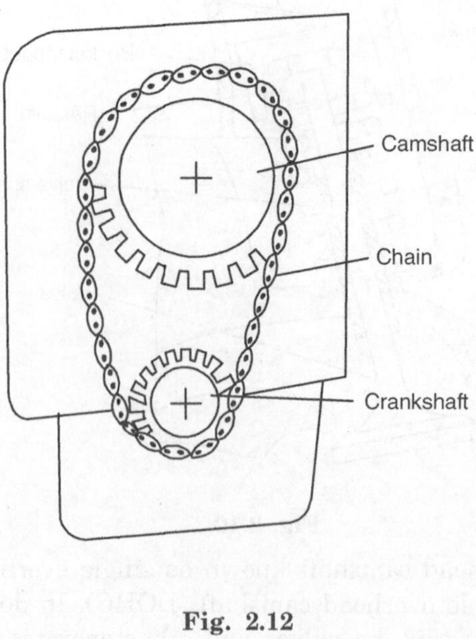

Fig. 2.12

Double overhead camshafts (DOHC) are driven by a variety of methods. Figure 2.13. shows a DOHC V-6 engine using a single timing belt to drive all the four camshafts, two for intake values and two for exhaust valves. In most of the V-type engines the intake camshafts are towards the inner side and exhaust camshafts are towards the outer side of the engine.

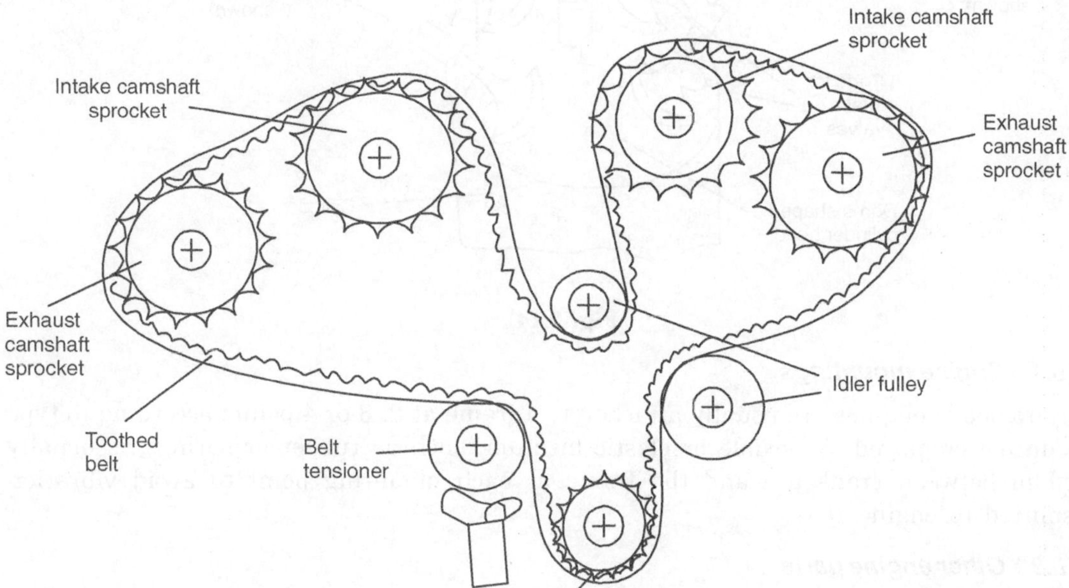

Fig. 2.13

2.2.10 Multivalve Engines

Using more than one valve for intake or exhaust is also a lately introduced modification in engine design. With a single valve for intake or exhaust the time period available for operation is very small particularly in a high speed engine. It intake or exhaust are incomplete the working of the engine is adversely affected.

In multivalve engines the number of values per cylinder could be three, four, five or even six. In case of engine with odd number of valves these are more valves for intake than exhaust. For example, in engine with three valves, two are for intake and one for exhaust. Similarly, in the engine with five valves three are for intake and two for exhaust. In case of engines with even number of valves the number of valves for intake and exhaust are equal. The additional valves cause an easy intake of air fuel mixture and easy exhaust of outgoing gases. Due to this the volumetric efficiency of engine is enhanced. The diameter of valves and their weight is also reduced. This causes a reduction in the force exerted by valve spring to close a valve.

Though the working and efficiency of the engine improves with increasing the valves but it makes the design of cam shaft more complicated. There may be some problem in lubrication also. In case of engines with four valves per cylinder the easiest way is to operate them with two overhead camshafts (Fig. 2.14).

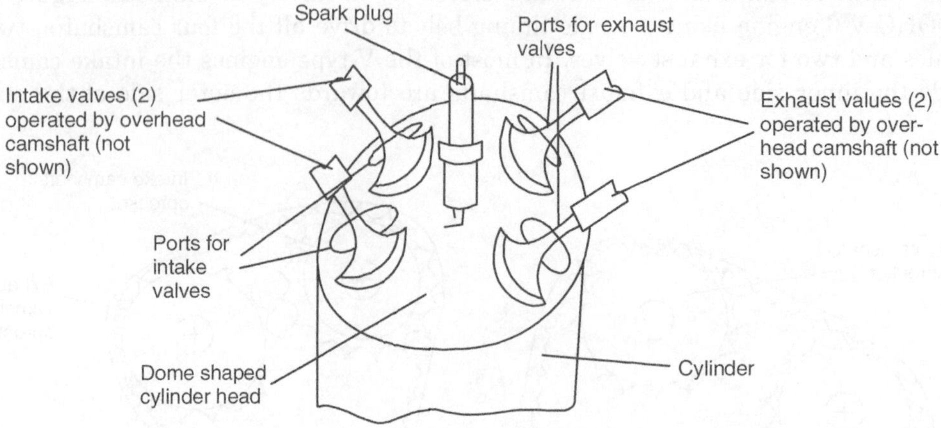

Fig. 2.14

2.2.10.1 Engine mountings

Automobile engines are usually attached to a frame at 2, 3 or 4 points according to type of mounting employed. A flexible or elastic medium, such as rubber or springs, is usually placed in between crankcase and the frame at each mounting point to avoid vibration transmitted to engine.

2.2.10.2 Other engine parts

In addition to above, we have in an engine spark plug, ignition devices, carburetor and manifolds. Some supplementary parts like air cleaner, oil filter, automatic choke and heat controls are used for better engine working. For detailed study of these components book on Internal combustion Engine can be referred.

2.3 WORKING OF FOUR-STROKE PETROL ENGINE

The working cycle of the engine consists of four strokes which are completed in two revolutions of crankshaft. The cycle consists of following strokes:

2.3.1 Suction Stroke

Considering a vertical engine and assuming the piston to be at top dead centre (TDC), the suction stroke occurs during the downward movement of the piston till it reaches the bottom dead centre (BDC). During this movement, the intake valve remains open and fresh charge in the form of fuel-air mixture enters the cylinder. When the piston reaches at bottom dead centre the inlet valve closes.

2.3.2 Compression Stroke

During this stroke, both the valves (intake and exhaust) are closed and piston moves upwards to top dead centre. The charge is compressed as piston moves upwards. As the piston reaches top dead centre, the mixture is ignited by spark. The combustion process begins, hot gases are produced which expand. This expansion of gases causes the downward movement of piston.

2.3.3 Expansion Stroke

The downward movement of piston occurs due to expansion of gases therefore this is known as expansion stroke. During this stroke the output is delivered by the engine, therefore it is also known as Power Stroke. At the end of stroke, as piston approaches bottom dead centre the exhaust valve opens.

2.3.4 Exhaust Stroke

During this stroke the piston moves from bottom dead centre to the top dead centre. The exhaust valve remains open and used gases move out of the cylinder and are disposed off. As the piston approaches top dead centre the intake valve opens and from top dead centre the piston again starts moving downwards for intake stroke.

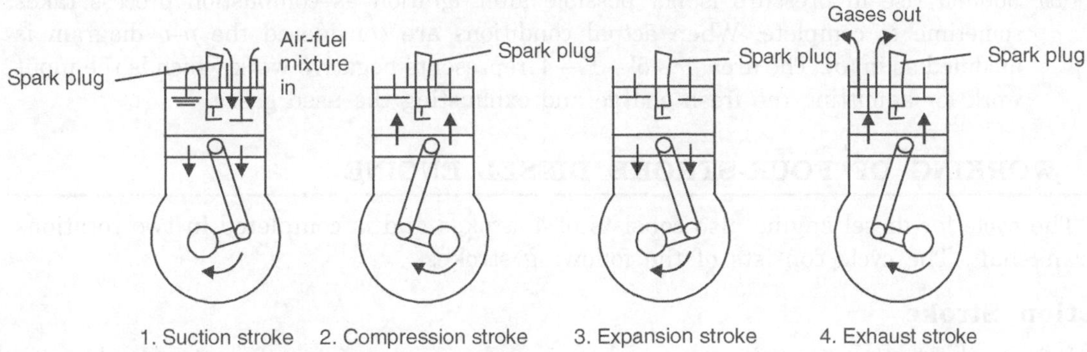

Fig. 2.15

***p–v* Diagram for 4-stroke petrol engine.**

While considering the working of 4-stroke petrol engine, following assumptions have been made :

1. Suction and exhaust take place at atmospheric pressure.
2. Compression and expansion are isentropic.
3. The combustion occurs instantaneously at constant volume.
4. Pressure suddenly falls to atmospheric pressure at the end of expansion stroke.

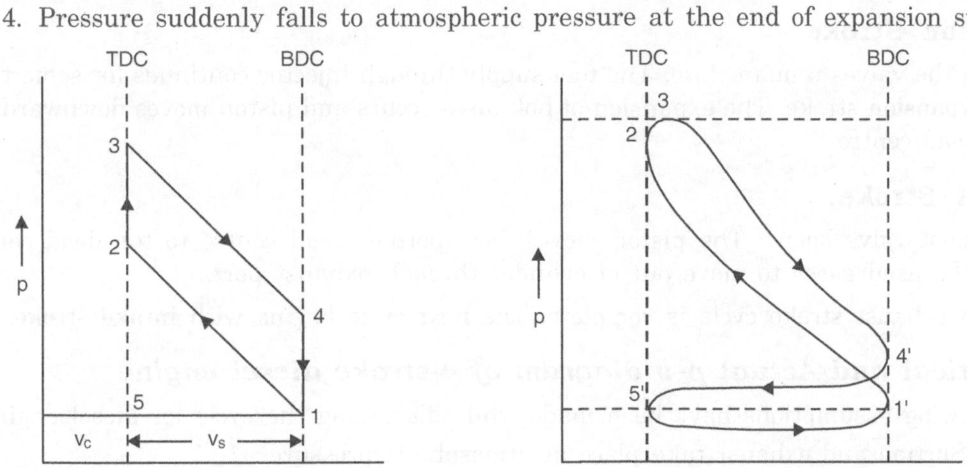

Fig. 2.16

The p–v diagram is the same as that of Otto cycle as shown in the figure 2.16. In figure, 5–1 represents suction stroke, and 1–2 represents the compression stroke. Similarly expansion and exhaust strokes are represented by 3–4 and 4–5. Process 2–3 represents instantaneous combustion.

Ideal conditions are assumed while describing the cycle. Actual conditions are different as described below:

(*a*) Pressure inside the cylinder is below atmosphere to make possible the suction of mixture.
(*b*) During exhaust the pressure of inside the cylinder must be above the atmospheric pressure.
(*c*) The compression and expansion processes are not truely isentropic.
(*d*) Sudden rise in pressure is not possible after ignition as combustion process takes sometime to complete. When actual conditions are considered the p–v diagram is modified as in (*b*). The area 4' – 5' – 1' – 4' represents negative work which is the input work for admitting the fresh charge and exhausting the used gases.

2.4 WORKING OF FOUR-STROKE DIESEL ENGINE

The cycle for diesel engine also consists of 4-strokes and is completed in two rotations of crankshaft. The cycle consists of the following strokes.

Suction Stroke

It is similar to that occurring in petrol engine when piston moves from top dead centre to bottom dead centre with intake valve open. The only difference being that only air is socked in and not air-fuel mixture as in petrol engine.

Compression Stroke

The piston moves upwards from bottom dead centre to top dead centre with both the valves closed. At the end of stroke the pressure is about 60 bars and temperature is about 600°C. At the end of compression stroke the fuel is injected inside the cylinder. Due to high temperature and pressure the fuel ignites on its own and combustion occurs.

Expansion Stroke

Both the valves remain close. The fuel supply through injector continues for some time during expansion stroke. The expansion of hot gases occurs and piston moves downwards to bottom dead centre.

Exhaust Stroke

Exhaust valve opens. The piston moves from bottom dead centre to top dead centre causing the used gases to move out of cylinder through exhaust port.

With exhaust stroke cycle is complete. The next cycle begins with intake stroke.

Theoretical and Actual p–v diagram of 4-stroke diesel engine

Following assumptions have been made while discussing the cycle for diesel engine:

(*a*) Suction and exhaust take place at atmospheric pressure.
(*b*) Compression and expansion are isentropic process.
(*c*) The combustion occurs at constant pressure during a part of expansion stroke.
(*d*) The duration of suction, exhaust, compression and expansion are 180° each.

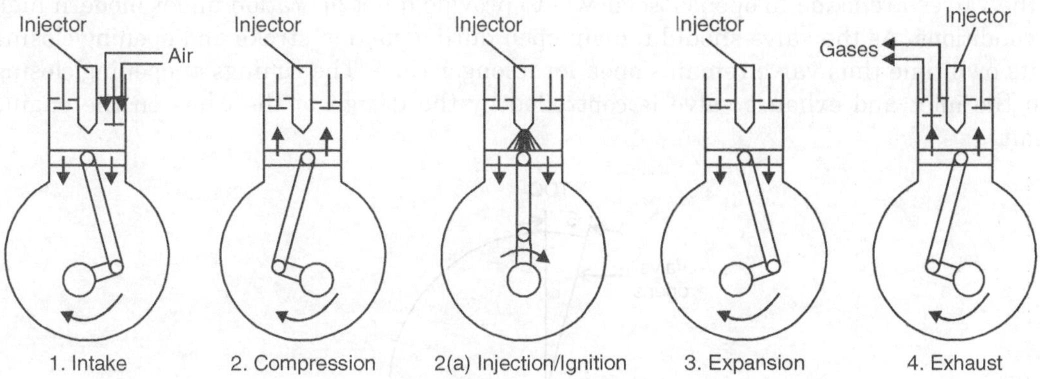

Fig. 2.16(c)

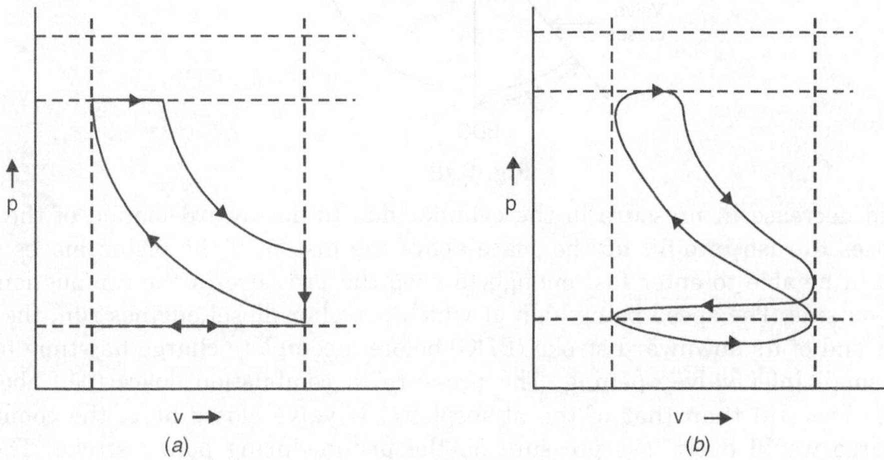

Fig. 2.17

The theoretical cycle is shown in Fig. 2.17 (*a*) and actual cycle in Fig. 2.17 (*b*). In actual conditions the difference from ideal conditions is as follows (*a*) the suction will occur only when pressure inside the cylinder is below atmospheric pressure (*b*). The exhaust will occur when pressure inside the cylinder is above the atmospheric pressure (*c*). The compression and expansion processes are not truely isentropic (*d*). Practically combustion does not occur at constant pressure.

2.5 VALVE TIMINGS

In 4-stroke cycle engine the inlet valve opens during suction stroke and exhaust valve during exhaust stroke. It is desirous that the opening and closing of valve(s) occurs at a particular piston position. There are some other practical aspects and assuming valves open and close instantaneously is wrong. Considering practical aspects let us take up operation of inlet and exhaust valves.

2.5.1 Inlet Valve

Practically, the valve opens slightly before the piston starts moving downwards from top dead center position (suction stroke) and closes after the piston has started moving upwards following the completion of suction stroke (*i.e.,* compression stroke). The reason for this being the valves are made to open/close slowly to provide quiet operation under modern high speed conditions. As the valve should remain open during suction stroke and opening/closing takes its own time thus valve remains open for a longer time. The timings of opening/closing of both the inlet and exhaust valve is controlled by the design of the cams on the engine camshaft.

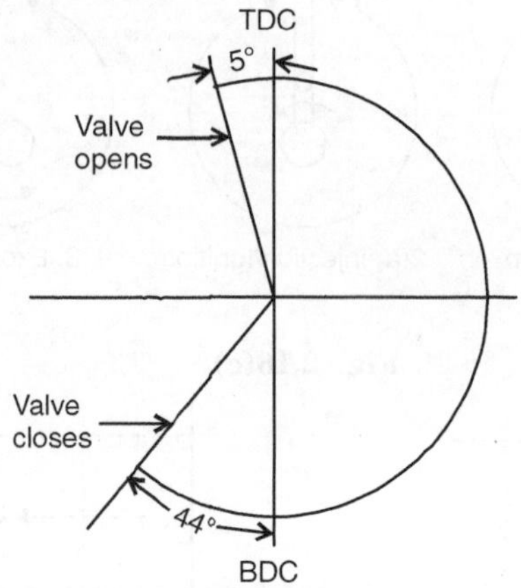

Fig. 2.18

The rapid decrease in pressure in the cylinder due to downward motion of the piston causes the gases to rush into fill up the space above the piston. If the piston moves slowly, the mixture will be able to enter fast enough to keep the pressure in the combustion space equal to that outside. The speed being high at which petrol or diesel engines run, the piston will reach the end of its downward stroke (BDC) before a complete charge has time to enter through the small inlet valve opening. The pressure in combustion space (*i.e.,* above the piston) will be low still than that of the atmosphere. If valve closes here, the combustion of partial charge would exert less pressure on the piston during power stroke. The inlet valve is thus permitted to remain open till the pressure is equal to outside pressure which occurs during the compression stroke (during up motion of piston). This period may vary from 28°–71° (of crankshaft turn). The movement of piston upwards reduces the space in the cylinder and compresses the charge ahead of it. When under compression, the gas is ready to be ignited and burned.

Figure 2.18 shows the valve timing data for the inlet valve popularly used. The total rotation of crankshaft when the inlet valve remains open is 229°.

2.5.2 Exhaust Valve

As the piston is forced downwards by the expanding gas, it has been found necessary to open the exhaust valve before the piston reaches the end of the stroke. Though it causes some wastage of force of expansion, but reduces the amount of work required during exhaust stroke. It is wrong to keep exhaust valve closed up to the very moment when piston is about to move upwards, for then, when piston starts moving upwards it would be confronted, may be just for in instance, with the force which had driven it downwards. It 'll mean power loss. Therefore, to prevent this loss, valve is opened slightly earlier.

During the next upward stroke, the remaining gases go out of the exhaust valve as the pressure inside the cylinder is more than atmospheric (the outside pressure). This also causes some gases (used) at higher pressure trapped in the clearance space above the piston after the exhaust valve closes. The best results are obtained, not by closing the exhaust valve at the end of exhaust stroke but a short time after the suction stroke has begun.

It may seem that it will draw exhaust gases back into the cylinder however it does not happen because (*i*) gases inside the cylinder are at higher pressure and (*ii*) piston's translatory movement is very less during first 0–15° movement of crankshaft. This does not increase the combustion space significantly.

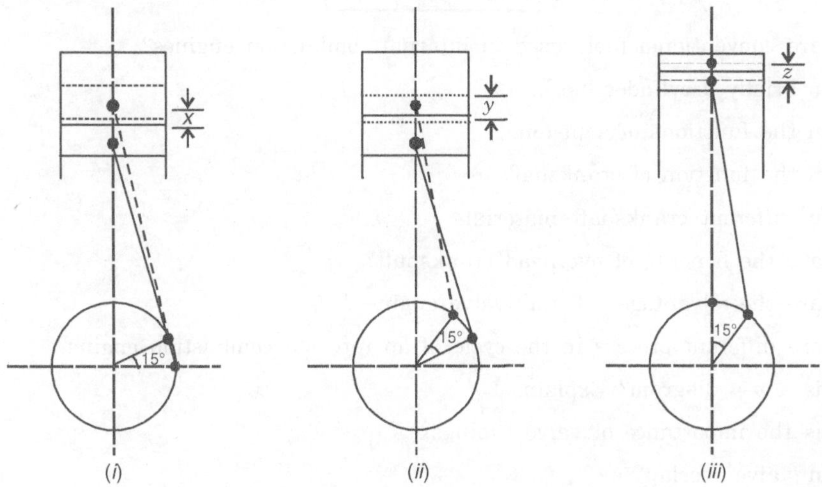

Fig. 2.19

Figure 2.19 (*i*), (*ii*), (*iii*) clearly shows the same. Movement of 15° for crankshaft in the middle of stroke (*i*) as shown in (*i*) means greater translatory movement of piston. The translatory motion is reduced when piston is in the middle of half stroke as shown in (*ii*) and becomes minimum at the end of stroke as shown in (*iii*). In third when practically there is no movement, this region is called the 'rock of the piston' Usually in this region the exhaust valve is closed after the T.D.C.

Figure 2.20 shows the exhaust valve-timing data. It opens 47° before the B.D.C. remains open throughout exhaust stroke (180°) and further 12° during suction stroke (*i.e.*, after T.D.C.). So it remains open for 239° of crank movement.

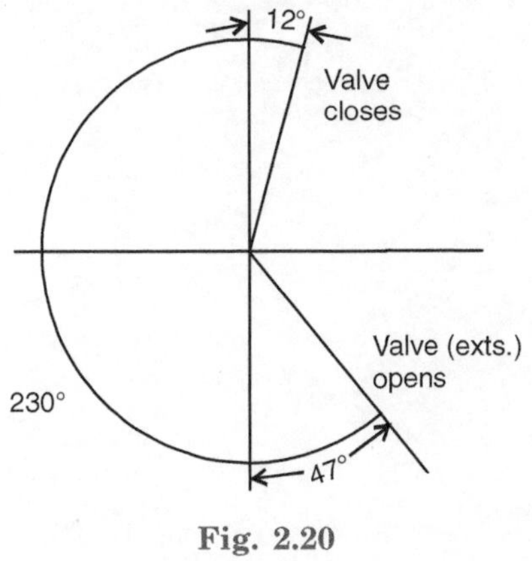

Fig. 2.20

2.5.3 Valve Overlap

Comparing the two figures for inlet and exhaust valve timing it is apparent that both the valve remain open for sometimes. This is called valve overlap. The opening of the inlet valve 5° before T.D.C. and closure of exhaust valve 12° after T.D.C. means it overlaps the inlet valve opening by 17°. Hence the valves overlap each other during 17° movement of crankshaft.

QUESTIONS

1. What are conventional fuels used in internal combustion engines?
2. Explain briefly a cylinder block.
3. Explain the functions of a piston.
4. Explain the function of crankshaft.
5. Explain different crankshaft materials.
6. What are the benefits of overhead crankshaft?
7. What are the advantages of multivalve engines?
8. What are different process in the cycle of an internal combustion engine?
9. What is a *p–v* diagram? Explain.
10. What is the importance of valve timings?
11. Explain 'valve overlap'.

3

FUEL INJECTION SYSTEMS

3.1 PETROL FUEL INJECTION SYSTEMS

Fuel injection has been utilised in compression ignition engines since beginning but has been introduced in spark ignition engine lately. These are developed for more efficient use of fuel. To make them more dependable electronic control systems were introduced in the eighties.

Fuel injection system supplies the engine with a combustible air-fuel mixture. The richness of the mixture varies depending upon various operating conditions. When cold started, the engine needs very rich mixture with higher fuel contents. Afterwards, as engine warms up lean mixture with less fuel is sufficient. During acceleration and at high speeds again the increased quantity of fuel is needed.

An electric fuel pump supplies the fuel to injectors under pressure. As soon as injector opens, the fuel sprays out. There can be single spray injector or dual spray injector as shown in Figs. 3.1 and 3.2.

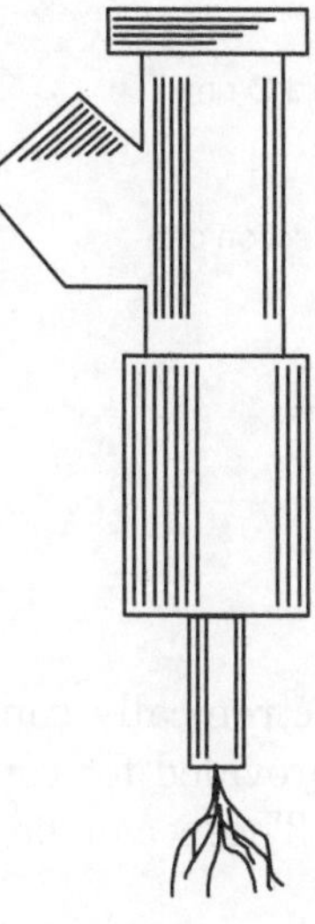

Fig. 3.1

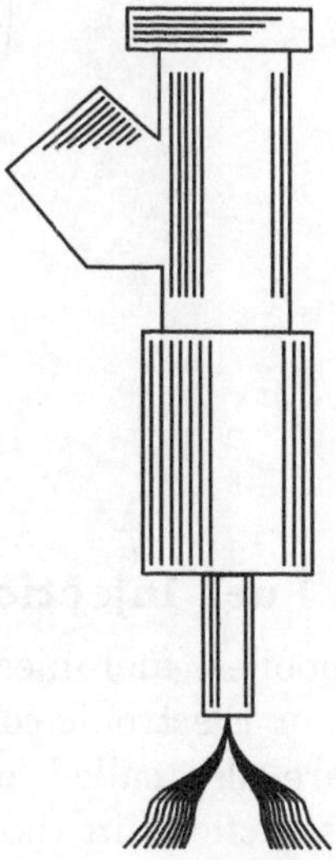

Fig. 3.2

An electric solenoid in the injector opens and closes the valve. The solenoid has a coil of wire which gets magnetised when electric current is supplied. The magnetic force lefts the armature which raises the needle valve or pintle off its seat and the fuel sprays. When the supply of electric current is stopped, the magnetic force does not exist and needle valve or pintle occupies back its seat. This stops the fuel spray. Figure 3.3 represents the details of fuel injector.

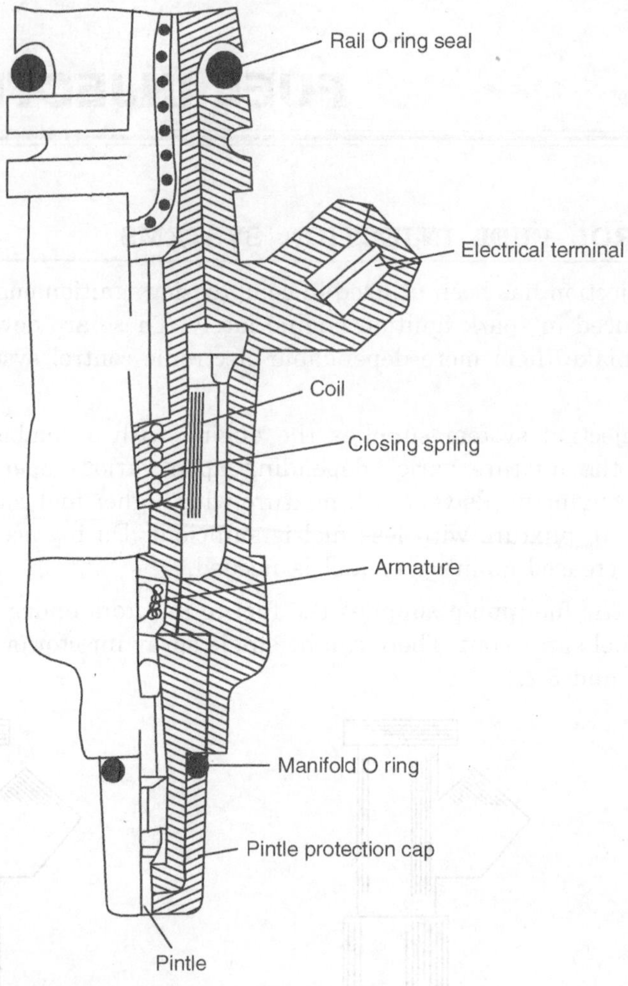

Fig. 3.3

3.1.1 Electronic Fuel Injection

In modern automobiles, fuel-injection systems are electronically controlled. Electronic control module (ECM) or Electronic control unit (ECU) is provided for control. These are so advanced that they are also called 'on-board computers'. These can be programmed and acted according to instructions in the programme.

As shown in Fig. 3.4, the ECM receives inputs such as engine speed, coolant temperature, air intake temperature, different parameters in intake manifold and exhaust through sensors. ECM processes these input signals and sends information in the form of output signals to various components of engine.

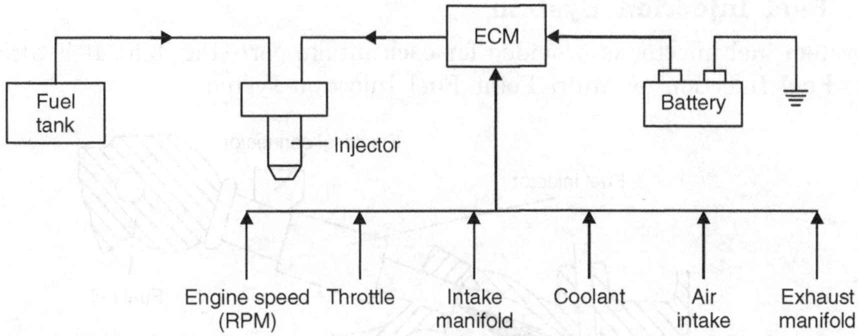

Fig. 3.4

3.1.2 Engine Sensors

To provide information regarding correct amount of fuel for different operating conditions the electronic control module processes the information from the following sensors:

(*a*) *Mass air flow sensor:* It is located in intake manifold and conveys the mass of air entering the engine.

(*b*) *Oxygen sensor:* It is located in exhaust manifold. It monitors the amount of oxygen in the exhaust so that ECM can determine the amount of fuel in the air-fuel mixture.

(*c*) *Throttle position sensor:* It monitors the throttle valve position and the amount of air going to the engine.

(*d*) *Coolant temperature sensor:* It conveys the temperature of engine to ECM which operates the engine when proper temperature has been attained.

(*e*) *Voltage sensor:* To monitor the system voltage so that ECM can raise the idle speed if voltage is dropped.

(*f*) *Manifold absolute pressure sensor:* It is located in the intake manifold and monitors the pressure of the air. The amount of air being drawn into the engine indicates the power output from engine.

(*g*) *Engine speed sensor:* It monitors the engine speed which is used in finding out the pulse width.

The electronic control module being a electronic processor is very accurate. It is programmed and can calculate the different parameters very quickly and very accurately. It is highly reliable. It is also capable of storing all the readings of different parameters in its data storage space. This data can be retrieved at the time of servicing and is highly useful. In some automobiles these parameters are available to driver 'live' at the time of driving and he can use it to improve the performance of engine.

3.2 FUEL-INJECTION SYSTEM FOR SPARK IGNITION ENGINE

Can be classified as:

(*a*) Post fuel injection system and

(*b*) Throttle body fuel injection

3.2.1 Post Fuel Injection System

In this system fuel injector is provided for each intake port (Fig. 3.5). It is also termed as Multi Port Fuel Injection or Multi Point Fuel Injection system.

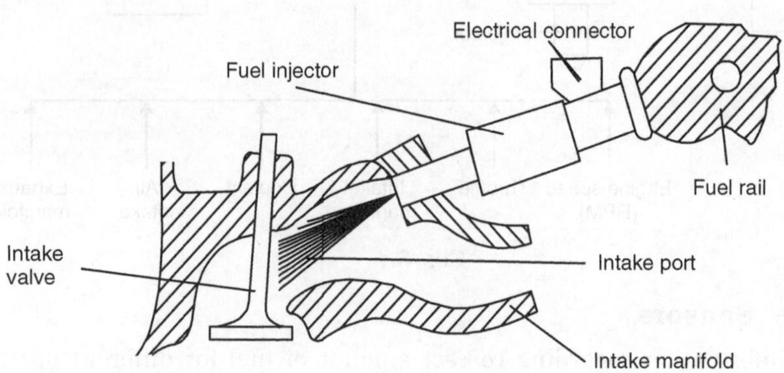

Fig. 3.5

In Multi Point Fuel Injection system, the injectors and paired or grouped together and operate together (Fig. 3.6.). If there are two groups, these operate alternately in each revolution.

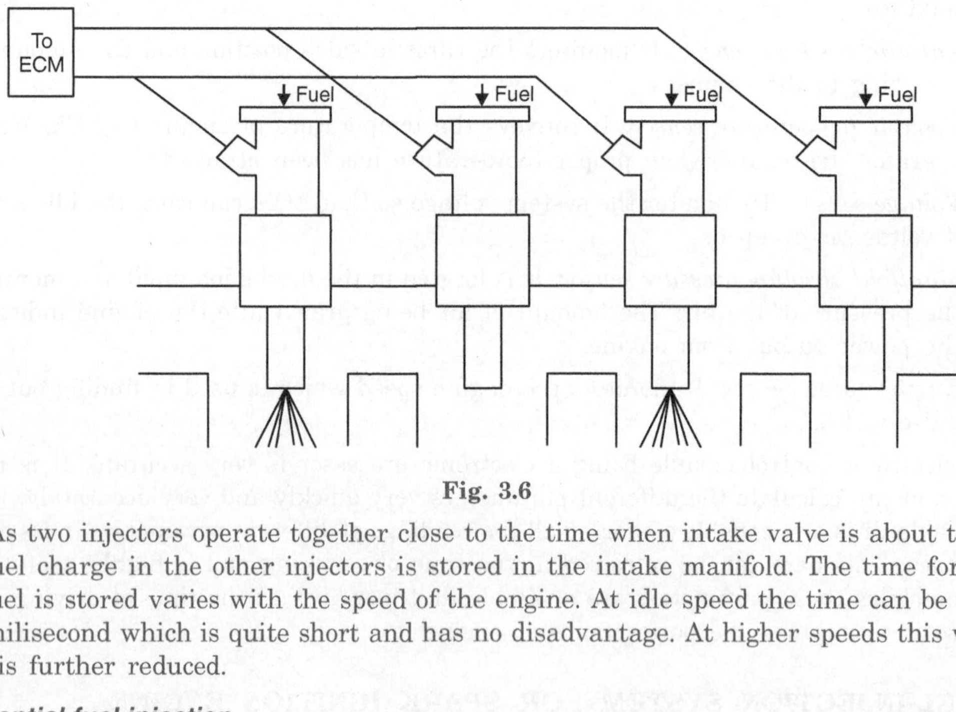

Fig. 3.6

As two injectors operate together close to the time when intake valve is about to open, the fuel charge in the other injectors is stored in the intake manifold. The time for which the fuel is stored varies with the speed of the engine. At idle speed the time can be around 150 milisecond which is quite short and has no disadvantage. At higher speeds this waiting time is further reduced.

Sequential fuel injection

In this system each injector is controlled individually by ECM so that it opens just before the intake valve. There is no waiting period for fuel in the intake manifold and thus adjustments to the mixture can be made instantaneously between firing of injectors. It is highly accurate method of regulating port injection.

3.2.2 Throttle Body Fuel Injection System

In this case fuel injector is located above the throttle valve (Fig. 3.7.). These have a throttle body assembly mounted on the intake manifold in the position usually occupied by a carburettor. The throttle body assembly has one or even two injectors.

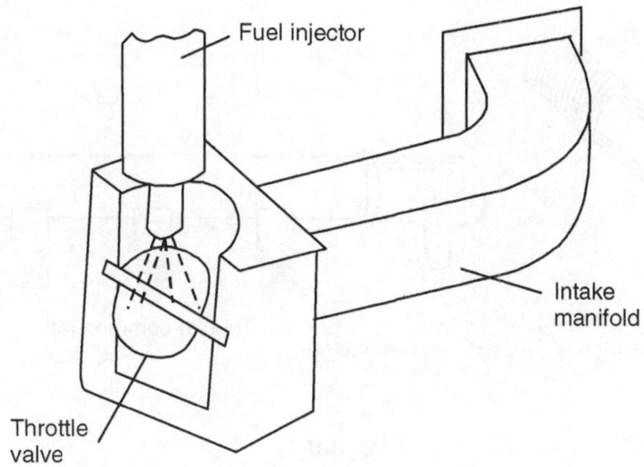

Fig. 3.7

3.3 COMMON RAIL DIRECT INJECTION (CRDI) SYSTEM

As we have discussed fuel injection in spark ignition engine, the fuel injection in compression ignition engines have also been improved. It is in the form of Common Rail Direct Injection System (CRDI) (Fig. 3.8). This is a reliable system and can be adopted for most existing diesel engines after suitable changes.

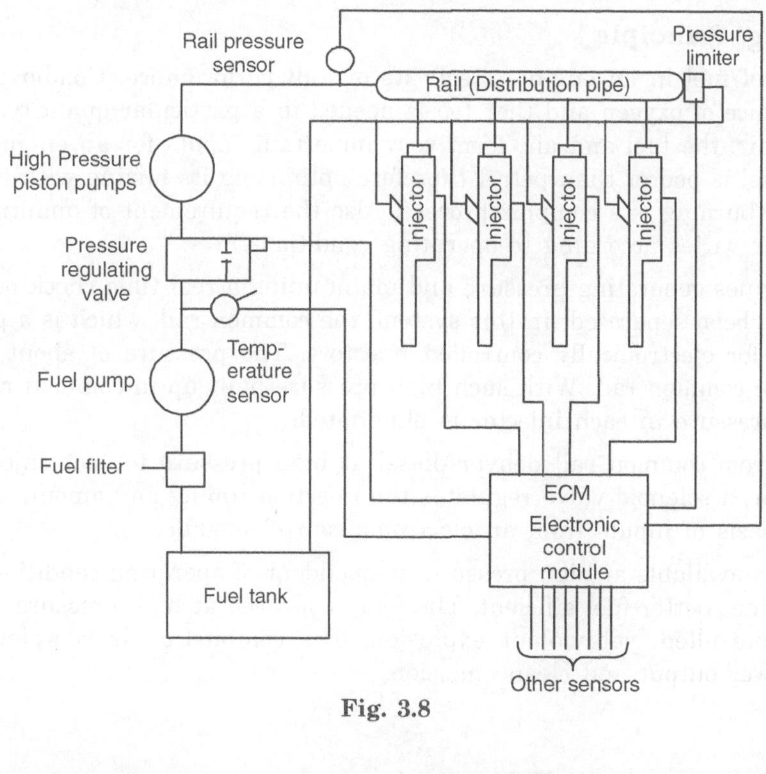

Fig. 3.8

The common rail is a large manifold and is continuously fed by fuel under pressure with the help of a pump driven by engine. The injectors are fed by pipes connected to rail (Fig. 3.9). The injector is opened electronically. The injection timing is independent of engine cycle.

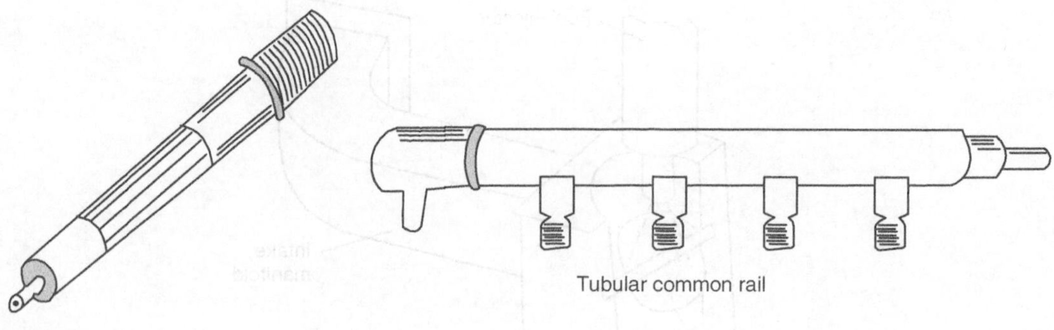

Fig. 3.9

Diesel is the lower quality fuel and has particles larger and heavier than petrol. It is difficult to pulverise them. If pulverisation is not proper the combustion of fuel leaves behind unburnt particles, and more pollutants. This means lower fuel efficiency and less power. Common rail technology improves the pulverisation of fuel. Here, a separate pump is used and fuel at high pressure is fed to individual fuel injector through a common pipe (Common rail). Fuel always remains at high pressure and therefore whenever the injector opens, high pressure fuel can be injected into combustion chamber quickly. As a result, alongwith improvement in pulverisation timing of fuel injection can be precisely controlled.

3.3.1 Working Principle

Combustion of fuel in an engine affects its overall performance. Combustion can not occur in the absence of oxygen and that too is needed in a particular quantity. The source of oxygen being air, the fuel and air, form very important inputs for an engine. The combustibility of diesel is poorer than petrol therefore optimising its mixing with air, injection into cylinder and burning is a complex process. Also the requirement of quantity of fuel to be mixed with air varies according to operating conditions.

In CRDI engines generating pressure and maintaining a real time check on amount of fuel injected have been separated. In this system, the common rail, which is a pipe, acts as shared reservoir for electronically controlled injectors. The pressure of about 1500 bar is maintained in the common rail. With such high pressure built up in common rail the need for building up pressure in each injector is eliminated.

Connectors from common rail deliver diesel at high pressure to each injector. At the end of the injector, a solenoid valve regulates the injection timing and amount of fuel to be injected on the basis of inputs from an electronic control module.

As the fuel is available at high pressure independent of operating conditions of engine it makes the engines better fuel efficient. The fuel is sprayed at high pressure, it ignites in the form of a controlled, yet violent explosion. The combustion is complete and that enhances the power output and clean emission.

The violent explosion, accompanying the combustion process, produces noise and vibration. To avoid this, most CRDI engines employ 'Pilot Injection' or 'Pilot Burn'. A small amount of diesel is injected just before the main fuel injection. This starts the process of combustion before the injection of main fuel. This helps to make the explosion less violent. The rise in temperature and pressure is staggered and makes the operation of engine less noisy.

In modern diesel engines, ECM plays a very important role. Each injector has a solenoid valve that adjusts the injection timing as well as the amount of fuel to be injected. The electronic control module is speedy, reliable and can be programmed to take into account a large number of operating conditions. A mechanical device does not have all these qualities.

QUESTIONS

1. What is the difference between Single Spray and Dual Spray Injector?
2. What are the functions of 'electronic control module'?
3. Explain the following sensors (*a*) Oxygen sensor (*b*) Throttle position sensor (*c*) Manifold absolute pressure sensor.
4. With the help of diagram explain port fuel injection system.
5. Explain throttle body fuel injection system.
6. What is CRDI? Explain.
7. What are advantages of CRDI system?

4

MULTICYLINDER INTERNAL COMBUSTION ENGINES

Internal combustion engines are the source of power in an automobile. Automobiles became more usable with introduction of internal combustion engine. An automobile requires good amount of power to move efficiently. There are certain problems in employing a single cylinder engine. One of these being that a single cylinder engine will become too heavy. It will also have problem of mechanical balance. Therefore, multicylinder engines are better option.

In an automobile providing comfort to its user is of prime importance. One of the sources of discomfort can be vibrations which are produced in an engine. These vibrations are transmitted to the body of the automobile and cause discomfort to its users. It is very essential that these vibrations are not transmitted to the body of the automobile.

The vibrations may be produced in engine itself due to improper mechanical balancing. The mechanical imbalance occurs due to movement of parts. It becomes more complicated as the nature of movement is different for different moving parts. The piston has reciprocating movement while the crankshaft has rotary motion. Connecting rod has quite complicated motion. Its small end reciprocates with piston while big end rotates along with crank pin. The different parts of engine have to be balanced individually and in sub-assemblies. Finally after the assembly, mechanical balancing of engine as a single unit is done.

4.1 GENERAL CONSIDERATIONS OF ENGINE BALANCE

Balancing of an engine includes power as well as mechanical balance. An engine is said to be in power balance when the power strokes occur at regular intervals with relation to the revolution of crankshaft and each power stroke exerts the same force. Mechanical balance is obtained in an engine when the moving parts, reciprocating as well as rotating are so arranged that these cause least vibrations by counterbalancing vibrations produced by each other. The rotating parts can be balanced mechanically by bringing them into 'static' and 'dynamic' balance. The crankshaft and flywheel are principal parts to be balanced mechanically.

To balance reciprocating parts is quite difficult. The weight of the pistons and connecting rods moving, one way then other, produces considerable vibrations. The crankshaft is subjected to 'shocks' in bringing these parts to a stop at the end of each stroke. These shocks on the crankshaft are called 'primary inertia forces' and are increased in intensity by the pressure on the pistons at the end of the cycle. Also there is *second inertia force* due to angularity of the connecting rods, which if not balanced, causes 'secondary vibrations'. In a well designed engine every piston and connecting rod is of the same weight within limits, and the flywheel and crankshaft assembly has a perfect dynamic balance. Through this practice, vibrations are minimised.

The dynamic balancing of a complete engine is also relevant as a slightly unbalanced part on one side of the engine may cause unwanted vibrations.

Vibrations produced in an engine

Every well-balanced engine must eliminate all sort of vibrations either through original design or subsequently added expedients. Vibrations are undesirable as these cause discomfort to the user of the automobile. Vibrations may occur due to different causes.

Torsional vibration is caused by twisting/untwisting of the crankshaft due to force exerted by power stroke. Its magnitude will be more if crankshaft has small diameter in-proportion to its length. Also the crank throws near the flywheel transmit their forces to flywheel with little twisting or crankshaft 'wind up'. On the other end, there is a considerable length between the crank throw and flywheel hence winding/unwinding has greater magnitude. Every crankshaft has an inherent natural period or frequency of vibration. If the frequency of this torsional vibration due to winding/unwinding corresponds to its natural frequency the vibration would become excessive, with serious results, like breakage of crankshaft. The phenomenon is commonly known as 'resonance'. Speeds at which resonance occurs is called 'critical speed'.

One of the methods to minimise the torsional vibration is to design the crankshaft such that its critical speed is above the maximum speed of the engine. The torsional vibration which occur at lower speed can be damped by special means. Most in line 6 or 8 cylinder automobile engines, however, because of their length and high speeds, depend upon torsional vibration dampers to neutralise torsional crankshaft vibration. One type of vibration damper is provided with an inertia weight which is set in vibration by the torsional vibration in a direction opposite to that of crank pin and thus tries to neutralise the vibrations (Fig. 4.1).

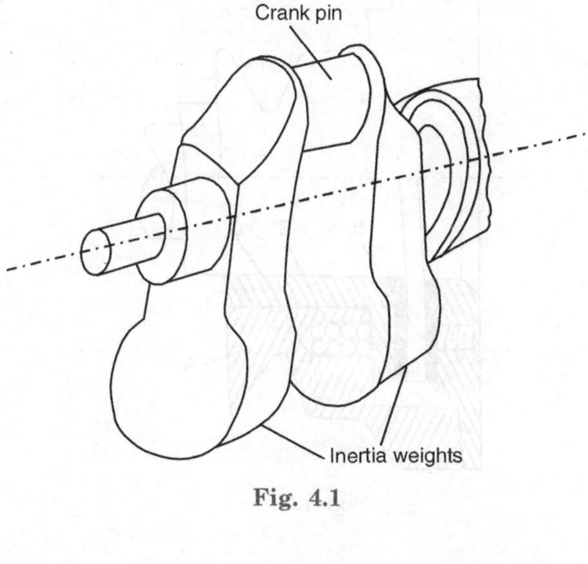

Fig. 4.1

Another type of vibration damper is known as harmonic balancer (Fig. 4.2.). The damper mounts on the front end of the crankshaft and forms the hub. On this hub the crankshaft pulley is attached. The inertia ring is bonded through the rubber ring to the pulley. The inertia ring provides the damping effect.

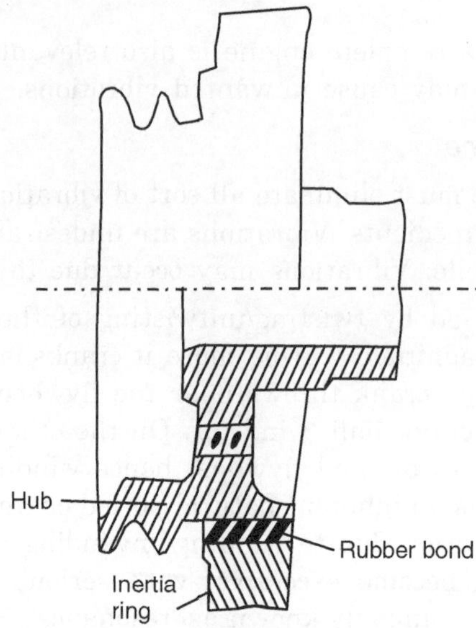

Fig. 4.2

In another variant, the vibration damper is mounted on the front end of the crankshaft (Fig. 4.3.). It consists of a flywheel A driven through a friction disc clutch B. The frictional force is adjusted through springs S. These springs keep pressure plate C pressed. The friction does not permit the transmission of torsional vibrations due to high torsional acceleration which occurs because of winding/unwinding of crankshaft. Thus the friction absorbs the torsional vibrations.

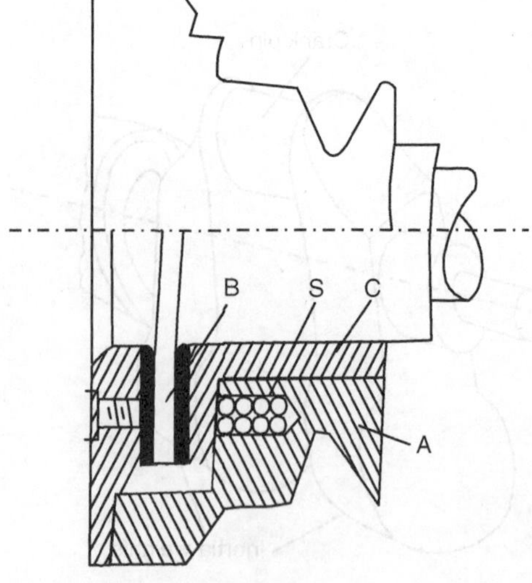

Fig. 4.3

Third type of damper is viscous type. (Fig. 4.4). There is an annular flywheel seated in a metal casing. The space between the two is filled with silicone fluid. The fluid has high viscosity. The casing rotates and oscillates with the crankshaft and the flywheel tends to maintain steady motion. Due to this, there is a viscous drag between the two. This causes the dissipation of energy of vibration as heat.

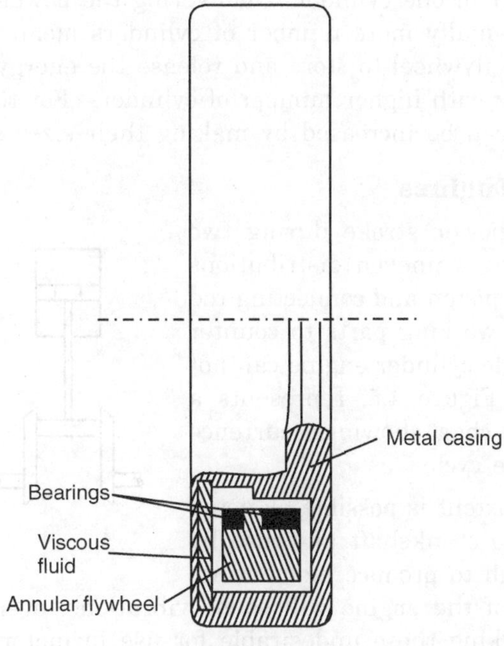

Fig. 4.4

Another type of vibration is caused by the 'torque reaction' of the connecting rods on the cylinder block as they push against the crank pins to cause the rotary motion of the crankshaft. The torque reactions of the connecting rod impulses tends to rotate the cylinder block in the opposite direction. As the impulses are fluctuating in nature so are the reactions and so cylinder block tends to vibrate. To minimise it either number of cylinder are increased or mounting of engine is done on sort of elastic material (like rubber) which can take the vibrations.

4.2 FIRING ORDER

The sequence in which the power impulses occur in an engine is called the firing order. When the cylinders are in line, the cylinder nearest to radiator is designated as number 1. The one directly behind it is number two and so on. In V-type engine the numbering of cylinders is not uniform.

4.3 BALANCE AND FIRING ORDER OF VARIOUS ENGINES

Though generally an automotive engine has four or more cylinders yet to understand the engine balancing properly knowledge of balancing one and two cylinders engine is important.

In case of a single cylinder engine there is one power stroke for every two revolutions of crankshaft and thus inspite of having a large flywheel it may not run smoothly and quietly. Due to relatively large cylinder size and large time between power strokes, the parts must be made large and heavy to withstand the resultant rough operation. If there are two, four, six or eight cylinders in engine there will be 2, 4, 6 or 8 power strokes for two revolutions of crankshaft meaning there by uniform application of torque and thus smoother operation. In a four-cylinder engine, there is no time period when a cylinder is not delivering power during the cycle. In an engine with more than four cylinders, there will be a time period during which more than one cylinder is delivering the power simultaneously. If the power strokes are spaced equally more number of cylinders mean less vibration and less work has to be done by the flywheel to store and release the energy. Hence flywheel could be lighter in case of engines with higher number of cylinders. For the same power requirement number of cylinders can be increased by making their sizes smaller.

4.3.1 One Cylinder Engines

Here because of one power stroke during two crankshaft revolutions there is uneven distributions of power. Since there is one piston and connecting rod which reciprocates with no working parts to counter balance their weight, a single cylinder engine can not be balanced mechanically. Figure 4.5. represents a single cylinder engine with a chart showing occurrence of different processes in the cycle.

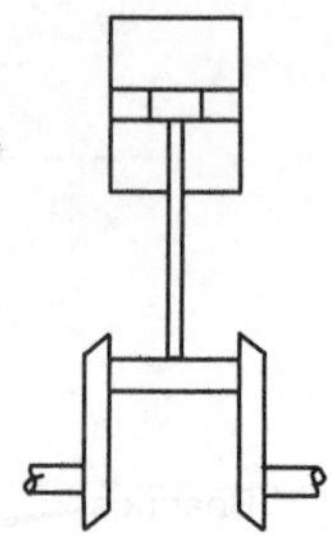

Cylinder	1
1st revolution	P
	E
2nd revolution	S
	C

Fig. 4.5.

Balancing up to some extent is possible by using counter weights attached to crankshaft and also by using flywheel heavy enough to produce momentum. Fluctuations in the speed of the engine will cause vibrations, even in the best designed single cylinder engines, making these undesirable for use in motor vehicles.

4.3.2 Two Cylinder Engines

Figure 4.6 shows two-cylinder vertical engine with 180° crankshaft. Here, the pistons move in opposite directions. Hence the engine is well balanced as far as primary inertia forces are concerned. Power balance as shown in the chart (Fig. 4.6). In (A) both power strokes occur in 1st revolution of the crankshaft while there are no power strokes during second revolution. In (B) power stroke occurs at the beginning of 1st revolution and then in the end of 2nd revolution. In both the cases, there is an irregular production of power which sets up vibrations and causes the engine to run unevenly. A two-cylinder engine has a better balance if the cylinders are horizontal and arranged on the opposite sides of the crankshaft.

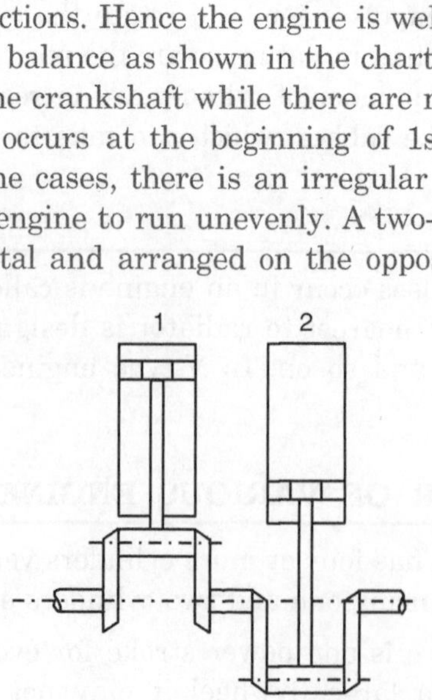

(A)

Cylinder	1	2
1st revolution	P	C
	E	P
2nd revolution	S	E
	C	S

(B)

Cylinder	1	2
1st revolution	P	E
	E	S
2nd revolution	S	C
	C	P

Fig. 4.6

4.3.3 Four-Cylinder Engine

Figure below represents a four cylinder engine with piston positions and charts displaying the processes occurring in different cylinders during the two revolutions of cycle.

In four cylinder engine the pistons are positioned at opposite ends. If in cylinder number 1 and 4 the pistons are positioned at the top dead centre in cyulinder number 2 and 3 these are at bottom dead centre (Fig. 4.7). The pistons move in opposite direction. These will move downwards in cylinder number 1 and 4 and upwards in cylinder number 2 and 3. This arrangement tends to neutralise primary inertia forces because piston subassemblies, i.e., piston, piston pin and small end of the connecting rod, are of the same weight. This provides good primary mechanical balance.

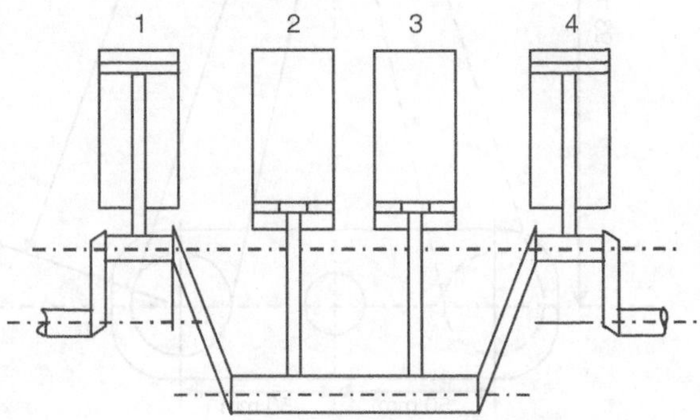

(A)

Cylinder	1	2	3	4
1st revolution	P	E	C	S
	E	S	P	C
2nd revolution	S	C	E	P
	C	P	S	E

(B)

Cylinder	1	2	3	4
1st revolution	P	C	E	S
	E	P	S	C
2nd revolution	S	E	C	P
	C	S	P	E

Fig. 4.7

As far as secondary inertia forces are concerned in a four-cylinder engine, these are not balanced. An example given below well demonstrate it. The pistons in cylinder number 1 and 4 are at top dead centre (T.D.C.) and pistons in cylinder number 2 and 3 at bottom dead centre (B.D.C.). If the stroke is 100 mm and connecting rod is 200 mm long, two pistons (2 and 3) are 150 mm above crankshaft centre and two pistons (1 and 4) are 250 mm above, the crankshaft centre. In view on right hand side, when crank arms are horizontal with all the four pistons near (but not exactly at), the middle of stroke. The connecting rod forms hypotenuse of 200 mm and horizontal side of 50 mm in the right angle triangle. Computing the third side, it comes out 193 mm approximately (Fig. 4.8).

Hence in the centre of gravity of body comprising of 4 pistons, piston pins and smaller ends of connecting rods is about 7 mm lower when the crank becomes horizontal from verticle. The total weight of these parts (approximately 6 kg) must be moved up 7 mm and down 7 mm during each 1/2 crankshaft revolution. The vibration has frequency twice the engine speed. Hence a four cylinder engine is subjected to undesirable vibrations. In six or eight cylinder in line engines there is no such problem.

Charts show various firing orders for power balance. With either arrangement power strokes are evenly distributed. Table A gives firing order 1-3-4-2 while Table B gives firing order 1-2-4-3. American four-cylinder cars have adopted 1-3-4-2 as standard firing order.

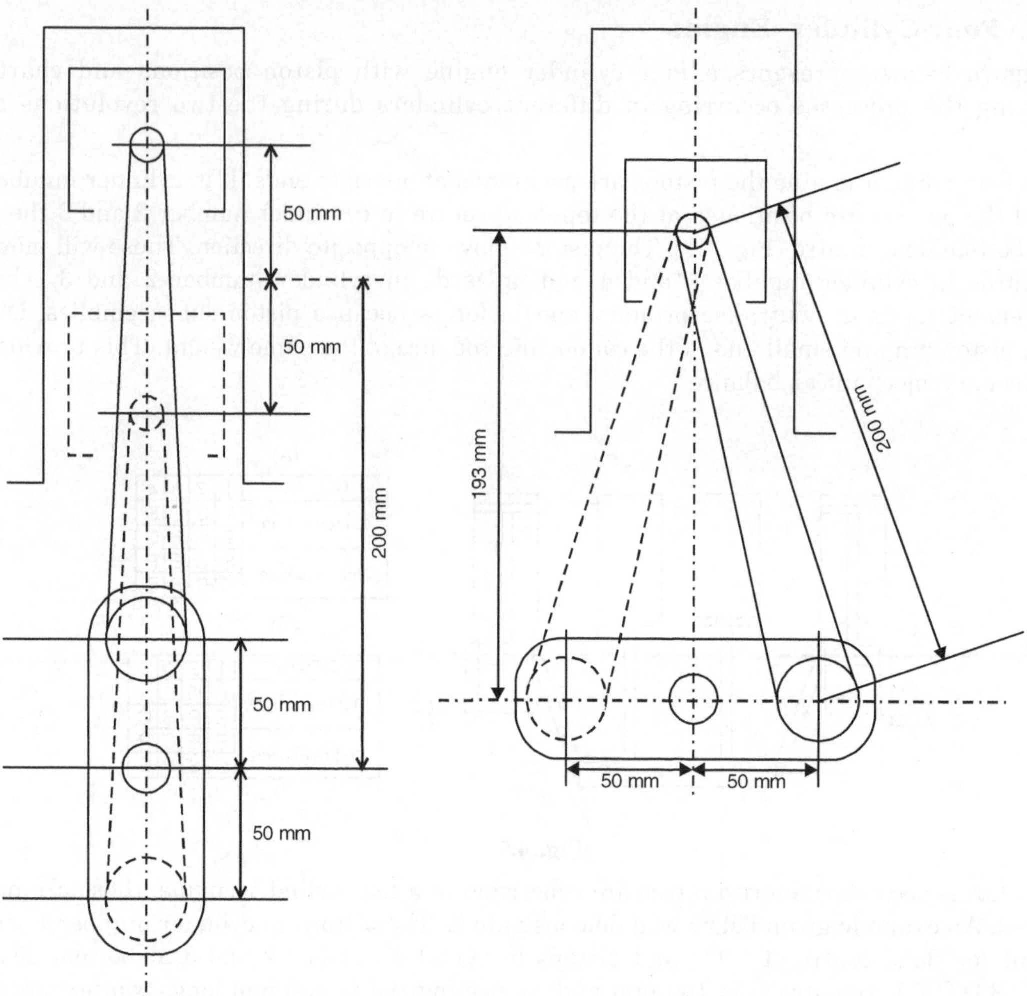

Fig. 4.8

4.3.4 Six Cylinder Engines

Six cylinder in line engines are built with 3 pairs of crankpins 120° apart as shown in Fig. 4.9. The arrangement is such that crank throws of number 1 and 6, 2 and 5 and 3 and 4 are in the same radial plane. The angle between the planes being 120°. Hence there are six power strokes during two revolutions (*i.e.,* 720°) of crankshaft or one power stroke every 120° of crankshaft rotation. Because of this, the engine with proper firing order has primary as well as secondary inertia forces perfectly balanced.

Viewing from front away from flywheel, crankpins 3 and 4 are to the left of 1 and 6 and as 1 and 6 start to more downward 2 and 5 are completing downward strokes. Crankpins 3 and 4 are on their upward stroke. With this arrangement, four different firing orders with good engine balance are possible. However, 1-4-2-6-3-5 is the firing order used with this arrangement [Fig. 4.9 (a)].

In another arrangement, crankpins 3 and 4 are to the right of 1 and 6 when viewed from front [Fig. 4.9 (b)]. Which is standard for all American in-line 6-cylinder passenger cars. Here 3 and 4 are finishing their downward stroke; 1 and 6 are starting downwards and 2 and 5 are on their upward stroke. Although four different firing orders are possible, the standard one is 1-5-3-6-2-4 which is adopted.

The V-6 engines have a firing order 1-6-5-4-3-2 where left bank is numbered as 1, 3, 5 and right bank is numbered 2, 4, 6 from the front of the engine. The passenger car has 90° angle between banks whereas in heavy vehicles it is 60°. The opposed 6-cylinder engine (with 180° between the banks) uses six-throw crankshaft. The firing order 1-4-5-2-3-6 is adopted where cylinders 1, 3, 5 are on the right bank and 2, 4, 6, are on the left bank.

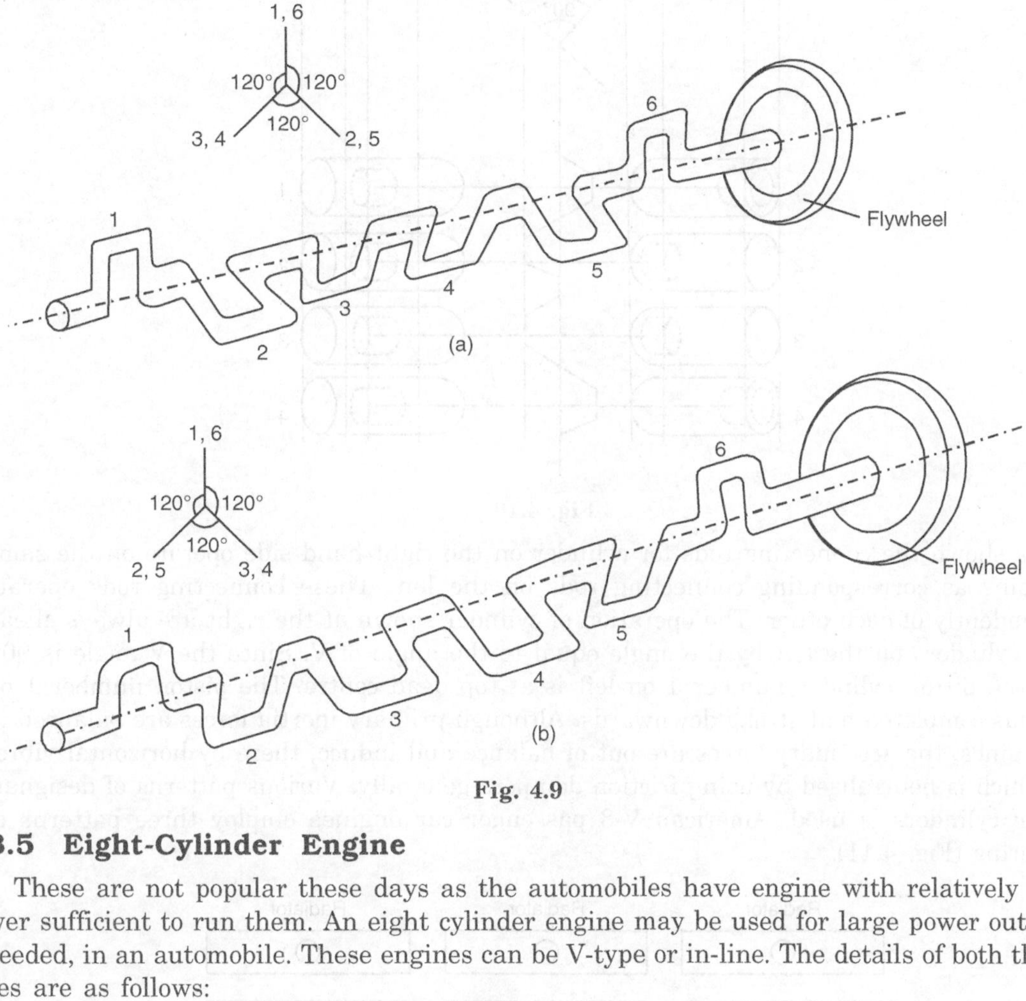

Fig. 4.9

4.3.5 Eight-Cylinder Engine

These are not popular these days as the automobiles have engine with relatively less power sufficient to run them. An eight cylinder engine may be used for large power output, if needed, in an automobile. These engines can be V-type or in-line. The details of both these types are as follows:

4.3.6 V-type Eight Cylinder Engine

This type of eight cylinder engine has two banks known as left and right banks (Fig. 4.10). Each bank has four cylinders numbered 1, 2, 3, 4. It is a combination of two 4-cylinder engines operating from a single crankshaft. The cylinders are so arranged that their centre lines make an angle of 90° forming a V with crankshaft at the junction point.

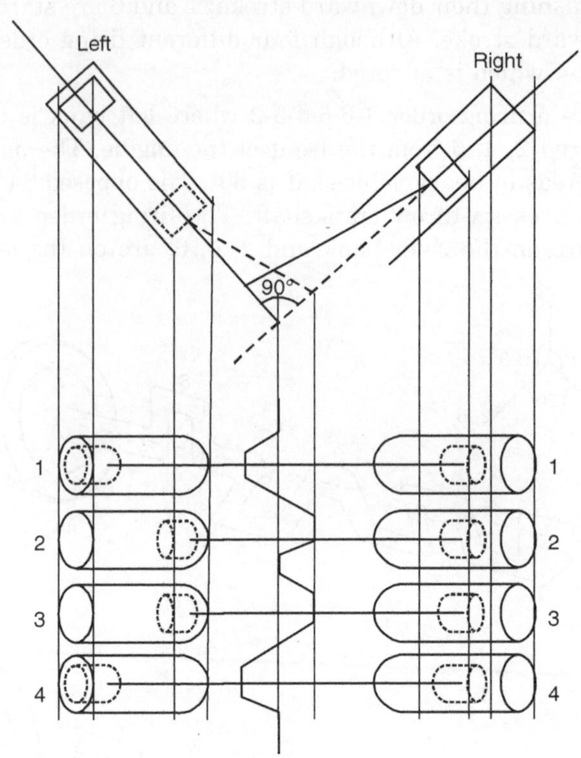

Fig. 4.10

As shown the connecting rods for cylinder on the right hand side operate on the same crankpins as corresponding connecting rods on the left. These connecting rods operate independently of each other. The operation of cylinder shown at the right are always ahead of the cylinders on the left by the angle equal to the angle of V. Since the V angle is 90°, when the piston cylinder number 1 on left is at top dead centre. The piston number 1 on right has completed half stroke downwards. Although primary inertia forces are balanced in V-8 engines, the secondary forces are out of balance and induce, there by horizontal vibration which is neutralised by using friction dampers generally. Various patterns of designating the cylinders is used. American V-8 passenger car engines employ three patterns of numbering (Fig. 4.11).

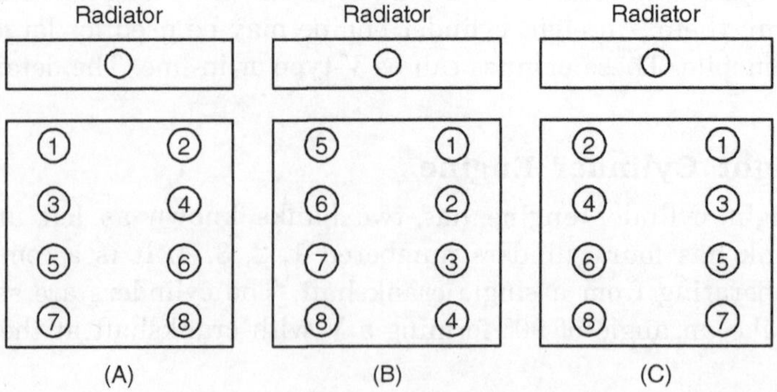

Fig. 4.11

A – General Motors and Chrysler Products adopt this pattern. The standard firing order adopted is 1-8-4-3-6-5-7-2. Another manufacture olds mobile also adopts the same pattern with standard firing order as 1-8-7-3-6-5-4-2.

B – Ford, Mercury or Lincoln V-8 adopt this pattern. The firing order which has been adopted as standard is 1-5-4-8-6-3-7-2. Alternatively another firing order 1-5-4-2-6-3-7-8 has also been adopted as standard.

C – Buick has adopted this pattern of designating the cylinders. Firing order 1-2-7-8-4-5-6-3 has been adopted as standard in this case.

Straight-eight engines with cylinders in line use a crankshaft with throws set at 90° from each other. The crank pins for cylinder 1, 8; 2, 7; 3, 6; and 4, 5 are in the same radial plane. There are two possible arrangements (shown in the Fig. 4.12.) 1, 8 may follow 3, 6 as shown in (*a*) or 4, 5 as shown in (*b*).

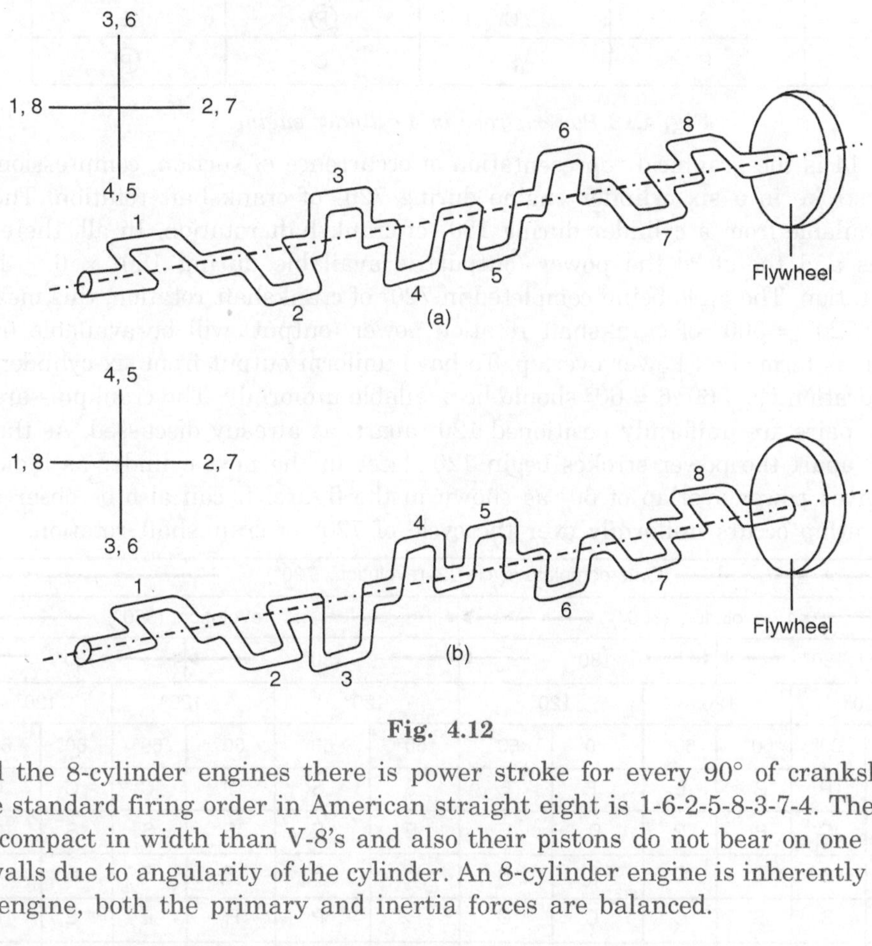

Fig. 4.12

In all the 8-cylinder engines there is power stroke for every 90° of crankshaft movement. The standard firing order in American straight eight is 1-6-2-5-8-3-7-4. These engines are more compact in width than V-8's and also their pistons do not bear on one side of the cylinder walls due to angularity of the cylinder. An 8-cylinder engine is inherently a perfectly balanced engine, both the primary and inertia forces are balanced.

4.4 POWER OVERLAP

As already stated more number cylinders mean smoother flow of power and so the operation. It's because of power overlap which is the availability of power (output) from two cylinders simultaneously.

In a 4-cylinder engine, each piston, during power stroke, pushes the crankshaft downwards during 180° of crankshaft rotation. There are four power strokes and therefore the total duration when power (output) is available from engine is 180° × 4 = 720°. The duration of cycle is also 720° of crankshaft rotation. Therefore in a 4-cylinder engine, there is uniform output available during the cycle but there is no overlap. Figure 4.13 represents the same graphically in table form.

Firing order	One complete cycle or 720° of crankshaft rotation			
	1 revolution (360°)		1 revolution (360°)	
	180°	180°	180°	180°
1	Ⓟ	E	S	C
3	C	Ⓟ	E	S
4	S	C	Ⓟ	E
2	E	S	C	Ⓟ

Fig. 4.13 *Power stroke in 4-cylinder engine.*

Figure 4.14 is the graphical representation of occurrence of suction, compression, power and exhaust stroke in a six cylinder engine during 720° of crankshaft rotation. The power (output) is available from a cylinder during 180° of crankshaft rotation. In all, there are six power strokes and therefore the power (output) is available during 180° × 6 = 1080° of crankshaft rotation. The cycle being completed in 720° of crankshaft rotation, this means that during 1080°–720° = 360° of crankshaft rotation power (output) will be available from two cylinders. This is termed as Power overlap. To have uniform output from six cylinders of the engine this duration *i.e.*, 360°/6 = 60° should be available uniformly. The crankpins are paired and the three pairs are uniformly positioned 120° apart, as already discussed. As the crankpins are 120° apart the power strokes begin 120° later in the next cylinder as indicated in figure. This gives power overlap of 60° as shown in the figure. It can also be observed from figure that overlap occurs uniformly over the cycle of 720° of crankshaft rotation.

	One complete cycle (2 revolution, 720°)											
	1 revolution (360°)						1 revolution (360°)					
	180°			180°			180°			180°		
	120°		120°		120°		120°		120°		120°	
	60°	60°	60°	60°	60°	60°	60°	60°	60°	60°	60°	60°
1	P_0	P	P_0	E	E	E	S	S	S	C	C	C
5	C	C	P_0	P	P_0	E	E	E	S	S	S	C
3	S	C	C	C	P_0	P	P_0	E	E	E	S	S
6	S	S	S	C	C	C	P_0	P	P_0	E	E	E
2	E	E	S	S	S	C	C	C	P_0	P	P_0	E
4	P_0	E	E	E	S	S	S	C	C	C	P_0	P_0

Fig. 4.14

If we consider an eight cylinder engine, there are eight power strokes of 180° duration. In all, the power output is available from engine during 180° × 8 = 1440° of crankshaft revolution. Cycle being completed in 720° of crankshaft revolution. In this case, the power (output) is available from two cylinder simultaneously *i.e.*, power overlap, during 1440° – 720° = 720° of crankshaft rotation. This means the overlap for each cylinder will be 720°/8 = 90°. The power (output) from two cylinder will be available during 90° uniformly during the cycle.

On the same lines, the power overlap duration, for 12 cylinder engine, would be 120° of crankshaft revolution and for a 16 cylinder engine the duration of overlap would be 135°.

4.4.1 Engine Balance and Firing Order Service

Whenever replacements of connecting rod or piston become essential, care should be taken in choosing it. It must be within the tolerance limit as far as weight is concerned. In V-type engine similar care should be taken, rather matching of weight need to be done parts by parts. If not done so, excess weight of one part or the other may cause vibration and throw the engine out of balance. Mostly flywheel can be mounted on the crankshaft in one position only. A punch mark on flywheel as well as shaft should be put so that the wheel can be remounted in the same position exactly.

4.5 IN LINE ENGINES WITH 3-CYLINDERS

In case of engines with odd number of cylinders, having crankpins pitched at equal intervals around the crankshaft, primary couples, secondary forces and secondary couples are induced in the plane of the cylinders axes. 3-cylinders with a crankshaft having cranks at 120° intervals provide better working in two stroke engines. The reason being three power strokes revolution makes torque output smooth. In case of 4-stroke diesel engines having three cylinders instead of four (for the same power output) provides a better layout and at the same time avoiding the problems of fuel injection into a cylinder of smaller diameter.

In a 3-cylinder engine, primary forces are balanced because the upward inertia force on any piston at the top dead centre is balanced by the downward forces on the other two pistons. However, when piston in cylinder 1 is at the top dead centre these forces tend to rotate the engine nose up, and when piston in cylinder 3 is at top dead centre couple acts in opposite sense. Only when piston in cylinder 2 is at the top dead centre the pitching couples are in balance. When the pistons are at bottom, the directions of the couple are in each instance reversed.

Counter weights, positioned opposite to the crankpins can be used to offset these couples. However, since the centrifugal force, they produce, act about the axis of the crankshaft, while the piston inertia forces act only vertically, these introduce yawing couples.

There are two ways of overcoming the problem. It is most economical to balance the pitching couples only partially so that the yawing couple is not of large magnitude. The engine mountings are so designed that they absorb both *i.e.*, pitching couple and yawing couple.

The second method, which is costly one, is to use, in conjunction with the balance weights, a balancer shaft rotating at crankshaft speed but in opposite direction, to counteract the yawing couples. To keep the weights on the balancer as small as possible, the shaft should be as long as possible. They are usually set at 60° and 240° relative to the crankshaft, so that they apply maximum yawing correction couples when the effective arms of the crank balancer weights are horizontal, and zero yawing couples when they are vertical. The couples due to secondary forces, which add to the primary forces when pistons are at top dead centre and subtract from the primary force when pistons at bottom dead centre, remain unbalanced, regardless of whether a balancer shaft is used or not.

The arrangement of pistons in a three cylinder inline engine is shown in Fig. 4.15. The crank pins are 120° apart. As shown, the piston in cylinder 1 is at the top dead centre. The piston in cylinder 2 is 120° ahead in terms of crankshaft rotation. It is moving downwards and has finished 2/3 of the stroke. The piston in cylinder 3 is again 120° ahead. It is moving upwards and has completed 1/3 of the stroke.

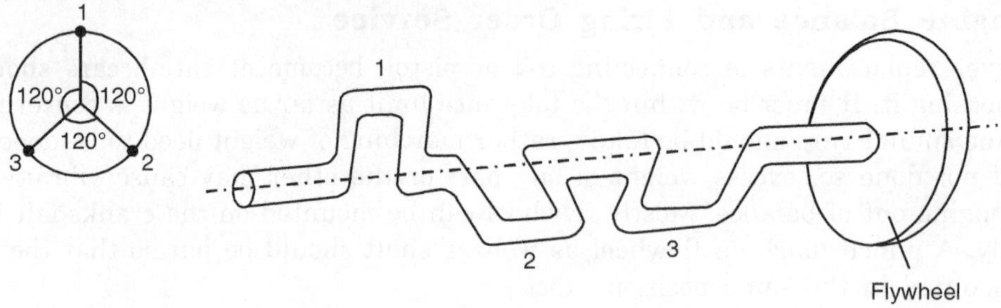

Fig. 4.15

The movement of pistons in the three cylinders is considered. The complete rotational movement of all the three pistons is considered in six different stages. This also gives the couples being generated during the movement of pistons.

(*a*) The upward inertia force due to piston in cylinder 1 is equal to combined downward inertia force due to pistons in cylinder 2, 3 (both being 60° from bottom dead centre). Due to these opposing offset forces along the crankshaft a vertical pitching couple in clockwise direction with front end lifted upward and rear end pressed downwards.

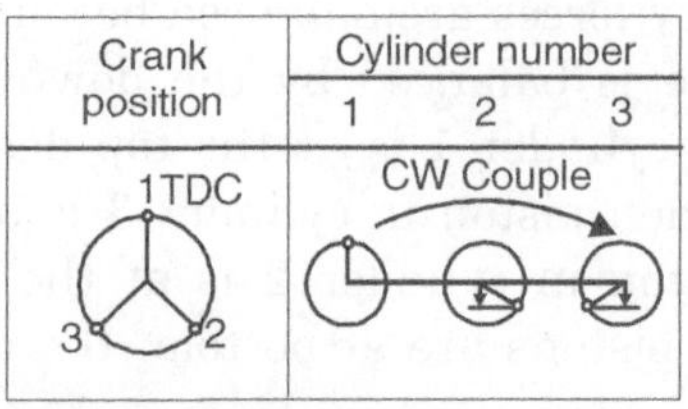

Fig. 4.16

(*b*) Here the piston in cylinder 2 is at bottom dead centre and piston in cylinder 1 and 3 is at 60° from top dead centre upward inertia force components are equal to downwards inertia force produced in there will be no pitching couple.

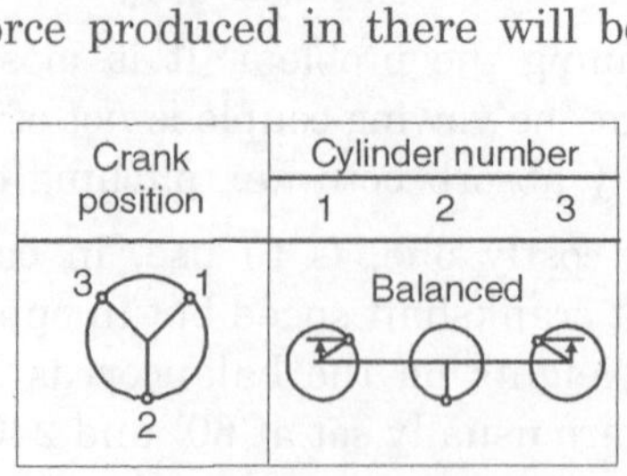

Fig. 4.17

(*c*) Rotating further piston in cylinder 3 reaches at top dead centre, pistons in cylinder 1, 2 move to 60° either side of bottom dead centre. The pistons in cylinder 1, 2 therefore, produce downwards inertia force components equal to that produced by piston in cylinder in 3 upward direction. But as all the pistons are in different planes, an anticlockwise pitching couple is produced. It tends to lift the engine downward at the front and upward at the rear.

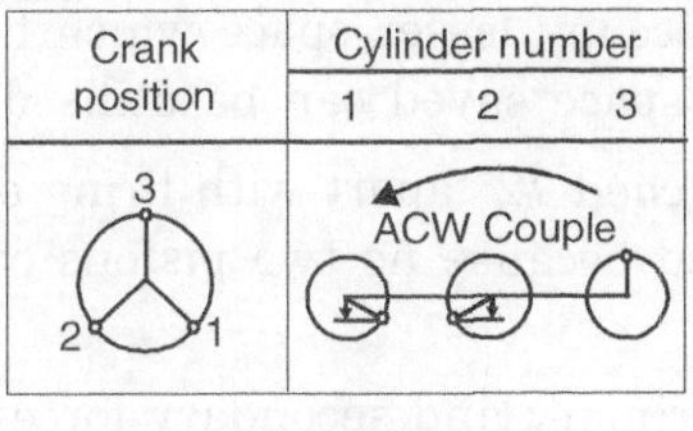

Fig. 4.18

(*d*) The next stage is the piston in cylinder 1 at the bottom dead centre. The piston in cylinder 2 approaching the top dead centre and piston in cylinder 3 moving downwards. Pitching couple is formed as in (a) but in anticlockwise direction.

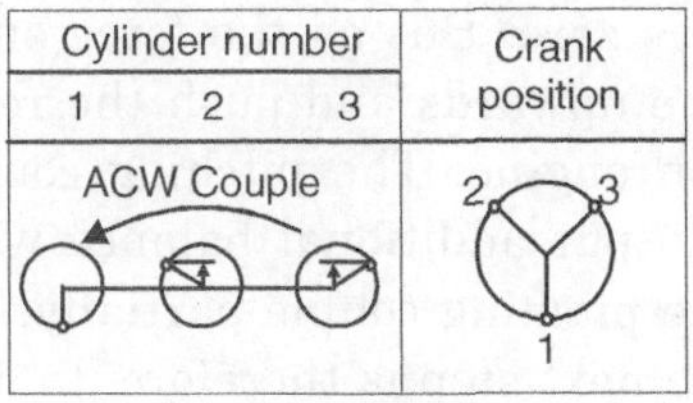

Fig. 4.19

(*e*) Considering further 60° movement of crankshaft. The piston in cylinder 2 reaches at top dead centre. The piston in cylinder 3 is moving downwards and in cylinder-1 is moving upwards. The inertia forces upwards and downwards balance each other and there is no pitching couple.

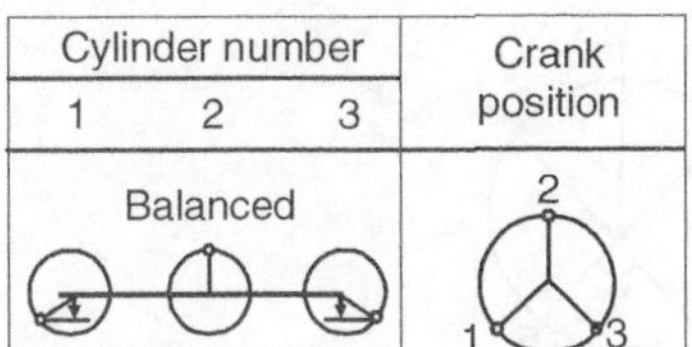

Fig. 4.20

(*f*) The next 60° movement places the piston in cylinder 3 at bottom dead centre and piston in cylinder 1 move upward and in cylinder 2 downwards. This gives rise to clockwise pitching couple due to inertia forces acting in opposite directions.

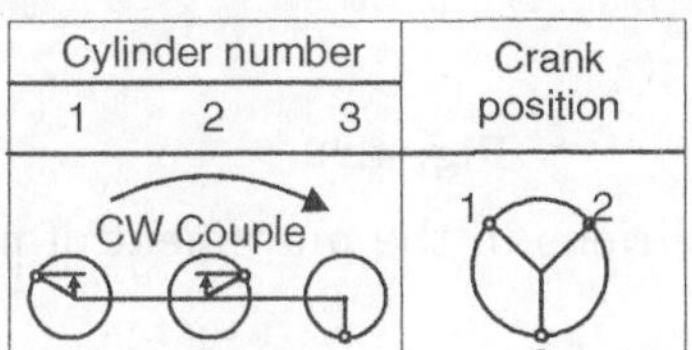

Fig. 4.21

The situations in (*d*), (*e*) and (*f*) are similar to that in (*a*), (*b*) and (*c*) except that directions of the pitching couples are opposite and their effect on engine is also opposite in nature.

4.6 ENGINES WITH FIVE CYLINDERS

A five cylinder engine may be more suitable for buses and coaches as these require lesser power as compared to goods carrying vehicles. They are compact also as compared to six cylinder engines and thus occupy lesser space which becomes useful particularly in a rear engine chasis because the space saved can be utilised to accommodate passengers.

The crankpins are usually pitched 72° apart with firing order of 1-2-4-5-3. This means primary forces are slightly reduced because no two pistons come together to either at the top or bottom dead centre.

In five cylinder engine, the primary and secondary forces are in balance. Considering cylinder 1, where piston is at top dead centre. If the primary force is one, the secondary force is 0.28. Total upward force is 1.28. At the same time, in cylinder 2 and 3 the sum of primary and secondary forces is 1.44 acting downwards. In cylinder, 4 and 5 the sum of primary and secondary forces is 0.16 acting upwards. Summing these together the net force acting on the whole engine is zero. But as the force on cylinder 1 is quite large, it tends to lift the front of the engine upwards and push the rear of the engine downwards. This imposes a pitching couple on engine. The pitching couples are confined to vertical plane, a compromise solution is to put additional balance weights on the front and rear webs of the shaft to counteract the pitching couple partially, but this introduces a yawing couple in the horizontal plane. The next step is therefore, to design engine mountings that will accommodate the yawing motions without transmitting the vibrations to the vehicle structure.

The position of pistons in different cylinder and couples imposed on engine are as discussed below. The crankpins are equally spaced and therefore these are 72° apart (Fig. 4.22).

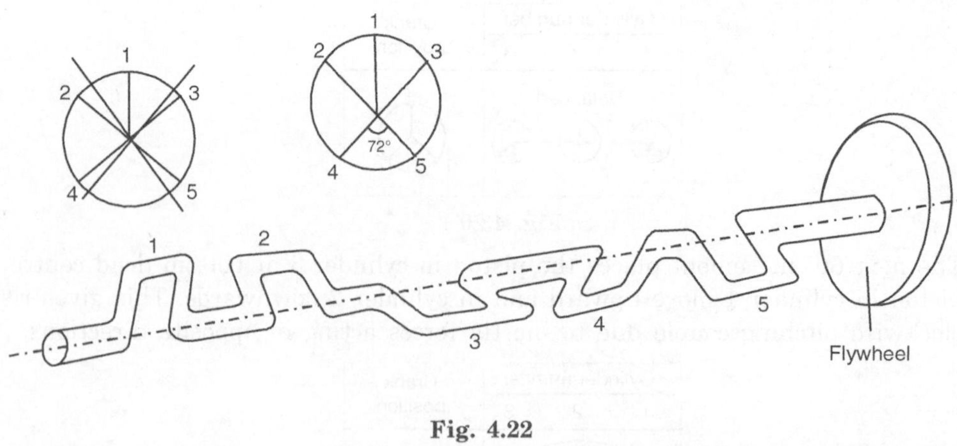

Fig. 4.22

Considering five different conditions in the movement of pin through 360° following are the details of forces and couples.

(*a*) As shown the piston in cylinder 1 is at top dead centre. The crank pin 1 is vertical. The other crank pins are 72° apart and forces and couple on cylinders is as shown below.

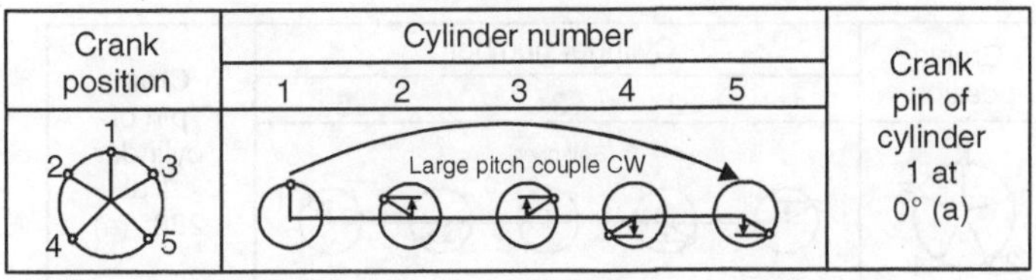

Fig. 4.23

The forces in cylinder 2, 3 and 4, 5 are equal and opposite. The force on cylinder 1 causes a large pitching couple in clockwise direction on the engine due to location of cylinder 1 at extreme end.

(*b*) After the piston in cylinder 1 has moved 72° downwards the forces and couples are as shown below. The piston in cylinder 2 being at top dead centre causes a clockwise couple of small pitch due to location of cylinder 2.

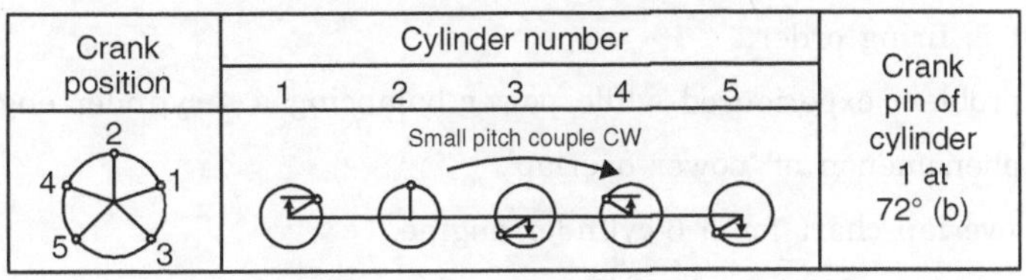

Fig. 4.24

(*c*) The piston in cylinder has moved further by 72° and piston in cylinder 4 is at top dead centre, causing a small pitch couple but in anti clockwise direction as shown in the figure.

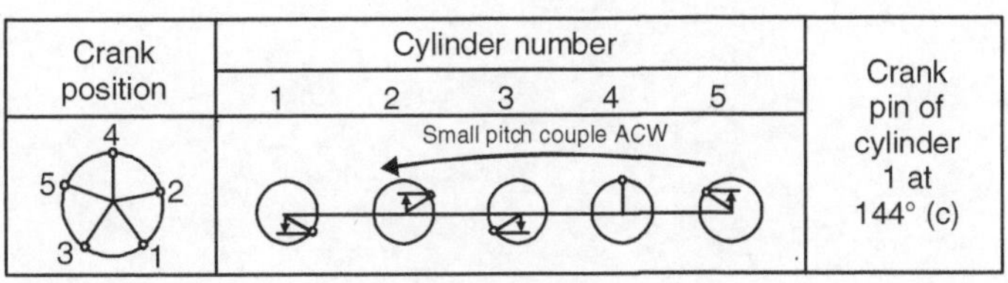

Fig. 4.25

(*d*) The piston in cylinder 1 moves 72° further and is now moving upwards. The piston in cylinder 5 is at top dead centre cause anti clockwise couple as shown.

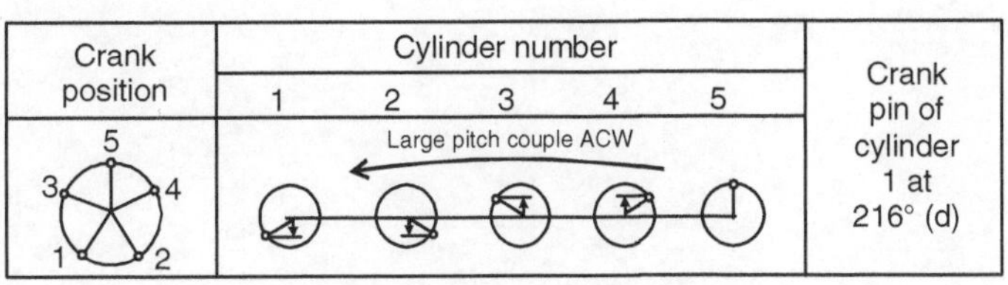

Fig. 4.26

(*e*) The piston in cylinder 1 has moved up further by 72° of crank pin rotation. The piston in cylinder 3 is at top dead centre. The cylinder being located in the middle of engine provides balanced forces and couple as shown.

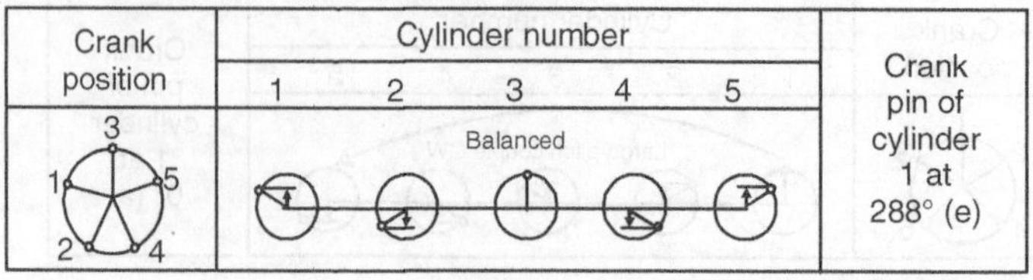

Fig. 4.27

QUESTIONS

1. What is 'mechanical balance' and 'power balance'?
2. How the vibrations are dampened? Explain.
3. What is a viscous damper? Explain.
4. Explain what is firing order.
5. Explain the problem experienced while power balancing a 4-cylinder engine.
6. Explain the phenomenon of 'power overlap'.
7. Draw power overlap chart for a 6-cylinder engine.
8. Explain the working of V-8 engine.
9. Explain balancing of 3 cylinder inline engine.
10. Explain the advantage of 5 cylinder in line engine.
11. Explain the unbalanced forces in the 5 cylinder in line engine.

5

PERFORMANCE OF INTERNAL COMBUSTION ENGINES

Internal Combustion engine is the source of power in an automobile. The performance of automobile is dependent upon the performance of its other components to a large extent. It depends upon the performance of engine. The performance of engine is measured on the basis of input which is given to engine and output that is obtained from the engine. The input to engine is in the form of fuel which is due to its properties containing chemical energy. The combustion of fuel, which is a form of chemical reaction, causes the conversion of chemical energy into thermal energy. As a product of combustion, hot gases are obtained at high pressure which force the piston to move downwards. Thus the thermal energy is converted into work. This motion is transferred to crankshaft and can be considered output of the engine.

The performance of engine is affected by a number of parameters. The parameters may be evaluated inside the cylinder where thermal energy is converted into work. Evaluation of performance on the basis of these parameters may not be true performance of engine. The losses occur as the work output is transferred to crankshaft through connecting rod and other components. The power developed inside the cylinder is termed as indicated power whereas that available at the crankshaft is brake power. The power, earlier known as horse power, is termed as power (in SI units). Some Indian manufactures are using German unit PS (Pferdestärke) which is no longer a legal unit.

The other parameters measured are mean effective pressure, specific fuel consumption, thermal efficiency, mechanical efficiency. Exhaust emission also affects the performance of engine.

5.1 EVALUATION OF PERFORMANCE

To evaluate these parameters it is essential to know about the following terms used in internal combustion engine.

5.1.1 Bore and Stroke

Bore is the diameter of the cylinder measured in millimeters and the stroke is the distance of a piston travels between top dead centre and bottom dead. If the bore of the engine is bigger than its stroke, it is said to be over square *viceversa* it is known as under square. Generally the engines are over square for automobiles. For vehicles such as tractor or trucks under square engines are preferred.

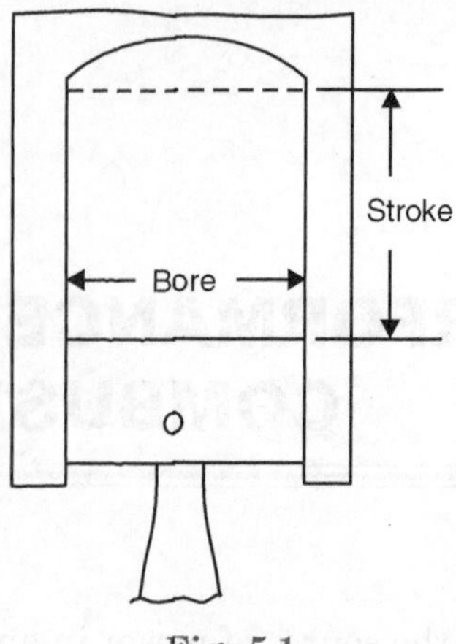

Fig. 5.1

5.1.2 Crank Throw

It is the distance from axis of the crankshafts main bearing to the axis of the crank pin. The stroke of the engine is twice its crank throw.

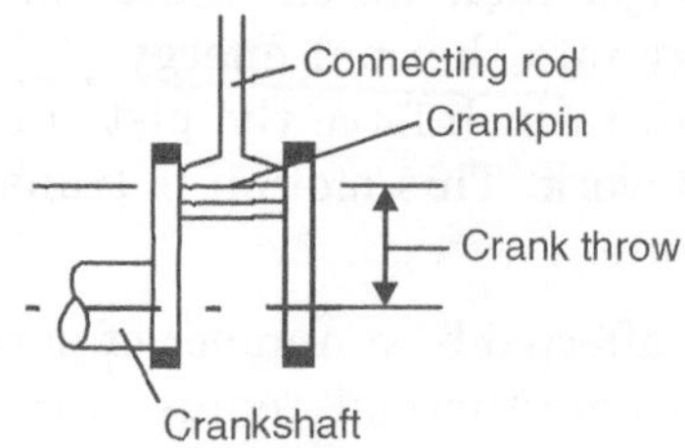

Fig. 5.2

5.1.3 Displacement

It is the volume of cylinder between top dead centre and bottom dead centre. It is usually measured in cubic centimeters or litres. The total displacement of the engine *i.e.*, sum of displacement in all the cylinders indicated the approximate power output from engine.

5.1.4 Compression Ratio

It is the ratio of the volume in the cylinder above the piston when the piston is at the bottom dead centre to the volume in the cylinder above the piston when it is at the top dead centre.

Ideally, higher the compression ratio more power will be produced in an engine. Also, with increase in compression ratio more heat will be produced when charge is compressed. Fuels with low octane rating burn fast and may explode instead of burning when compression ratio is high. This causes preignition which is not a desirable.

5.1.5 Torque

It is the twisting force. The force exerted on the piston is transmitted to the crank pin through connecting rod. The eccentricity between crank pin and crankshaft (described as crank throw previously) cause the torque transmitted to crankshaft when it rotates. This torque is transmitted to the road wheels of the automobile through the transmission system.

5.1.6 Power

It is the rate at which torque is produced. Power is the output from the engine. It is measured inside the cylinder or at the crankshaft. The power inside the cylinder is more than that at crankshaft as some of it is lost in between. The loss of power is due to friction mainly. The power developed inside the cylinder is known as indicated power whereas the power available at crankshaft is known as brake power. Indicated power is more than brake power. The difference between the indicated power and brake power is known as friction power.

5.1.7 Engine Efficiency

It is the ratio of amount of available energy from the engine to the amount of energy put into the engine. It is expressed in percentage and is always less than 100 as output is always less than input.

$$\text{Efficiency} = \frac{\text{Output energy}}{\text{Input energy}} \times 100$$

Efficiency could be mechanical efficiency if output and input is being measured in the form of mechanical energy. The efficiency is less than 100 as there is some loss due to friction. Similarly efficiency could be thermal efficiency when output and input are being measured in the form of thermal energy. Due to combustion of fuel, chemical energy possessed by fuel is converted into thermal energy. This thermal energy is converted into mechanical energy. Due to limitations of system whole of this thermal energy is not converted into mechanical energy rather major part of thermal energy is lost to surroundings or in the exhaust. Therefore the internal combustion engines have quite low thermal efficiency.

5.2 PERFORMANCE CURVES

With the increase in speed of the engine, the mean effective pressure and mechanical efficiency also change. The mean effective pressure at low speed is less than its maximum value due to carburation effects and valve timings. Valve timings, in majority of engines, is designed at a particular speed. Below this speed the mean effective pressure is less. At engine speeds, higher than designed speed, the mean effective pressure again decreases due to lower volumetric efficiency.

The mechanical efficiency of the engine also varies with the speed of the engine. At high speed, due to inertia, high stresses and high bearing loads are set up, which may ultimately lead to fracture or bearing seizure.

When the mean effective pressure falls with rise in engine speed the power output remains constant but when the mean effective pressure falls more rapidly the power output reduces with increase in engine speed.

If brake mean effective pressure is considered the reduction at high engine speed is more significant as shown in the diagram. This is due to reduction in mechanical efficiency because of higher friction losses. The curve representing brake power departs from ideal straight line more rapidly than does the indicated power curve (Points A, B, C, D in the Fig. 5.3). The figure represents a particular engine with 75 mm bore and 120 mm stroke.

As it can be observed that brake mean effective pressure is maximum at engine speed (E). The maximum indicated power occurs at engine speed (F) which is 2.2 times of OE and maximum brake power occures at G which is approximately 2.3 times of OE.

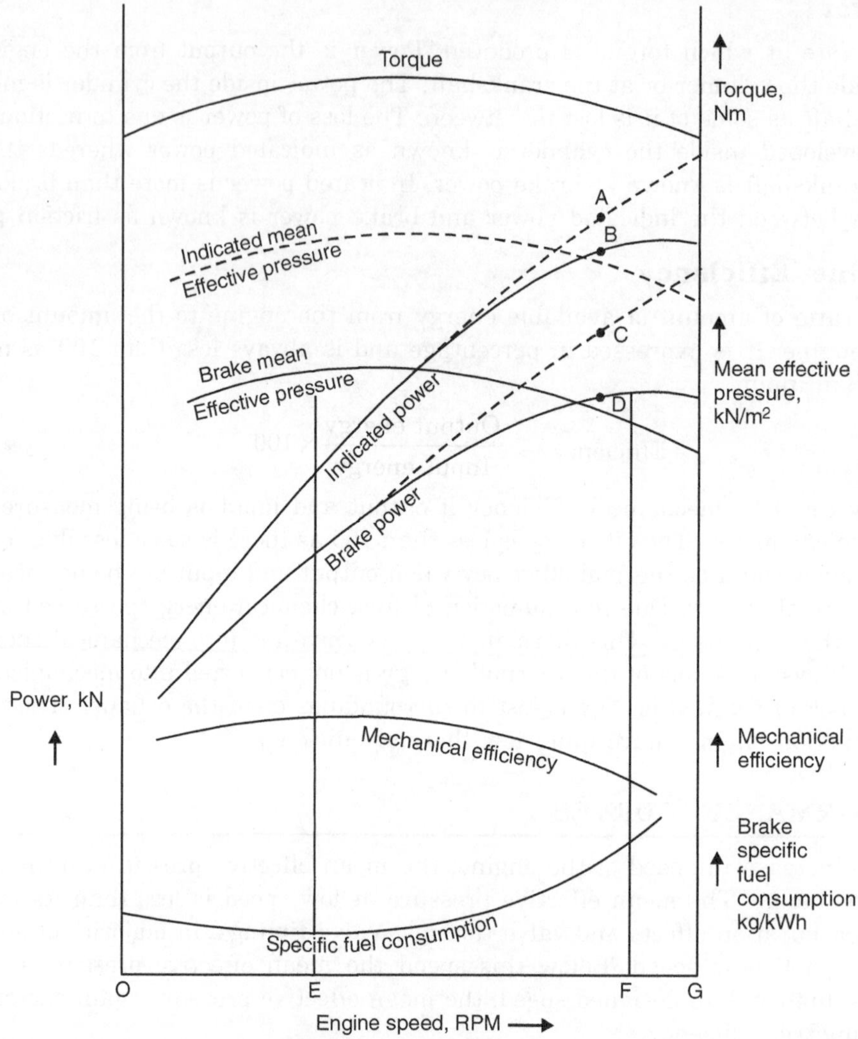

Fig. 5.3 Performance curves.

The curve representing the variation of torque with engine speed can also be drawn. The brake mean effective pressure, *bm ep* = $T \times \frac{4\pi}{V}$ where V is the total stroke volume. The factor, π / V is only a numerical constant for an engine which means that torque curve is similar to brake mean effective pressure curve as shown in the diagram. While drawing the curves full throttle conditions have been assumed.

Brake Specific fuel consumption is the fuel consumed per unit brake power. The quantity of fuel can be represented in volume or in weight. Generally, the quantity is measured by weight because the calorific value of a fuel varies more widely when measured per unit volume. While drawing the curve it has been assumed that the engine is running at maximum load over the whole range of engine speed.

Super charging enhances the power output. This improvement also depends upon degree of supercharging. Let us consider two different cases (Fig. 5.4.).

(*a*) When super charging is barely sufficient to maintain the volumetric efficiency. The brake mean effective pressure and power enhance in the higher range of engine speed and this can be utilised to increase the maximum road speed.

(*b*) When degree of super charging is high the power and brake mean effective pressure are enhanced through the whole range of engine speed. Piston loads and crank shaft torque also increase. These can be reduced by modifying compression ratio and ignition timing. This also has an adverse effect on specific fuel consumption and causes excessive heating which may cause waste heat disposal problem.

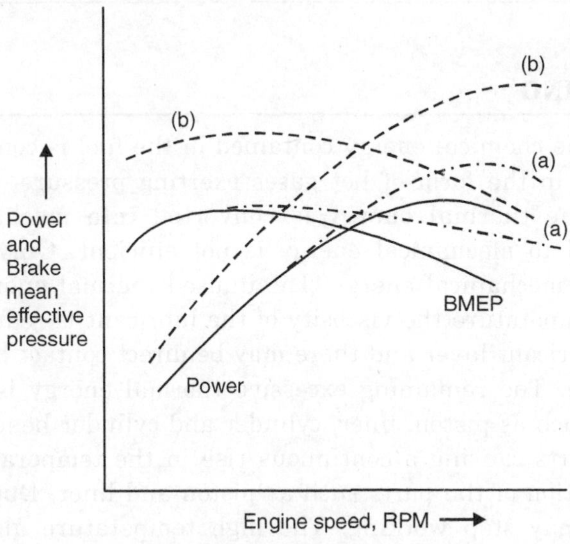

Fig. 5.4 Effect of super charging.

QUESTIONS

1. Explain 'bore' and 'stroke'.
2. What is compression ratio?
3. Write the expression for engine efficiency and explain it.
4. What are performance curves?
5. Explain the 'brake power' 'indicated power' and 'friction power'.
6. Why brake power is less than indicated power?
7. What is specific fuel consumption? Explain.
8. What is the effect of supercharging on brake power.

6

COOLING AND LUBRICATION

6.1 ENGINE COOLING

Upon combustion, the chemical energy contained in the fuel is converted into thermal energy. Thermal energy in the form of hot gases exerting pressure on piston causes its movement. This way the thermal energy is converted into mechanical energy. The conversion from thermal to mechanical energy is not efficient. Only a part of thermal energy is converted into mechanical energy. The unused thermal energy causes a rise in temperature. At high temperature the viscosity of the lubricant is reduced. The change in viscosity deforms the lubricant layer and there may be direct contact between metal parts causing damage to them. The remaining excessive thermal energy is transferred to the adjoining engine parts such as piston, liner, cylinder and cylinder head. The heat may get accumulated in these parts causing a continuous rise in the temperature of these parts. This may cause deformation of the parts such as piston and liner. Due to this, the piston may seize and engine may stop working. The high temperature also causes excessive stress in parts reducing their life. Some heat may be carried away by lubricating oil and some may get lost to the surroundings through radiation. But heat taken by lubricating oil and that lost to surroundings is only about 12% of total heat supplied. A part of heat is taken away by exhaust gases. This may be about 30% in case of four stroke engine. Useful conversion into mechanical energy is about 28%. Hence about 30% of unused heat remains inside the engine. To avoid accumulation of this heat in the different parts it becomes essential to dissipate it efficiently and continuously. To achieve this, an engine cooling system is employed.

The transfer of heat occurs from higher to lower temperature. During intake stroke, the fresh charge enters the cylinder at lower temperature. The temperature inside the cylinder is high due to outgoing exhaust gases. Considering a particular case of four stroke engine, the temperature may get reduced to below 500 K from more than 1000 K. During compression stroke the temperature starts rising. During combustion there is sudden rise in temperature and it may attain a value of around 2000 K. During expansion thermal energy is converted into mechanical energy. This causes a reduction in temperature to about 1200 K. The gases go out during exhaust stroke at about 1200 K. The heat retained inside the cylinder affects most the parts like liner, piston and rings and cylinder head. The cylinder head contains intake, exhaust ports and valves therefore these are also affected.

6.2 COOLING SYSTEMS

The heat that remains unused in an engine affects adversely its working. Therefore it is essential that this heat is dissipated properly. To achieve this, cooling system is provided in the engine. The cooling system can use water/coolant or air for cooling purpose.

6.3 COOLING SYSTEM WITH WATER/COOLANT

This is an old and reliable cooling system. In recent past, chemical compounds, commonly known as coolant have replaced water. Water or coolant is circulated around the cylinder where it absorbs the heat. The heat is taken away by water or coolant as it moves and is replaced by fresh water or coolant. The water or coolant is made to flow through radiator. The heat absorbed by the water or coolant is transferred to atmosphere and water or coolant return back to atmospheric temperature. The cooled water or coolant is sent back to engine again to absorb heat. The cycle is repeated again and again with the water or coolant taking away the heat from engine parts continuously. The system is so designed that it is capable of keeping the temperature of engine parts within permissible limits. Components of the system are as described below.

6.3.1 Coolant

Water was being used for a long time as coolant. Its evaporation rate being high, particularly in summers, required frequent filling of radiator. In recent past, chemical compounds have been introduced as coolants. One such compound being used is ethylene glycol-based coolant. This acts as anti-freeze agent also. When this compound is mixed with water (in ratio 2:1) the mixture has a higher boiling point (113°C) and lower freezing point (–69°C). Generally a 50:50 mixture is used that gives the best results. The anti-freeze compound must not be less that 44%. If that happens, the engine parts may get eroded. Corrosion may severely damage the cooling system components. Also if water is not mixed this compound may cause serious damage to engine parts as tremendous heat would be produced.

6.3.2 Expansion Tank

This tank holds coolant that passes through the pressure cap when the engine is hot. As the engine attains a higher temperature, the coolant expands and causes the release of pressure cap. The coolant passes through the expansion tank. When engine is stopped, its temperature is reduced and coolant from expansion tank is drawn back into cooling system.

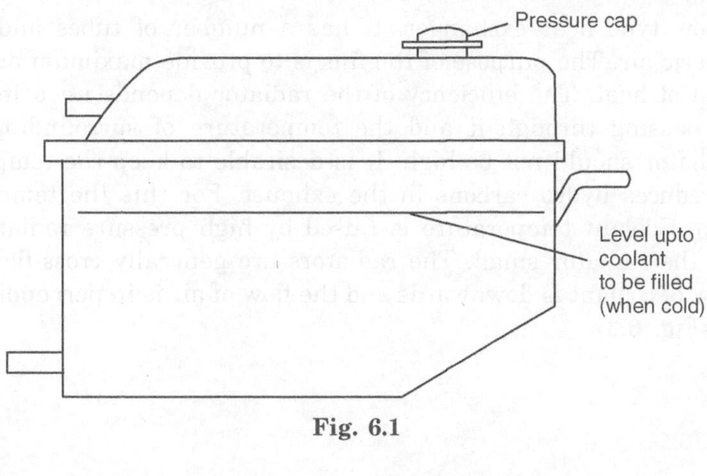

Fig. 6.1

6.3.3 Pump

The function of pump is to circulate the coolant through cooling system. It is a centrifugal pump. To run the pump, the output from the engine is used. The crankshaft is connected to pump through pulleys and v-belts. It has impeller at one end of the shaft. The shaft is mounted in the housing. Bearing is provided for support and smooth movement. At the other end of the shaft pulley is provided that connects the pump to crankshaft through belt. Proper seals are provided to prevent any leakage.

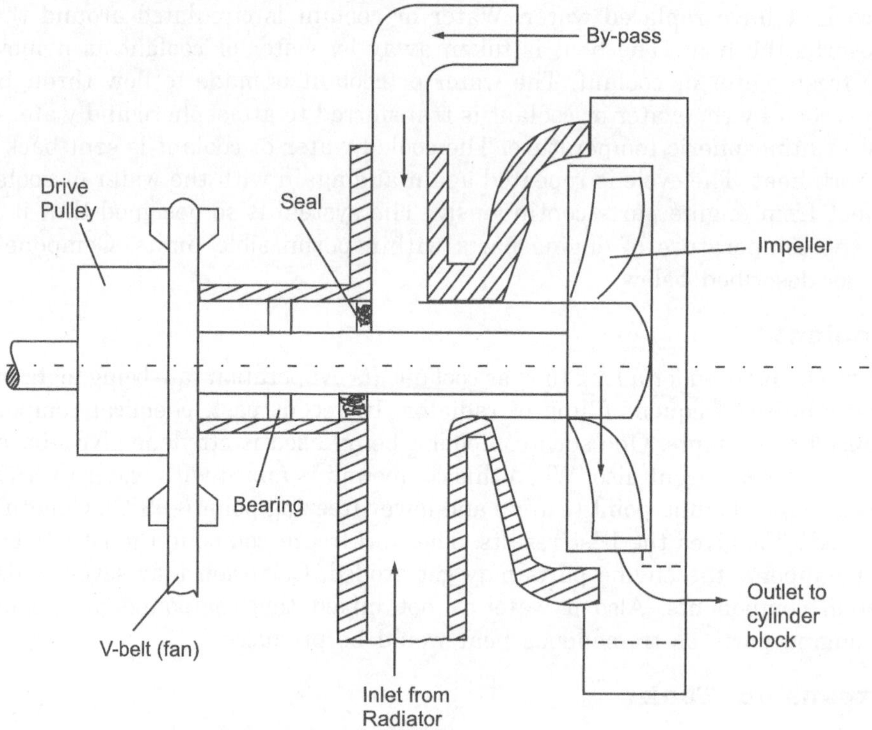

Fig. 6.2

6.3.4 Radiator

It is a cross-flow type heat exchanger. It has a number of tubes and fins that are exposed to atmospheric air. The purpose of the fins is to provide maximum area of exposure for quick dissipation of heat. The efficiency of the radiator depends upon its basic design, amount of coolant passing through it and the temperature of surrounding air. But the efficiency of the radiator should not be high. It is desirable to keep the temperature of the engine high. This reduces hydro carbons in the exhaust. For this the temperature of the coolant is kept high. Coolant temperature is raised by high pressure radiator cap and by keeping the size of the radiator small. The radiators are generally cross-flow type of heat exchangers. The flow of coolant is downwards and the flow of air is in perpendicular direction across the coolant (Fig. 6.3).

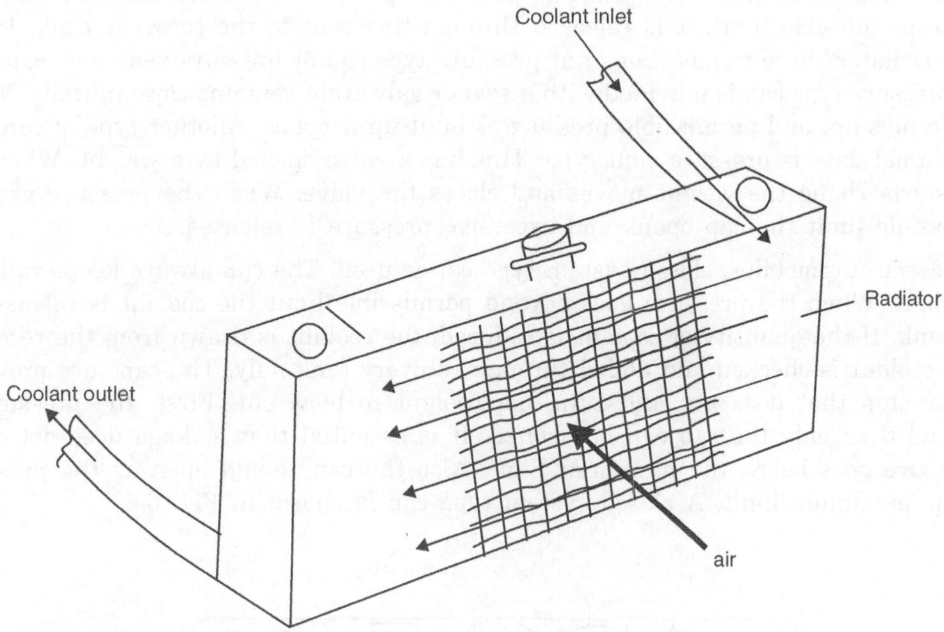

Fig. 6.3

The radiators are provided with plugs at the bottom to drain out the coolant. After the coolant has completed its life and has become ineffective it is to be replaced with fresh coolant. Fresh coolant is added through the radiator cap. Sometimes recovery tank is provided which provides fresh coolant (Fig. 6.4).

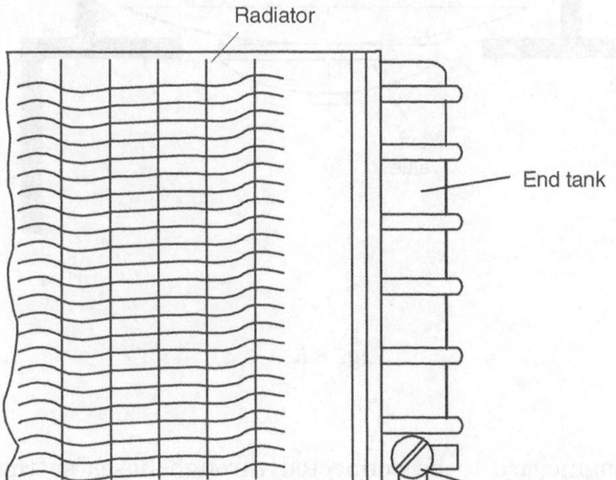

Fig. 6.4

6.3.5 Pressure Cap

Boiling point of a substance can be raised by increasing the pressure. It is desirable that the boiling point of coolant is raised. To achieve it, pressure cap is used to cover the radiator. The cap is provided with vent and pressure spring. When pressure in the radiator crosses the permissible limit, it is released through this vent to the recovery tank. In old times, the radiators used to have constant pressure type cap or pressure vent type cap. The constant pressure type cap is provided with a seal or valve that remains close initially. When coolant becomes hot and permissible pressure is built up it opens. Another type of cap that was used in old days is pressure vent type. This has a valve opened by a weight. When the pressure starts rising the weight moves and closes the valve. When the pressure exceeds the permissible limit the cap opens and excessive pressure is released.

In present automobiles, closed system type cap is used. The cap always keeps radiator full of coolant. When the pressure goes beyond permissible limit the coolant is released to recovery tank. If the quantity of coolant is reduced, the coolant is drawn from the recovery tank. The coolant is checked and added through recovery tank only. The caps are provided with safety stop that does not allow the hot coolant to blow out. First, the pressure is released and then only the cap can be opened. It is essential that leakage does not occur when pressure goes below the minimum limit. Also the cap should open as the pressure exceeds the maximum limit. A closed system type cap is shown in Fig. 6.5.

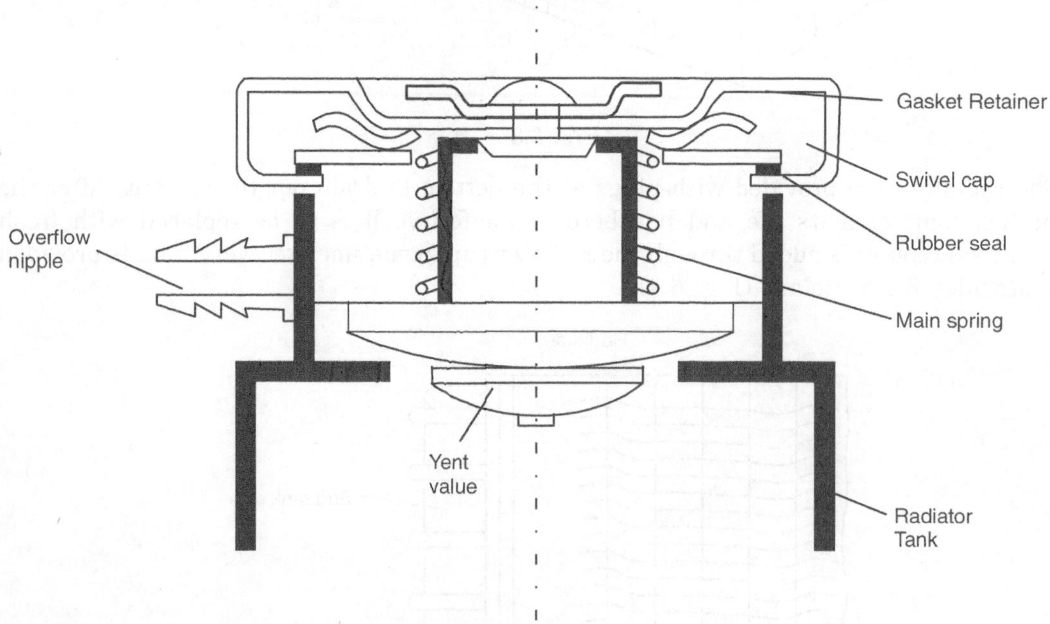

Fig. 6.5

6.3.6 Hoses

The radiator and engine are to be connected through hoses so that coolant can flow from engine to radiator. Normally, there are two heater hoses and an upper and lower radiator hose. A by-pass hose is also provided in some systems. The hoses are made up of butyl rubber. These are provided with wire reinforcement to provide strength. The hoses are provided with expansion bends. The upper hose is exposed to coolant at highest temperature. Also, it absorbs maximum vibrations and is exposed to toughest working conditions.

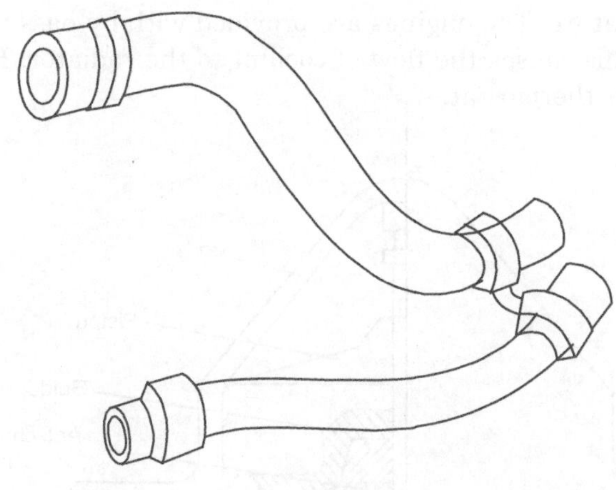

Fig. 6.6

The hoses get deteriorated from inside and pieces of rubber accumulate in the radiator core and cause clogging. The hoses become hard or spongy towards the end of their life. The hoses are clamped at the outlet and inlet of the radiator, pump and heater. Some of the clamps that are used are worm gear type, spring type or twin wire type (Fig. 6.7).

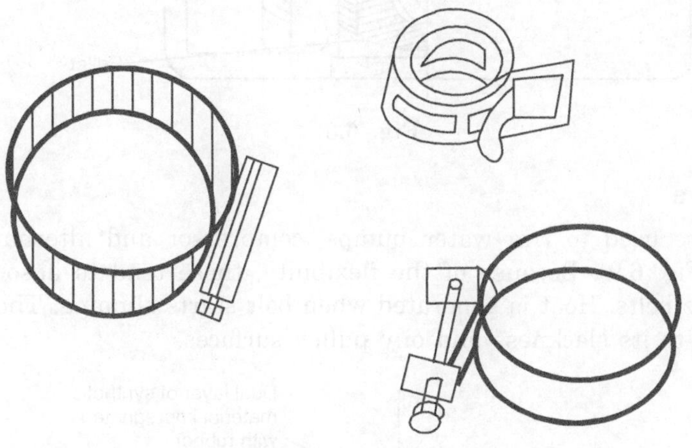

Fig. 6.7

6.3.7 Thermostat

Thermostat consists of wax and powdered metal pellet. These are accommodated in a copper cup with a piston. The pellet expands with rise in temperature. This causes outward movement of piston which in turn opens the valve of the thermostat. The pellet can sense the change in temperature and accordingly opens and closes the valve to control the flow of coolant and its temperature. The flow of coolant can also be slowed down.

Thermostat is generally located at the front top of the engine. The pellet is exposed to hot coolant and its top is covered by housing. The thermostat causes fast warming up of engine. It is essential to avoid the condensation in the combustion chamber. The condensation that occurs in combustion chamber causes sludge formation in the crank case. For efficient working of engine the coolant is kept above a particular temperature. Thermostat keeps the coolant above this temperature. The engines are provided with by-pass valve that closes, the by-pass after warm up. This causes the flow of coolant to the radiator. Figure 6.8 shows the constructional details of a thermostat.

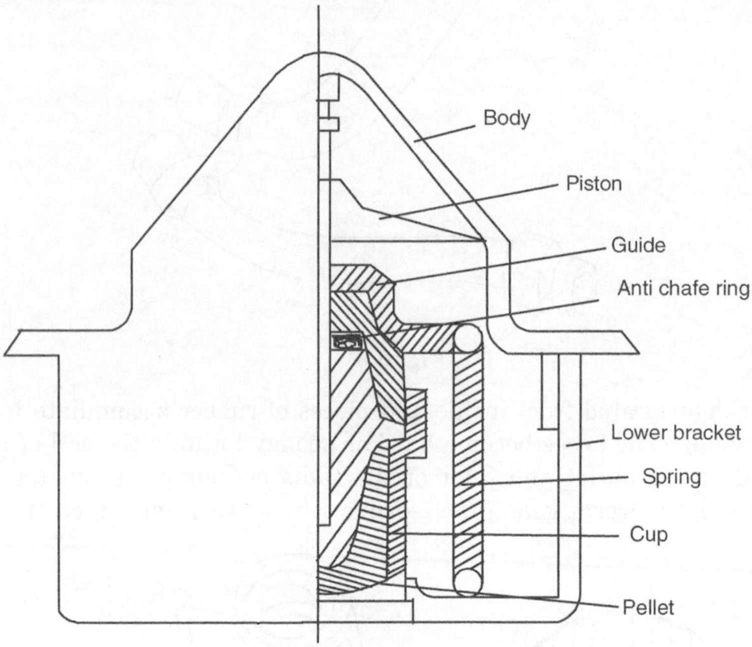

Fig. 6.8

6.3.8 Belt Drive

Belts drives are used to run water pumps, compressor and alternators. These are generally V-belts (Fig. 6.9). Because of the flexibility, these tend to absorb shocks. Heat adversely affects the belts. Heat is generated when belt starts slipping. The slipping of the belt may occur due to its slackness and oily pulley surfaces.

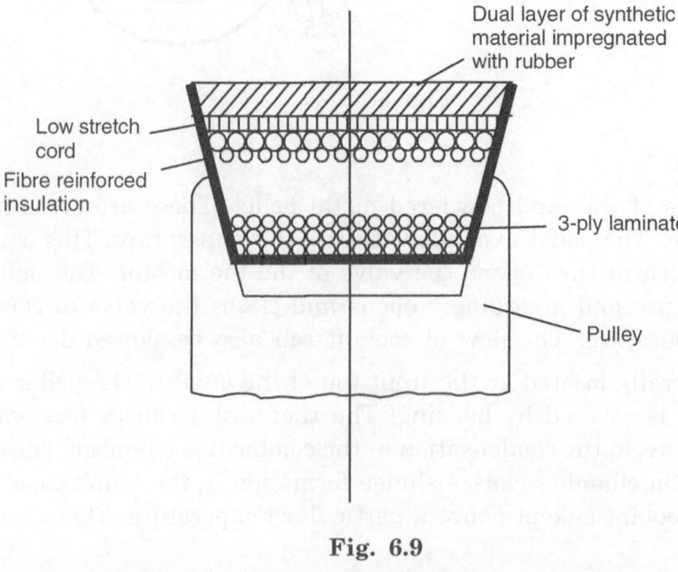

Fig. 6.9

6.3.9 Fan

The cooling system essentially removes the heat from the engine and dissipates it. For this purpose circulation of air is required. When a vehicle moves at high speed, the air passes through the radiator and is sufficient to dissipate heat. But at low speed or when the vehicle is stationary the air is to be provided through a fan. The size of the fan, its pitch and number of blades depend upon the air requirement. The fan can be two in numbers sometimes depending upon the cooling needed. The fans are masked and placed slightly away from the radiator. The fan is balanced properly and is made of steel generally. But now day's synthetic materials such as nylon and fibre-glass are also used. The fans use the output from the engine to run. Alternatively, the fan can be run on battery also. The fan can be switched off automatically when coolant temperature is low and heat to be dissipated is less. This is possible through the sensors that can sense the coolant temperature. The fans design has been improved by using flexible blades. These blades bend and pitch is reduced with increase in fan speed (Fig. 6.10).

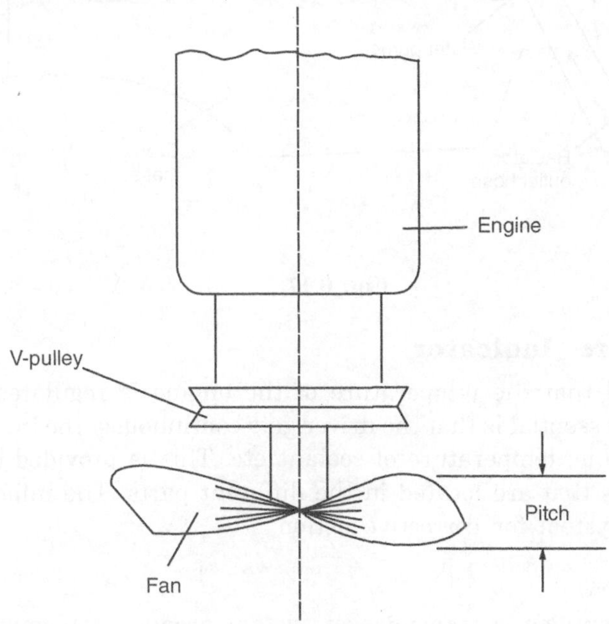

Fig. 6.10

6.3.10 Water Jacket

The cylinder block and cylinder head of the engine is provided with hollow space that acts as water jacket. The cylinder block surrounds the cylinders of the engine and cylinder head takes care of the top most part of the cylinder where combustion chamber is located. These parts are worst affected due to unused heat. The jacket is provided with inlet and outlet so that water can enter. After moving through the jacket and taking away the unused heat the water moves out through the outlet.

The cooling system representing the water jacket radiator, fan and other components is shown in Fig. 6.11.

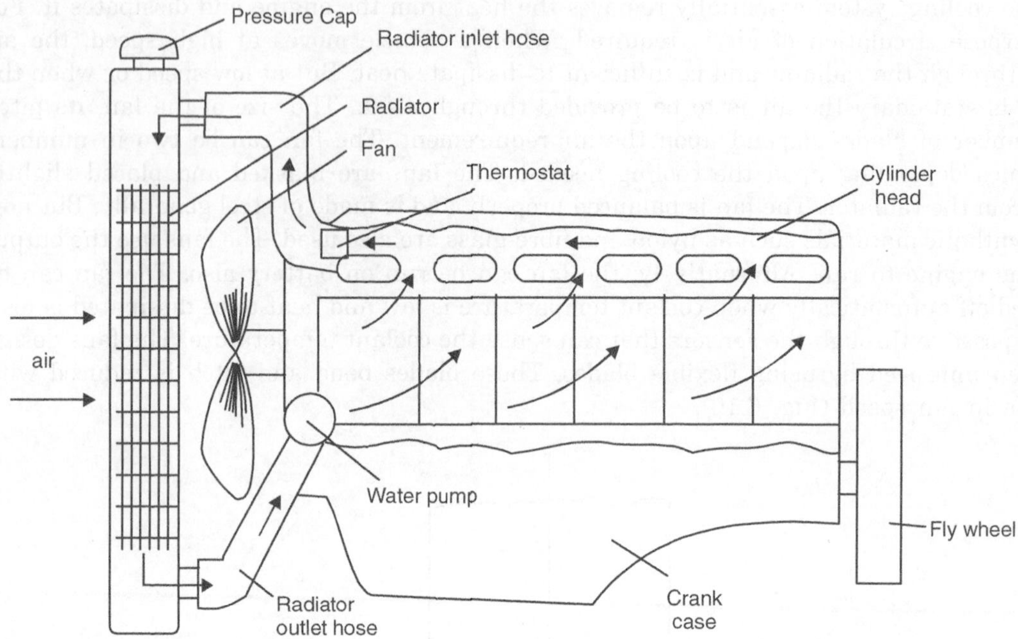

Fig. 6.11

6.3.11 Temperature Indicator

It is very essential that the temperature of the engine is regulated with the help of cooling system. Equally essential is that the driver gets continuously the information regarding the temperature of engine, temperature of coolant etc. This is provided on the dash board with the help of sensors that are located in the different parts. The information is also fed to the engine control system for corrective action.

6.3.12 Oil Cooler

Automobiles with automatic transmission system produce transmission fluid at high temperature. The temperature of this fluid is reduced by passing through sealed radiator located in the coolant tank. Transmission fluid is circulated back after its temperature is reduced.

6.4 AIR COOLING SYSTEM

Air is used as medium of cooling in this type of systems. The engine is kept open to atmospheric air. The air moving over the engine takes away the heat. To make the dissipation more effective area of heat transfer is increased. This is done by providing fins on the outer surface of cylinder wall (Fig. 6.12). The fins are at right angle to the cylinder axis. Small engines, particularly those used in motorcycles, are air cooled. Also the engines used in tractors are air cooled. Very few automobiles also use as an air cooled engine.

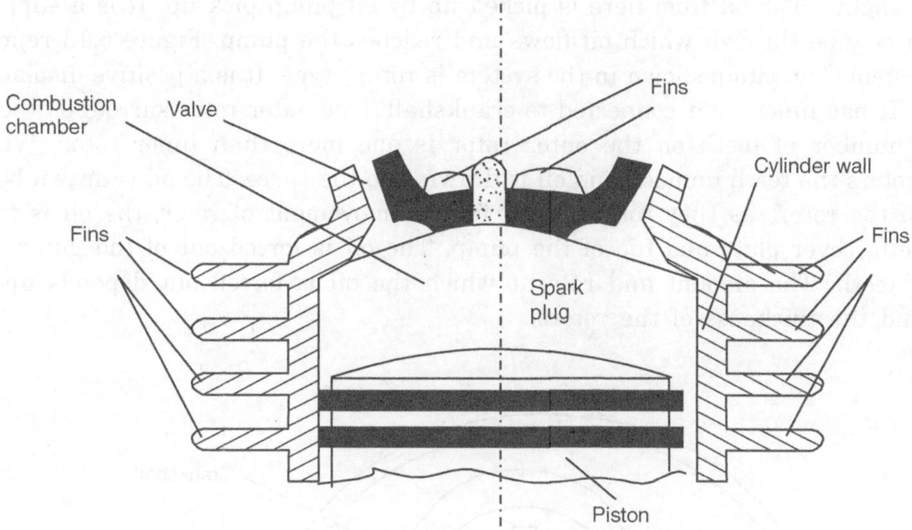

Fig. 6.12

6.5 ENGINE LUBRICATION

The engine of an automobile has several moving parts. These parts may have rotary as well as linear motion. Also the moving parts are in contact with each other. This causes friction between these parts. The friction tends to resist the motion of these parts. To overcome friction more input is needed that reduces the efficiency of the engine. Therefore, it becomes essential to remove the friction though practically it is not possible. The friction can be reduced to minimum level only by using lubricants. Engine lubrication system provides lubricating oil to all the moving parts of the engine. Lubricating oil has certain properties that enables it form a thin film between the surfaces so that direct frictional contact between them is avoided.

6.5.1 Lubrication System

It provides lubricating oil to all the moving parts of the engine. The lubricating oil is provided in the main bearing supporting rotating shafts, between piston rings and the cylinder walls, in the camshaft, crankshaft, valves and piston pin and several other parts where moving surfaces are in contact with each other. Lubricating system has the following main components:

Oil pump

The oil pump picks up oil from the oil pan and circulates it through different passages provided in the engine. It is located in the oil pan. It is essential that it delivers adequate amount of oil so that lubrication is proper. The flow of oil depends upon the viscosity of the oil. It is essential that properties of the lubricating oil are not adversely affected in hot environment of the engine.

Oil pan or sump acts as a reservoir for the lubricating oil. It is located in the lower most part of the engine. The oil from here is picked up by oil pump pick up. It is a supply line in the form of tube through which oil flows and reaches the pump. Figure 6.13 represents one such system. The pump shown in the system is rotary type. It is a positive displacement type pump. It has inner rotor connected to crankshaft. The outer rotor surrounds the inner rotor. The number of teeth on the outer rotor is one more than inner rotor. With the turning of rotors the teeth unmesh and oil is drawn into the space. The oil is drawn between the teeth of the rotors as they rotate. With further movement of rotor, the oil is trapped between teeth, cover plate and top of the pump. The oil is forced out of the pump by the meshing of teeth. The amount and rate at which the oil is forced out depends upon the diameter and the thickness of the rotors.

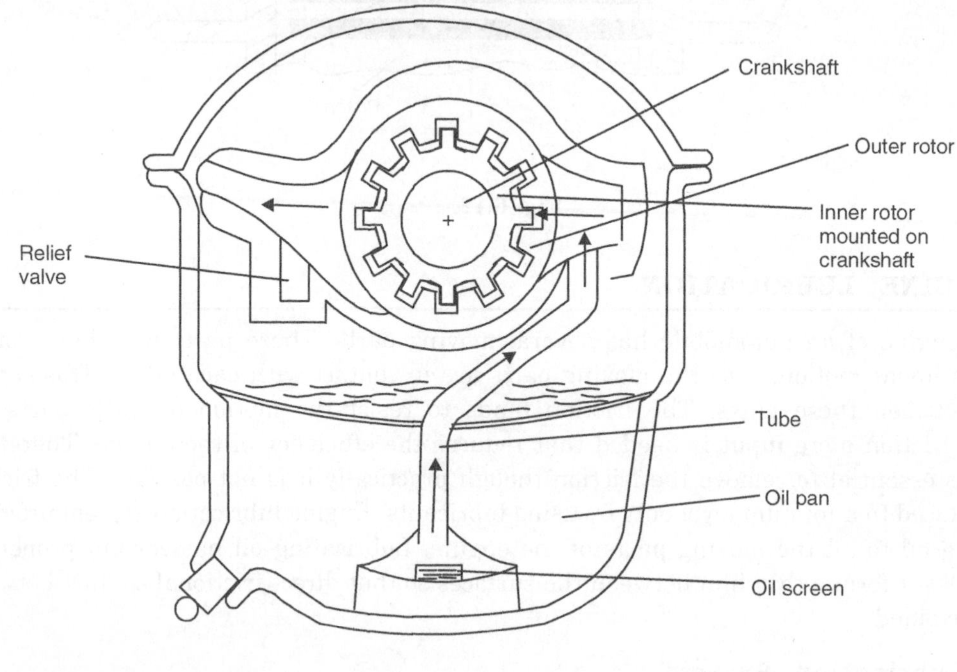

Fig. 6.13

Since the pump is positive displacement type, to avoid excessive pressure, pressure relief valve is provided. It is a spring loaded valve. It opens at the limiting pressure and allows the excess oil to go back to sump.

Another type of pump used is gear type. There is a pair of gear meshing. The oil is trapped between their teeth and the pump wall. The oil enters, moves between the teeth of meshing gears and comes out at the other end when teeth unmesh. The output volume of the oil depends upon length and depth of the teeth. The pump may be driven by the crankshaft directly or through the camshaft (Fig. 6.14).

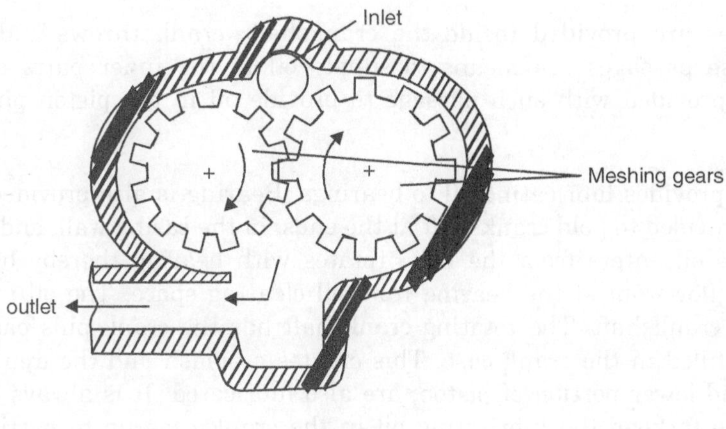

Fig. 6.14

Oil filter

The oil is made to pass through the filter as it comes out of the pump. This helps to retain any insoluble impurity. The oil that goes to engine must not have any impurity as that can damage the engine. Clogging of the filter may occur after continuous use. To avoid it, filter should be cleaned periodically. There is a valve provided in the filter. This valve prevents the drainage of oil when engine is not running. When the engine is started, the pressure builds up and the filter permits the supply of oil. Figure 6.15 represents an oil filter.

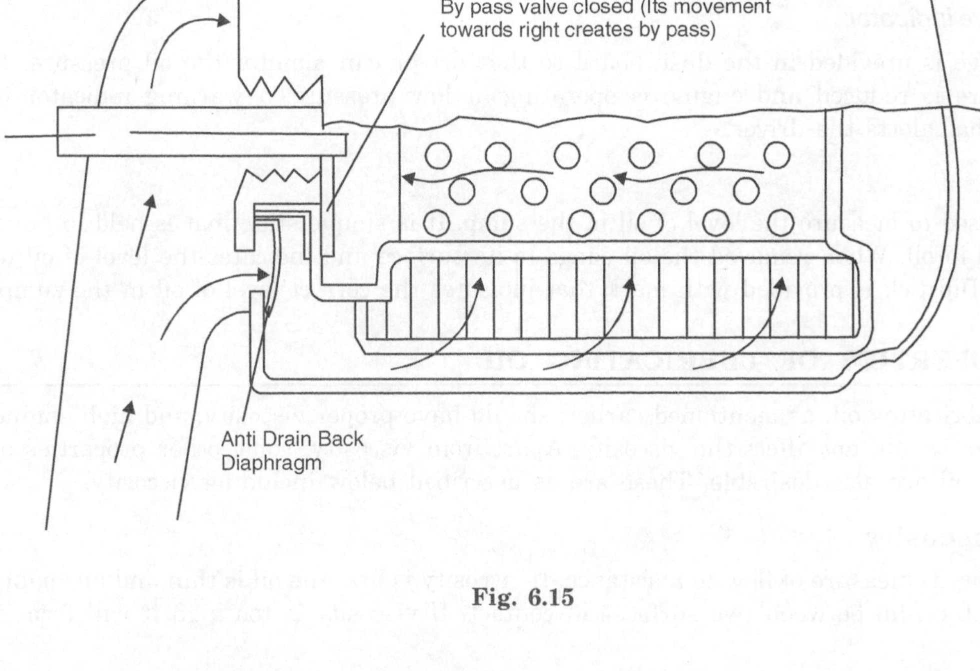

Fig. 6.15

Oil passages

These passages are provided inside the crankshaft, crank throws and through the engine block. These passages are means to supply oil to the inner parts of the engine. Connecting rod is provided with such passage to provide oil in the piston pin and rings.

Bearings

An oil gallery provides lubricating oil to bearings. Bearing is also provided with a hole. The bearings are provided to hold crankshaft at the ends, in the casing wall, and in connecting rod. In bearing the oil enters from the hole, rotates with bearing thereby lubricating the surfaces, and then flows out of the bearing from oil clearing space. The oil is then thrown off by the rotating crankshaft. The rotating crankshaft has the crank pins passing through the lubricating oil filled in the crank case. This creates a splash and the upper parts, such as cylinder wall and lower portion of piston, are also lubricated. It is always recommended by the manufacture to keep the lubricating oil in the crank case up to particular level so that splashing occurs.

Oil seals

The oil seals are used to prevent the leakage of oil. These are provided at the shaft ends where they are supported in the crank case walls. Also seals may be used to prevent internal leakage. These are made of rubber, plastic, cork and fibre. Gaskets are used that act as oil seals. These gaskets are provided between the lower part and the lid.

Oil cooler

Lubricating oil must not attain high temperature. If the temperature goes beyond 120°C oil gets oxidized and breaks into carbon and varnish. The higher the temperature, the oxidation occurs at faster rate. Therefore, it is essential to provide oil cooler if the temperature is likely to go above this limit. It is a small radiator that is mounted in the front of the engine. This has air as coolant and keeps the oil at low temperature.

Oil pressure indicator

A gauge is provided in the dash board so that driver can monitor the oil pressure. If the pressure is reduced and engine is operating at low pressure, a warning indicator is provided that alerts the driver.

Dipstick

It is used to measure the level of oil in the sump. It is simple stick that is held in hand and dipped in oil. When removed the oil clings to its surface and indicates the level of oil in the sump. Dipstick is provided with mark that indicates the correct level of oil in the sump.

6.6 PROPERTIES OF LUBRICATING OIL

The lubricating oil, as mentioned earlier, should have proper viscosity, and high engine temperature should not affect the viscosity. Apart from viscosity, some other properties of lubricating oil are also desirable. These are as described below including viscosity:

6.6.1 Viscosity

Viscosity is measure of flow to resistance. If viscosity is low, the oil is thin and incapable of forming the film between two surfaces in contact. If viscosity is too high it will form a film that itself would produce resistance (and thus friction) to motion. Therefore, it is very essential to use lubricating oil with suitable viscosity. Viscosity is measured on a scale from 0 to 100 accepted as standard by SAE. Accordingly SAE numbers have been provided to lubricating oils from 0 to 100. Suitable oil is recommended by the manufacturer and that should be followed. Now-a-days multi-grade lubricating oils have been introduced. The viscosity of these oils remains relatively unchanged in a particular range of temperature.

Resistance to oxidation and carbon formation: The lubricating oils should not get oxidized at high temperature. Also carbon formation should not occur at high temperature.

Resistance to corrosion: The lubricating oils should have anti corrosion properties. These should not allow occurrence of corrosion in the engine parts.

Resistance to foaming: The movement of crankshaft in the oil to produce splash may cause the production of foam in the oil. This affects the lubricating properties of the oil. Due to foaming volume increases and overflow may occur. Baffle in the sump is provided to prevent foaming.

Pressure resistance: This property is essential to prevent breakage of oil film under extremely high pressure. The oil resists penetration and squeezing of film.

Energy conservation: This property reduces the consumption of fuel and provides better fuel economy. Additives, soluble and insoluble, are mixed in the oil. These additives are capable of altering the properties of oil.

These properties in the lubricating oil are achieved by mixing suitable additives. These additives are chemical compounds that are capable of imparting these properties in the oil. In recent times, some synthetic lubricating oils have also been introduced. These have better heat resistance, produce smaller amount of sledge and have longer life.

QUESTIONS

1. Explain the necessity of engine cooling.
2. What are different cooling systems employed in an engine? Explain briefly.
3. What is radiator? On which factor the efficiency of radiator depends?
4. Explain the functions of pressure cap.
5. Explain the function of thermostat.
6. What are the functions of water jacket?
7. What are the advantages of air cooling system?
8. Explain the importance of engine lubrication.
9. What are the functions of oil filter? Explain.
10. Why an oil cooler is needed in lubricating system? Explain.
11. Explain briefly the properties of the lubricating oils.
12. How are properties in the lubricating oils achieved?

7

ENGINE FUELS

The energy input in an internal combustion engine is the chemical energy possessed in the fuel. This is naturally provided in the fuel. Combustibility is another property naturally provided to fuels. Combustion of fuel causes conversion of chemical energy into thermal energy. Thermal energy is available which is further converted into mechanical energy that is the output from the internal combustion engine.

Almost all of the fuels for internal combustion engine are derived from petroleum, which is a complex combination of hydro carbon compounds. Petroleum is available from beneath the surface of the earth and is a product of process that has undergone beneath the surface of the earth for several hundred years. Unfortunately, it is not available uniformly throughout the earth. It is accumulated in some parts of the earth and some parts are totally deprived of it. Another important thing is that the quantity of fuel available from beneath the surface of the earth is limited.

7.1 THE NATURAL FUELS

The origin and subsequent evolution of petroleum in liquid and gaseous, beneath the surface of the earth, is not known exactly. However, petroleum is found in certain rock formations. These rock formations were the floors of ocean thousands of years back. It is believed that marine organic material on the bottom of the sea was enfolded by the layers of rock. Under these layers of rock, it was subjected to high pressure and a temperature of around 150°C. It remained under these conditions for a period of 1–2 million years. It cracked to a lower molecular weight bituminous material. The heavy oils or tar sands were first formed. These contained few light hydrocarbons, the younger rock formations contained heavier oils with little or no light portions and with appreciable amount of compounds containing Nitrogen and Oxygen. Nitrogen was eliminated later though it is not yet known how that happened. Some think that it was due to bacterial action while some consider it happened due to slight radioactive exposure that continued for ages from the neighbouring rocks.

Coal was produced by the similar process. The only difference here was that the original organic material was land vegetation. It is quite possible that oil distillated from the same vegetation and migrated to other locations.

The crude petroleum is found accumulated in porous rocks or sands or limestone and covered with a rock cap through which it could not pass. These underground rock traps also contained large amount of natural gas and salt water. It is believed that salt water has been retained in the pores of rocks from the time when organic matter was first deposited. 'Tar sand' deposits are also found consisting of common sand bonded together by a viscous tar of petroleum origin. As slate is also present, an 'oil shale' is formed that is a hard rock formation. These formations have veins or strata of an organic matter called 'Kerogen'. The Kerogen consists of 40% hydrocarbon and remaining are the compounds of Nitrogen, Sulphur and Oxygen.

Some earthquakes or other natural changes that occurred from time to time might have caused uplifting of land and exposure of buried oil. Natural refining or weathering took place and oil was converted into 'Asphalt'—the name given to a heavy, black, tar like substance of certain chemical structure but with relatively small amount of Hydrogen.

Geological evidences indicate that about 25 million square kilometers of the earth's surface, which was once under water, probably contained oil. Most oil originated in the shallow waters of land locked seas. This was because that shallow waters contained abundant organic material buried through the years. This organic material was provided by sediments deposited from the surrounding land. Petroleum was also formed without the necessity of burial ground. It is concluded that small amounts of hydrocarbons are continuously being evolved from the chemical action on the remains of aquatic organisms in both the fresh and salt water.

There are four great oil regions, each near a land locked sea. These are Eastern Mediterranean basin, the Caribbean basin, the Far East basin and the North Polar basin. The richest source lies at eastern end of Mediterranean Sea. It includes the land surrounded by Caspian Sea, Red Sea, Black sea and Persian Gulf.

If the geographical conditions are considered 'offshore' the earth are under water plains. These are also termed as 'continental shelves'. The shelves descend slowly for a very long distance to a depth of hundreds of meters before descending sharply (continental slope) to meet the 'continental rise' of the ocean's floor. Beneath the shelf (and slope) are large sources of oil and gas. This has started 'offshore' drilling throughout the world. India, too has some 'offshore' drilling projects near Mumbai.

In drilling for oil, the rock cover must be penetrated to reach the 'pool of oil'. It is known as pool of oil though it may be a bit misleading. The oil in the pool is dispersed through tiny pores and hairline cracks of the rock. When a well is bored, depending upon its location and underground stratum, oil flows into the well as primary recovery. This happens because of (*i*) expansion of gas dissolved in the oil, (*ii*) pressure on the oil from expansion of a gas cap above it, (*iii*) force of water on oil from below and (*iv*) the weight of the oil in steeply inclined formations. There are secondary recovery techniques that are used. These techniques are (*a*) Water Injection where water at high pressure is injected, (*b*) Gas Injection where gas is injected, (*c*) Thermal Drive where air is pumped and ignited (also known as fire flooding) and (*d*) A technique where a miscible fluid such as alcohol or Liquefied Petroleum Gas (LPG) is added to reduce the viscosity.

From scientific experience and calculations done that are based on test drillings, it is possible to find out the amount of crude oil in a pool. It is termed as 'proved reserves'. Sometimes the oil that can actually be recovered from such pool may be less than calculated. The 'proved' reserve figure is the actual amount of oil that can be recovered 'economically' from an oil pool.

In modern days, since the development of the satellite technology, it has become possible to know the availability of crude oil from beneath the surface of the earth. It is also possible to map the areas and to calculate the amount of crude oil that may be available. The technology is quite dependable and has eased the pre-exploration process. Earlier, pre-exploration process used to consume a lot of time and required a considerable amount of expenditure.

7.2 CRUDE PETROLEUM

Crude oil is a mixture of an almost infinite number of hydro carbon compounds. This ranges from light gases of simple chemical structure to heavy tar like liquids and waxes of complex chemical structure. The oil as it comes from ground also contains different amounts of sulphur, oxygen, nitrogen, sand and water. Although, the compounds or components of crude oil vary widely from pool to pool, the ultimate constituents are relatively fixed. The carbon generally varies from 83–87% and hydrogen varies from 11 to 14%. Many compounds of crude oil primarily belong to paraffin, naphthalene and aromatic families along with a considerable amount of asphaltic material of unknown chemical structure.

Primary families of hydrocarbons present in crude oil are (*i*) Paraffin (Alkanes) C_nH_{2n+2} (chain compound), (*ii*) Naphthalene - C_nH_{2n} (Ring Nucleus), (*iii*) Aromatic Benzene – C_nH_{2n-6} (Ring Nucleus) and (*iv*) Aromatic Naphthalene – C_nH_{2n-12}, (Ring Nucleus). However, the mixture of oil cannot be entirely divided into these separate families, because oil molecules can be made up of several families. A ring nucleus may be joined to a chain compound and, also, several rings of either the same or of different families, may be joined together to form a single molecule. Due to this complexity, the components and properties of the crude oil and products will exhibit extreme differences.

The crude oil is often classified by relative amount of Paraffin wax and asphalt residue in the oil. These are classified as Paraffin based, Asphalt based and mixed based type. This classification, though widely accepted, does not give the information about the other constituents in crude oil.

7.2.1 Conversion of Crude Oil into Petroleum Products

It is possible to convert crude oil into petrol and diesel completely but it is not economically feasible. The oil is released from the ground at high pressure. This allows the lighter boiling constituents to flash into gaseous state and the oil emerging from the well is a mixture of wet gases and liquids. These are separated with help of an absorber. The liquid droplets are absorbed and dry gases leave. The liquid petrol, when freed by heating the gas oil, requires relatively little treatment. It's known as natural petrol. The preliminary stage in petroleum refining is to pass hot crude oil into fractioning tower. The tower is about 35 meter high with trays at 0.65 meter interval. The vapours at high temperature bubble through condensed liquid in the trays and the crude oil is separated into several fractions. Compounds with higher boiling point are condensed in the lower trays and compounds with low boiling point are condensed in upper trays. This is because temperature decreases in the tower upwards. At arbitrary levels in the tower, the condensed compounds are withdrawn for specific treatments to yield the desired products.

The top fraction is called straight run petrol representing the lighter fractions of the crude oil. A low grade gasoline is known as Naphtha. At slightly higher boiling points the end product is kerosene oil. The heavier fractions are called 'gas oils'. The end products of paraffin based oil are 'cylinder stocks' for lubrication. If the crude oil has asphalt base or mixed base, the end products are lube distillates and an asphalt residue. When the asphalt is decomposed or cracked in a still, low quality gas oil is obtained. The carbonaceous residue from this operation is 'coke'.

Fractional distillation can be replaced by other method known as 'cracking'. In this method, molecules of hydrocarbons are decomposed into less complex compounds of lower boiling point. However, actual cracking reaction always involves recombination and decomposition. Therefore, the products of cracking are not only the low boiling point compounds but, also, a series of compounds with high boiling point. The latter can be recycled *i.e.*, returned to cracking chamber. The net result is the production of gases, petrol and a heavy fuel oil.

7.2.2 Refining and Octane Rating

Up to around 1910, the petrol had uncertain octane rating ranging between 10–80. At that time, there was no Octane Scale. Following World War I, natural petrol was added to straight run petrol to obtain validity necessary for starting.

General Motors (USA) experimented with several thousand compounds to prevent knocking and finally in 1923, emerged Tetraethyl Lead (TEL) as commercial product to prevent knocking. Taking petrol having octane number as 75 and then adding 3ml TEL high octane petrol (87) was achieved. This helped in increasing the compression ratio. On the other side, pure 2-2-4 Trimethyl Pentane (Iso-octane) was developed. Octane scale was devised. On this scale, *n*-heptane are considered with octane number 0 and 2-2-4 Trimethyl pentane (Iso-octane) are considered with octane number 100. Octane number 80 means the test fuel will yield the same knock reading in a standard engine under prescribed operating conditions as a solution (by volume) of 80 parts of Iso-octane and 20 parts of *n*-heptane with specific test methods. The scale is extended beyond 100 by adding Tetraethyl Lead (TEL) to Iso-octane.

With the advent of World War II, a high octane number fuel was acquired. This was considered good for combat operations. Present engines run at compression ratio less than optimum. This is to avoid air pollution.

Petrol, these days, may be made of straight run petrol, through fractional distillation. It can also be made from cracked petroleum through catalytic cracking or can be formatted through catalytic reforming. Alkylate and polymerized gasoline can be produced from gases. By adding butane or propane desirable properties can be imparted. Apart from these, some additives are added for different purposes.

To impart anti-knock characteristics in spark ignition engines, Tetraethyl Lead (TEL) and scavengers are added. Some additives are added to change the chemical character of combustion chamber deposits. This is done to reduce the surface ignition and spark plug fouling. Anti-oxidants are added to prevent the decomposition of Tetraethyl Lead (TEL). Detergents are added to prevent deposits in the carburetor. Metal deactivators are added to destroy the catalytic activities of copper traces. Anti rust agents are added to prevent rusting. Anti icing agents are essential when the engine is being used in extremely cold climatic conditions. Dyes and lubricants are used to identify the Tetraethyl Lead (TEL) and lubricate valve guides and upper region of cylinder.

Petrol

The petrol available in the market is a blend of a number of products. These are produced by different processes. Such blending helps in attaining the desired operating characteristics. The petrol should have the following characteristics:

1. *Anti-knock characteristics:* Octane number or presence of octane helps in attaining these characteristics.
2. *Volatility:* This has got so many components. First is starting characteristics— The fuel with low ignition temperature would enable a combustible mixture at the surrounding air temperature. Secondly there are vapour lock characteristics—A low vapour pressure is desirable as it prevents vaporization in feed lines and carburetor, which would otherwise prevent or reduce liquid fluid flow. Thirdly for proper running performance, fuel with lowest distillation temperature is required. Then to prevent dilution of lubricating oil when fuel condenses or does not vaporize in the engine, low distillation temperature is desirable.
3. *Gum and varnish deposits:* The fuel should not cause deposition of gum or varnish deposits in the engine.
4. *Corrosion:* The fuel and products of combustion should not cause corrosion in the components of the engine.
5. *Price:* Though price is sometimes beyond control but then the effort should be made that fuel is available at lower most prices.

Due to a number of variables, each characteristic of the fuel cannot be correlated to get best results from the engine. If each specification for fuel is made rigid the price of the fuel would become exorbitant. Therefore there should be enough flexibility so that price can be kept under control. The engine is designed and manufactured to use the fuel commercially available. If designed compression ratio is low and low octane fuel can serve the purpose it is wastage of money to use costly high octane number fuel. If carburetor can be provided with hot spot, highly volatile fuel may not be needed. It is, therefore, essential to optimize the characteristics to get a well performing fuel at reasonable price.

7.2.3 Requirements of Fuel for Compression Ignition Engine

These are even more complicated than the requirements of fuel for spark ignition engine. This is the complexity arising due to heterogeneous mixture. Also the combustion process is greatly affected by the injecting characteristics of the fuel injection system. However, a fuel for compression ignition engine should have the following characteristics:

1. *Anti-knock characteristics:* These characteristics are measured through cetane rating. Knock can be avoided by using fuel with high cetane rating.
2. *Starting characteristics:* This means easy starting demands high volatility, to form readily combustible mixture. The fuel should have high cetane number to attain low self-ignition temperature.
3. *Smoking and odor:* The fuel should not produce smoke when burnt. There should not be any odor produced when it burns. The combustion is required to be complete which is achieved by good mixing.
4. *Corrosion and wear:* The fuel should not cause corrosion before and corrosion and wear after combustion. These requirements are directly related to amount of sulphur present, ash and other residual contents produced after combustion.
5. *Handling ease:* The fuel should be in liquid form. It should flow in all the conditions of temperature and pressure that is determined by pour point and viscosity. It should have high flash point.

7.3 FUEL AND COMBUSTION

In fuel the combustible elements, predominately, are carbon and hydrogen. Other element present in small quantity is Sulphur. Liquid fuels used in automobiles are the mixture of complex hydrocarbons. Hydrocarbons have hydrogen and carbon as constituents. When combustion occurs, the presence of oxygen is essential. Air has about 20% oxygen and is the natural source of oxygen. The other major constituent of air is nitrogen which is about 80%. When the fuel burns completely it is considered to be 'perfect' combustion. The hydrogen in the fuel combines with oxygen in the air and forms water (H_2O) or water vapours while carbon in the fuel combines with oxygen to form carbon dioxide (CO_2). However, incomplete combustion may occur in the engine when fuel is partially burnt. This causes production of monoxide of nitrogen (NO_X) and monoxide of carbon *i.e.,* carbon monoxide (CO). These products of combustion along with partially burnt particles of fuel are emitted with the exhaust gases which are a health hazard for living beings. It is essential that harmful constituents do not emit from the automobiles.

To avoid emission of harmful constituents, control systems are provided in crank case. The gases emitted are sent back to the engine to be burnt again. Then there is evaporative emission control system. In this system, fuel vapours are trapped that are coming out from air cleaner fuel tank and other places. The vapours are returned back to engine to be burnt. Another type of control systems is exhaust emission control systems. These reduce the pollutants in the exhaust gases through engine management, using electronic control module, and emission control devices.

Unleaded petrol: To contain the menace of pollution, one of the devices used is catalytic converter. For the satisfactory working of catalytic converter, it is essential that unleaded fuel is used. Otherwise also, considering the poisonous nature of lead its use has been avoided. In India since 1999, only unleaded fuel is being used.

Some other combustible organic compounds have also been used as supplementary to main fuel. These are mixed with main fuel to impart desirable properties to the fuel.

Gasohol: This is the name given to a mixture of petrol and ethyl alcohol. It has 90% petrol and 10% ethyl alcohol. Ethyl alcohol is produced from sugar, grain and even from certain organic materials. It's production is not dependent upon conventional natural source and can be grown again and again like other crops. The fuel system of engine requires no major change. If amount of ethyl alcohol is increased beyond 10%, richer mixture is required and modification in the fuel system also becomes essential.

Methanol: Methanol or methane is another organic compound from the alcohol family. It has got higher boiling point and makes cold start difficult. To warm-up the engine in the cold conditions, and to ease cold starting, petrol is mixed with methanol in the ratio of 15:85. It also makes the mixture safe. Pure methanol burns with colourless flame which is undesirable. The mixing of petrol provides colour to flame. This makes it safer. The calorific value of methane is 15,900 kJ/kg which is about half of 32,300 kJ/kg — the calorific value of petrol. The advantage of methane is that it can be produced from coal, wood, manure, oil shale. Aluminium and plastics are attacked by methane and therefore the fuel system should be made of stainless steel. Another disadvantage is that water can mix with the blend of petrol and methanol. There it separates petrol and alcohol. This stalls the engine. Variable fuel sensor is used to overcome this problem. This sensor can detect the amount of methanol being used in fuel. The information is sent to electronic control module that adjusts the fuel injector timings to match the fuel. In India also blending of methanol with petrol has been started.

Liquefied petroleum gas: Commonly known as LPG, it is made from crude oil. Initially, it is stored under pressure and is in liquid state but when pressure is released it is converted into gas. LPG can be used as fuel in vehicle after modifying the fuel system. The LPG has octane rating more than 100 that means engines running on LPG can have higher compression ratio.

Compressed natural gas: Also known as CNG, it is another fuel that has replaced diesel. It produces less amount of carbon dioxide and slightly higher amount of nitrogen oxides. Still it is considered to be atmospheric friendly fuel. It has replaced diesel in the public transport system in metro cities of India.

Due to limited availability of fossil fuel, throughout the world the attempts are being made to develop alternate fuels. In present times, we are heavily dependent upon the fossil fuel. It fulfils several requirements of our daily life and it is nothing short of nightmare that one day it would not be available. Also measures are being found to conserve the fossil fuel so that it can last for longer. As far as alternate fuels are concerned, the use of hydrogen as fuel is being made in some developed countries. It is considered to be clean fuel. Certain problems regarding its handling are the hurdles. The attempts are on to remove them. Solar energy is another alternate source of energy being exploited. It has its own limitation one of these being availability. Solar energy is not available continuously. During night hours we have to do without it. Also the availability of solar energy depends upon the weather conditions. Another limitation is collection of solar energy. Collector with large area is required that creates practical difficulties. There are attempts to use electric current to run motor which acts as prime mover for an automobile. In this case the current is supplied by rechargeable battery. To charge the battery, electricity is needed which is produced from conventional sources. Charge available in battery limits the distance automobile can travel before recharging is needed.

QUESTIONS

1. Explain briefly the evolution of petroleum in nature.
2. Name the main oil regions.
3. How the availability of crude oil below the surface of the earth is estimated in present times?
4. What is crude petroleum? Explain.
5. Explain the process of conversion of crude oil into petroleum products.
6. What is octane rating?
7. What are characteristics of petrol? Explain briefly.
8. Explain the blending of petrol with other organic compounds.
9. Explain briefly the alternate fuels being explored.

8

TRANSMISSION SYSTEM

8.1 INTRODUCTION

The internal combustion engine generates power which is transmitted to the road wheels. The output from engine is available in the form of rotation of crankshaft. This rotary motion is transmitted to the road wheels. The friction between road and the surface of the wheel makes possible the movement of automobile. Transmission system performs this function. The transmission system consists of a number of components. These components work together to transmit the rotary motion at the crankshaft smoothly and efficiently to the road wheels.

Sudden change of state, from rest to motion or vice versa is not desirable. It may be uncomfortable, or even injurious, to the occupants of the automobile. Therefore, the rotary motion of crankshaft should be transmitted gradually and not suddenly. Another aspect of transmission is that the motion from the crankshaft should not be transmitted as soon as the engine starts. It is not desirable that as soon as the engine starts the vehicle begins moving. The motion is required to be transmitted only 'when desired'.

The rotary motion of the crankshaft gives rise to torque and transmission of this torque to road wheels give rise to a propulsive force or tractive effort causing the movement of wheels on the road. When starting from rest large tractive effort is needed. The engine produces almost the same torque. This torque has to be enhanced so that enough tractive effort is produced. This necessitates the introduction of 'leverage' between the engine and the road wheels.

A variation in the leverage is essential because if the same leverage is used for climbing as well as moving on the level road, the maximum possible speed would be unduly low. A large leverage implies a large reduction in speed between the engine and the wheels and at quite moderate road speeds the engine speed would be very high. But at high engine speeds the engine torque falls off so that tractive effort available would be less thereby reducing the road speed.

On properly maintained road comfortable cruising speed for car can be approximately 50 kilometer per hour and with wheel diameter of 30 cm it will have a rotary speed of about 1060 rpm. Considering the speed of the engine to be about 3500 rpm the transmission system will have to reduce 3500 rpm at engine to about 1060 rpm at wheel (ratio 3.3:1). This ratio may vary with the engine size and engine specifications.

While the nature of transmission is not affected greatly by the changes in the form of 'carriage unit', so that the transmission of a 4-wheel driven vehicle is similar in nature to that of rear wheel driven vehicle, the arrangement of transmission will be different in both the cases.

The axis of the rear road wheels, where the motion is usually transmitted is perpendicular to the centre line of the automobile. Therefore, drive between the engine and the road wheels is turned through 90 degrees. If the automobile moves on a circular path, the inner and outer wheels will traverse circles of different radii. Thus, the inner and outer wheels travel different distances. Because the automobile moves as a single unit they have to travel different distances during the same time period.

In majority of automobiles, the engines are fitted in the front portion on the frame of the carriage unit. Usually the motion is transmitted to the road wheels on the rear side. The distance between the two is quite considerable. The motion is required to be transmitted through this distance. Also, the rear axle is attached to the frame through springs. Due to uneven surface of the road the axle moves up and down and the springs flex. The relative positions of the engine and the axle changes and transmission system should be capable of taking it up.

The transmission system, therefore, should fulfill the following requirements:

1. Enable the engine to keep disconnected from the road wheels. These should be connected only 'when desired'.
2. Enable the engine, when running, to be connected smoothly and gradually without jerk-to the road wheels.
3. Enable the leverage between the engine and the road wheels. This leverage should be variable to cope with the different conditions such as starting from the rest, moving at uniform speed or climbing a hill.
4. Enable the reduction in the engine speed.
5. Turn the drive through 90 degrees.
6. It should enable the running of inner and outer road wheels at different speeds when the vehicle moves on a curved path.
7. It should provide the relative motion between engine and the road wheels when they move up and down due to uneven road surface.

8.2 COMPONENTS OF TRANSMISSION SYSTEM

The transmission system consists of the following components:

(*a*) Clutch
(*b*) Gear box
(*c*) Propeller shaft
(*d*) Differential
(*e*) Live Axle

(*a*) **Clutch:** This component enables the engine to keep disconnected from road wheels. The rotary motion available at the crankshaft is not transferred to road wheels. It allows the transfer of motion when desired by the driver of the automobile. It also allows the transfer of motion gradually so that the vehicle starts moving gradually. It works on the principle of friction.

(*b*) **Gear box:** It consists of a number of pairs of gear wheels. These transmit the motion available from the crankshaft, through clutch, at different speeds. This provides required leverage between engine and the road wheels. This leverage is variable to cope up the different conditions encountered during the movement of the vehicle.

(*c*) **Propeller shaft:** The third component of the transmission system which transfers motion from the gear box end to the differential end. The distance between the two can be large and therefore it is a shaft which is thin and long to connect the two.

(*d*) **Differential:** One of the requirements of the transmission system is to turn the motion through 90 degrees as the axis of the propeller shaft and live axle are at right angle to each other. This is performed by the differential through wheel and pinion arrangement. Another function performed by the differential is the variation in the speeds of inner and outer wheels when the vehicle is taking a turn.

(*e*) **Live axle:** The axle where motion from crankshaft of the engine is transferred is known as live axle. The other axle takes up only the load of the vehicle and therefore is termed as dead axle or simply the axle. The motion is generally transferred to rear axle but it can be transferred to the front axle or to both the axles. When the motion is transferred to both the axles it is known as four wheel drive.

Finally motion is transferred to the road wheels at the two ends of the live axle. The wheels rotate and friction between their surface and road surface makes possible the movements of the vehicle on the road. In the forth coming chapters the components of transmission system are discussed in details.

QUESTIONS

1. What is transmission system? Explain briefly.
2. What are requirements of transmitting motion to road wheels ?
3. Name and explain briefly the device in transmission system providing leverage between engine and the road wheels.
4. What are the functions of differential?
5. Explain the live axle. What are its functions ?

9

THE CLUTCH

The clutch enables the rotary motion of crankshaft to be transmitted to driven shaft when desired and gradually. These can be classified as:

1. Positive Clutch
2. Gradual Engagement Clutch

1. **Positive Clutch:** In this type of clutch there are two positions. Either it is 'in' when the two shafts are rigidly connected and revolve at the same speed or it is 'out' when the shafts are entirely disconnected and there is no transfer of motion and the driven shaft is not moving. This type of clutch is not suitable for use between the engine and the gear box as the motion will be suddenly transferred. Also the transfer of motion will suddenly stop. This means sudden movement of vehicle from rest and sudden stoppage of the movement. Both the situations are not desirable as sudden change of state from rest to motion or vice versa will be highly uncomfortable or even injurious to the user of the vehicle.
2. **Gradual Engagement Clutch:** In this type of clutch it is possible that driving shaft is rotating and the other shaft, driven shaft, is stationary. As the engagement of clutch proceeds the speed of the two shafts gradually becomes almost equal. When the clutch is fully engaged both the shafts rotate almost at the same speed. This type of clutch is used between the engine and the gear box. In this type of clutch the transfer of motion between the two shafts depends upon the friction between the surfaces of the two shafts when these come in contact with each other. Therefore, these are also known as friction clutch.

9.1 PRINCIPLE OF FRICTION CLUTCH

Consider two shafts A and B duly supported in bearing C and D, as shown in figure 9.1, and free to rotate about their common axes XY. Shaft A is driving shaft and shaft B is the driven shaft. Two discs E and F are attached at the ends of the shafts with the help of keys. Driving shaft A and E rotates along with it. Shaft B and disc F are at rest. If the two shafts are pressed together the surfaces of discs E and F come in contact with each other. The two surfaces are rough. Due to the movement of disc E some frictional force comes into existence. When the pressure is increased frictional force between two surfaces also increases. This causes transfer of motion and the disc F also starts rotating. This means the driven shaft also starts rotating. Initially the speed of driven shaft is low. As the pressure between the surfaces of disc E and F increases the frictional force between the two also increases and so increases the rotary speed of the driven shaft. The speed of the driving and driven shaft can be equal only when the coefficient of friction between the surfaces of discs E and F is one which is not possible practically. Therefore the speed of driven shaft is always less than the speed of the driving shaft. To make the contact possible between the two surfaces of the discs E and F, force should act along the axes of the shaft.

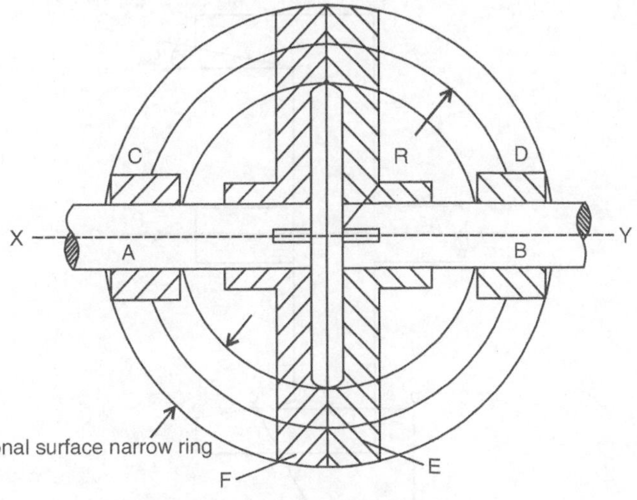

Fig. 9.1

The axial force is provided by the springs. Suppose the springs, used to press the discs together, exert a total force of P Newton normal to the surface of the disc. Upon the magnitude of this force depends the magnitude of frictional force which tends to prevent the transfer of motion to the driven shaft. The magnitude of this force is μP, μ being the coefficient of friction. This force is the sum of the large number of component forces acting all over the disc surface. These forces may be considered acting on the narrow rings which divide the circular surface of the disc into a number of parts. The resultant frictional force, μP can be considered acting tangentially at a radius of R which is the mean radius of disc. This force will cause a moment about the shaft axis equal to $\mu P \times R$ Newton-meter. This moment tends to stop the driving shaft A and to drive the driven shaft B. This is the torque transmitted.

The magnitude of torque depends upon the radius at which the friction force acts. The magnitude of the torque can be increased by increasing the radius of disc. The magnitude of torque also depends upon the co-efficient of friction, μ. The co-efficient of friction depends upon the nature of material of the surfaces in contact. The availability of space limits the radius of discs. The force P depends upon the springs. The torque required may be large in magnitude and it can be acquired only if certain modifications are made in the design of the clutch. The modification in the design has resulted in three principal types of clutches which are used in an automobile. These are cone clutch, single plate clutch and multi-plate clutch.

Cone Clutch: The friction surface of this type of clutch is in the form of a cone. The advantage of the cone clutch is that the normal force acting on the friction surface is greater than the axial force which is not so in case of single plate or multi-plate clutches.

Fig. 9.2 represents the geometry of a cone clutch, α is the semi cone angle and P is the force acting along the axis of the shaft. As a reaction to force P, there is an equal force H acting along the axis but in the opposite direction. Q is the force normal to conical surface and uniformly distributed over it.

The force P is exerted by the springs. The force which keeps the clutch engaged is Q. From the Fig. 9.2, it can be found that Q = H/sinα. The value of Q depends upon H (or P) and angle α. Smaller the angle α greater would be the value of Q. Though the force exerted to engage the clutch is higher than exerted by the springs but there is a limit to which the cone angle can be kept.

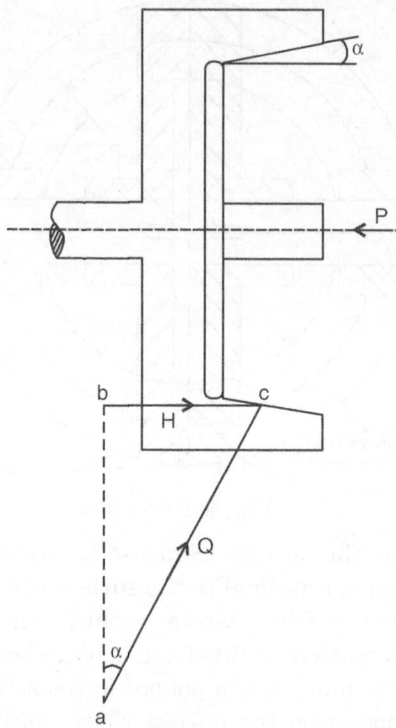

Fig. 9.2

Figure 9.3 represents a cone clutch. The male cone is mounted on the splined clutch shaft. Clutch shaft is the driven shaft and the motion is transferred to this shaft. The friction surfaces are mounted on conical portion. The male cone can slide on the splines provided on the clutch shaft. The splines allow sliding motion only along the shaft axis. There is no relative rotary motion between the shaft and male cone. The female cone is a part of the crankshaft of the engine. It rotates with the engine shaft. The spring exerts force along the axis of the shaft and keeps contact between friction surfaces of male and female cone. Due to the frictional contact the motion from engine shaft is transferred to clutch shaft. This is the 'engaged' position for the clutch. When the driver applies the force on clutch pedal it acts against the force of spring. It causes the movement of male cone away from female cone and the contact between the two is lost. In this position, the motion is not transferred from engine shaft to clutch shaft. This is the 'disengaged' position for the clutch. In practice, the clutch pedal is released gradually. This causes a gradual contact between the two conical surfaces and also gradual transfer of motion from engine shaft to clutch shaft.

The cone clutch is practically obsolete. There are some disadvantages associated with them. The disengagement of clutch becomes difficult with cone angle less than 20 degrees. If there is even small wear on conical surfaces, the axial movement of male cone becomes considerable which is not desirable.

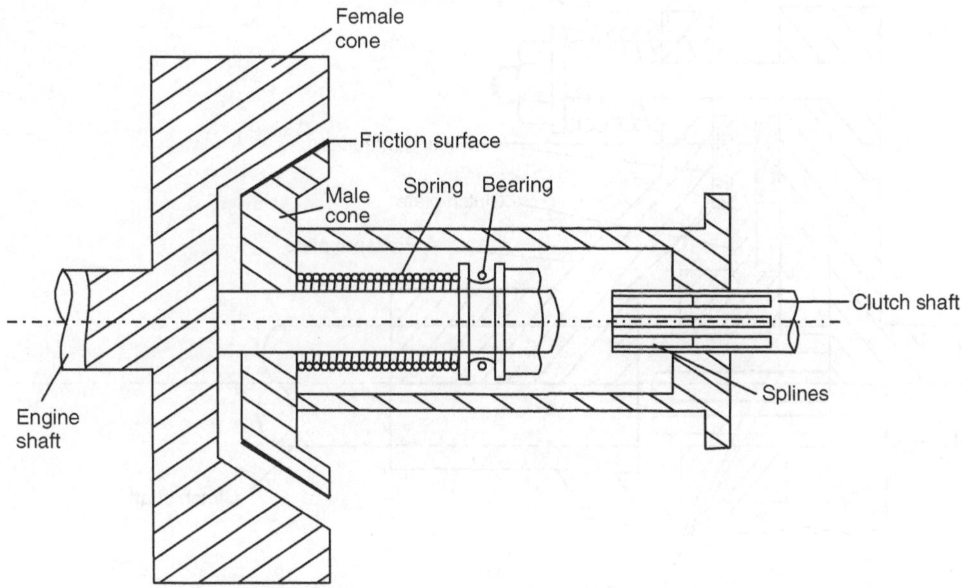

Fig. 9.3

9.2 SINGLE PLATE CLUTCH

This type of clutch has one clutch plate and works on the principle of friction. These are of two types: Helical spring type and Diaphragm spring type. In helical spring type clutches, the helical springs are used uniformly over the cross-sectional area of pressure plate to exert axial force. In diaphragm spring type clutch, diaphragm spring is used to exert axial force.

9.2.1 Helical Spring Type Single Plate Clutch

Figure 9.4 represents a single plate clutch of helical spring type. For simplicity sake, the clutch pedal and other links causing movement of pressure plate are not shown. The clutch plate is mounted on the splined shaft and can move along the axis of the shaft. There is no relative movement between plate and shaft as far as rotational movement is concerned. Both have same rotational movement due to splines provided on the shaft.

The flywheel is mounted on the engine crankshaft and rotates with it. The pressure plate is bolted to the flywheel through clutch springs. It can slide freely along the axis of the clutch shaft. The clutch is engaged due to force exerted by the clutch springs. This force causes contact between the pressure plate, clutch plate and the flywheel. The clutch plate is located in between the fly wheel and pressure plate. The clutch plate is provided with friction material on both the sides. The rotary movement from flywheel is transferred to the clutch plate and to the clutch shaft due to friction. The clutch shaft also acts as output shaft.

When the clutch pedal is pressed the clutch is 'disengaged'. The pressure plate moves back against the force of springs and the clutch plate becomes free between the flywheel and the pressure plate. Thus the flywheel continues to rotate as long as the engine runs but the speed of the clutch plate declines and becomes zero. In this situation, motion is not transferred to the clutch shaft.

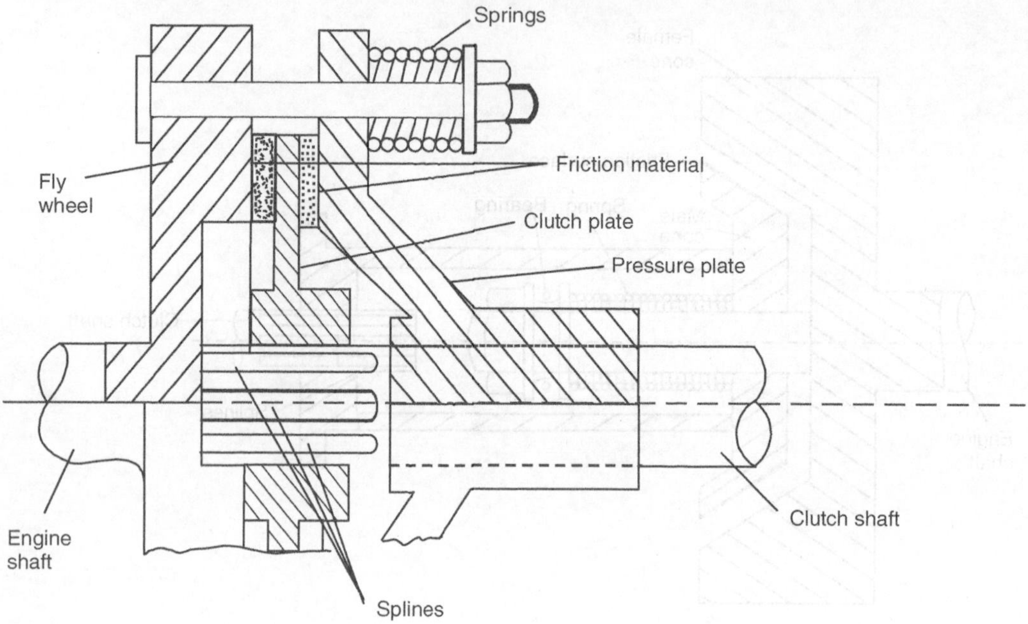

Fig. 9.4

9.2.2 Diaphragm Spring Type Single Plate Clutch

In this type of clutch, the helical springs are replaced by a single diaphragm spring which is a saucer shaped disc. The disc is provided with profile as shown in Fig. 9.5. The disc

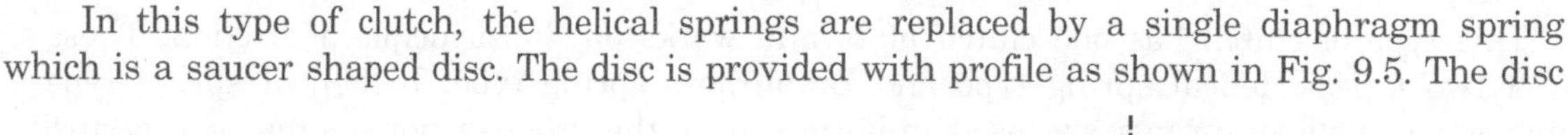

Fig. 9.5

adopts flat shape, as shown, when the clutch is engaged. In disengaged position, the disc adopts a buckled shape as shown. Figure 9.6 represents simplified view of the clutch assembly. The view shows the clutch in 'engaged' position. The force is exerted by the diaphragm spring on the pressure plate which causes the contact between pressure plate, clutch plate and flywheel. When force is applied through clutch pedal, the diaphragm spring is buckled and contact between pressure plate, clutch plate and fly wheel is lost. The clutch is 'disengaged' and motion from flywheel is not transferred to clutch shaft.

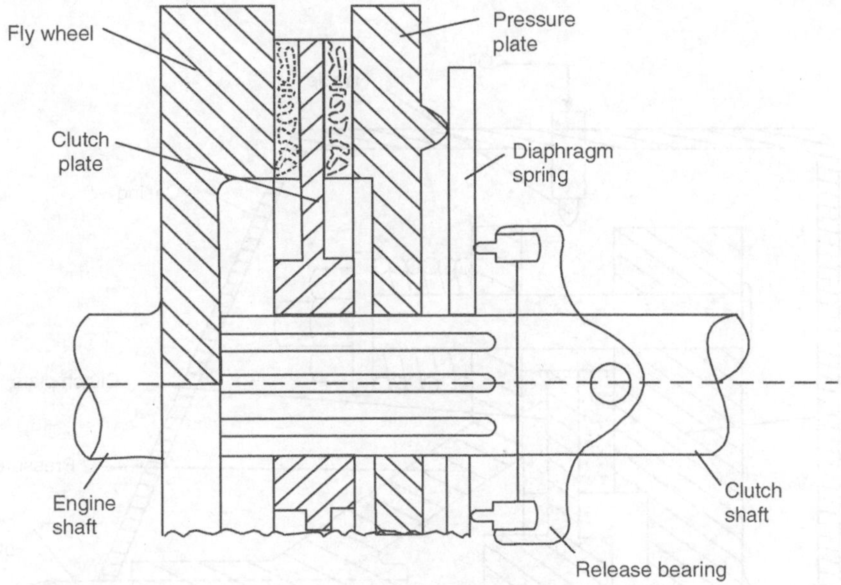

Fig. 9.6

9.3 MULTI PLATE CLUTCH

A number of times, single clutch plate can not transfer the required motion. This may be due to less friction force. The friction force can be increased by increasing the area of

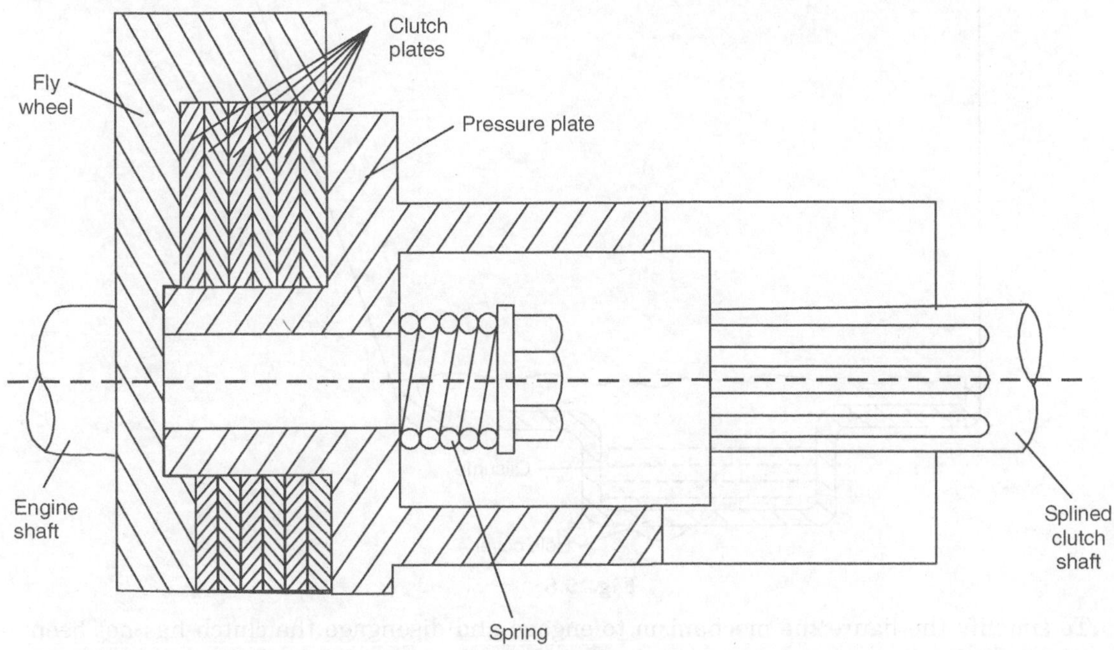

Fig. 9.7

contact. This increases the size of the clutch and due to limited space available; it may be difficult to increase the size. Therefore to increase the area of contact, the number of clutch plates is increased. The constructional details of multi plate clutch are as represented in the Fig. 9.7.

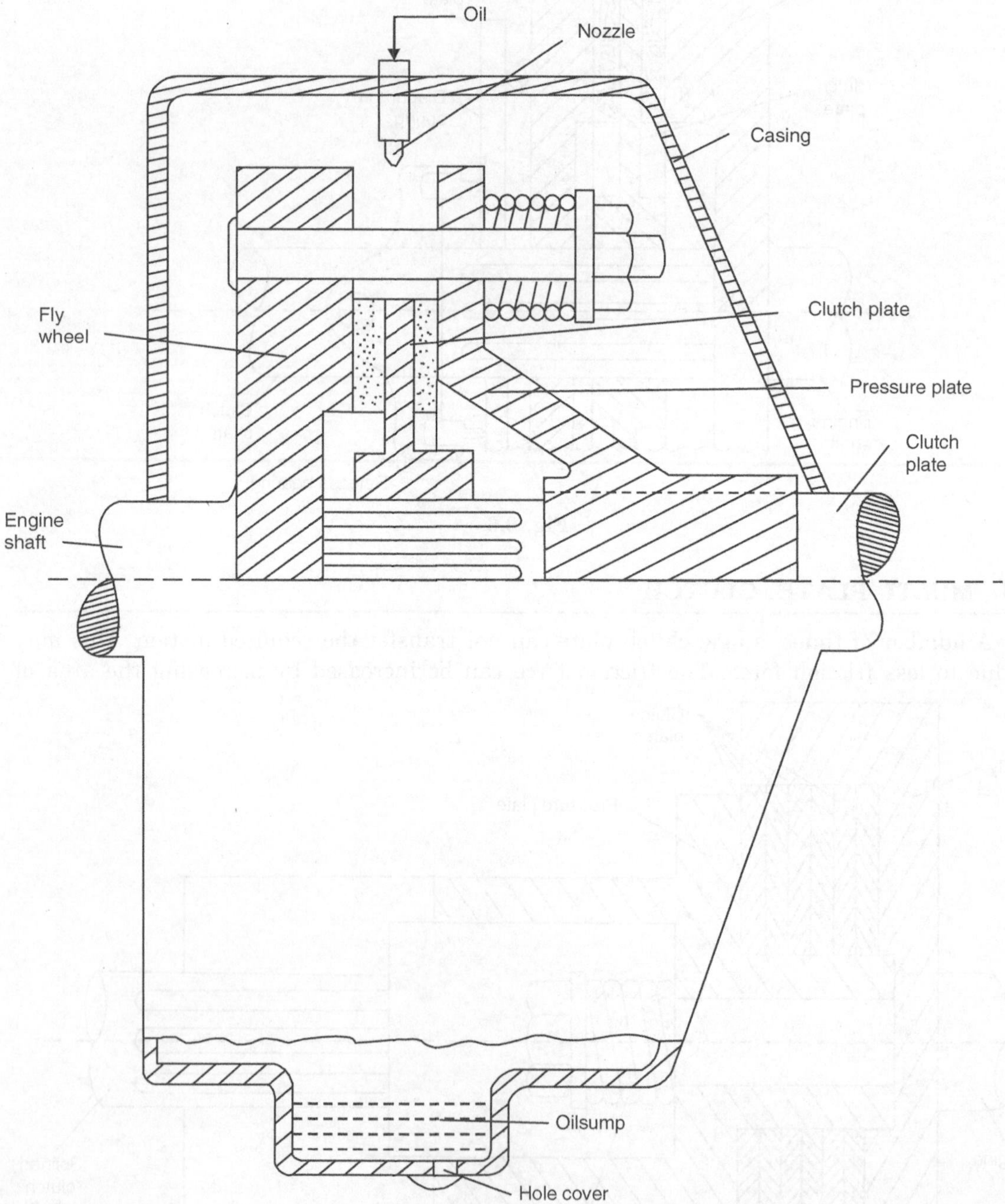

Fig. 9.8

To simplify the figure the mechanism to engage and disengage the clutch has not been shown. Internal splines are provided on the flywheel. The clutch shaft too is provided with splines. The clutch plates are assembled and are firmly pressed, with the help of pressure plate, by coil springs. These coil spring exert axial force, due to which there is contact between the clutch plates, flywheel and the pressure plate. The frictional surfaces on both the sides of plates helps transferring the motion from flywheel to the clutch shaft. This is 'engaged' position of clutch. By operating the clutch pedal, the force is exerted against the force of springs and contact between the flywheel, clutch plates and pressure plate is lost and no motion is transferred from fly wheel to clutch shaft. This is 'disengaged' position for the clutch. Presently, in all the automobiles multi plate clutches are being used.

Wet clutch is a variant of friction clutch. Here oil is sprayed on the plates with the help of nozzle. These are used in a variety of automobiles. The friction material used on clutch plates should have greater co-efficient of friction and these are perforated so that oil can pass through these. These clutches have intake for oil. A sump is provided at the bottom to collect the oil from where it is drained out (Fig. 9.8). These types of clutches have longer life than dry clutches due to better dissipation of heat.

9.4 CENTRIFUGAL CLUTCH

Many times attempts have been made to produce automobiles which are simple to operate. Pressing clutch and changing gears repeatedly is quite inconvenient for the user of automobile. Centrifugal clutch is a device which is not operated by the user. The axial force required to disengage the clutch is provided through centrifugal force. The centrifugal force depends upon the speed of the shaft. If the shaft speed is low the force is also low and clutch is disengaged. When the shaft speed increases, the centrifugal force increases and clutch becomes 'engaged'. As the shaft speed increases gradually, the centrifugal force also increases and the motion from driving shaft to driven shaft is transferred.

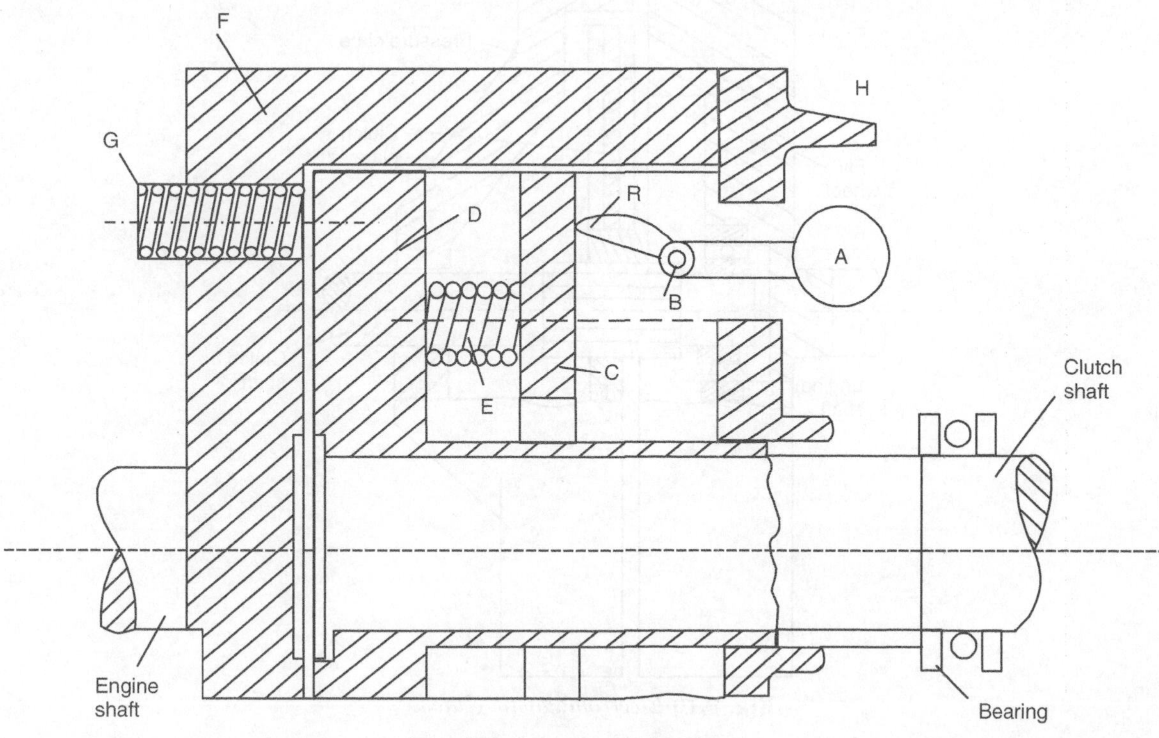

Fig. 9.9 *Centrifugal clutch.*

As this clutch uses centrifugal force instead of spring force to keep it in 'engaged' position no clutch pedal is required to operate it. It is operated automatically depending upon the engine speed. The vehicle can be stopped in gear without stalling the engine. Similarly, the vehicle can be started in any gear by pressing the accelerating pedal. Figure 9.9 shows the centrifugal clutch. It consists of a weights A pivoted at B. When the engine speed increases the weights fly off due to centrifugal force, operating the lever, R pressing the plate C. Movement of C compresses the spring E ultimately pressing the clutch plate D on the flywheel F against force exerted by the spring G. This makes the clutch engaged. The force exerted by spring G keeps the clutch disengaged at low speed. The stop H limits the outward movement of weight A due to centrifugal force.

9.5 ELECTROMAGNETIC CLUTCH

Here the flywheel consists of winding. Battery or dynamo provides current to the winding. Due to the current, electromagnetic field is produced which attracts the pressure plate creating contact between flywheel, clutch plate and pressure plate and thus the clutch is engaged. When supply is cut off, the electromagnetic field disappears, contact is lost and clutch is disengaged. The gear lever consists of a clutch release switch. The switch operation cuts off the current thereby disengaging the clutch. Sometimes at low speed, current supplied by dynamo is not sufficient to produce electromagnetic force. In this situation, force on the pressure plate is caused by the springs which keep the clutch in 'engaged' position at low speeds.

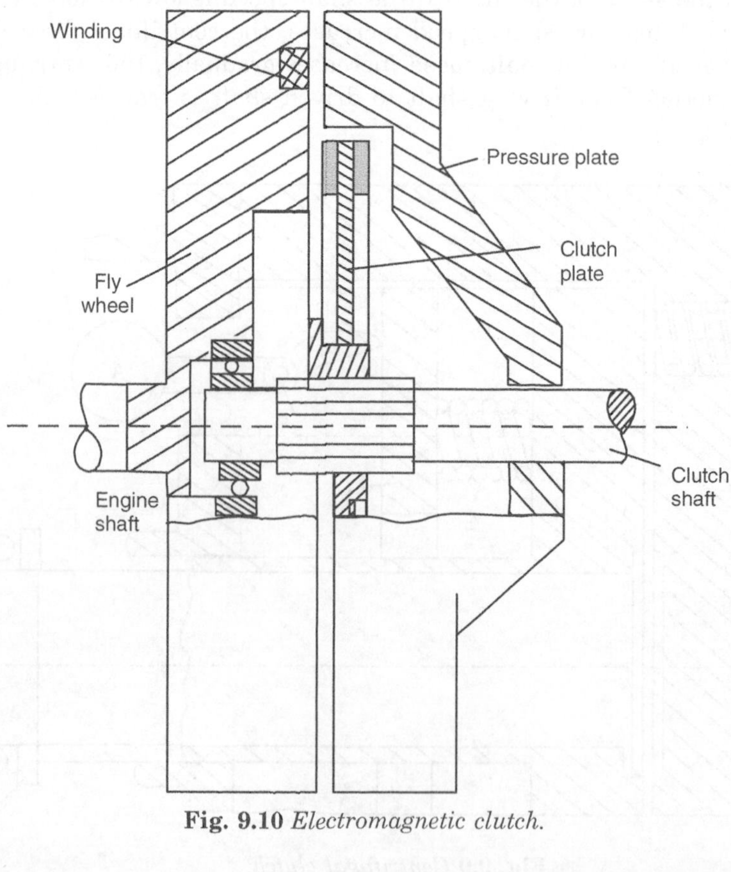

Fig. 9.10 *Electromagnetic clutch.*

9.6 CLUTCH LINING

In clutches, metal clutch plates engage with one another giving metal to metal contact. Such clutches may be enclosed so that oil could be kept in them to lubricate the engaging surface. Though this meant a reduction in frictional force. Alternatively, the clutch plate surfaces was covered with a friction fabric or a composite lining, so that frictional co-efficient was higher than metal to metal surfaces. Lining was secured to plate surface by means of copper or aluminium rivets. Their heads were sunk well below the lining surface so that these did not damage the lining on the adjacent plate.

9.7 FRICTION MATERIALS

The friction material acts as lining on the surface of clutch plate. As these are rough in nature, these offer higher co-efficient of friction. Higher co-efficient of friction causes higher force of friction and better transfer of motion.

These are of two types: (*a*) Woven friction material (*b*) Moulded friction material.

9.7.1 Woven Friction Material

This type of material is made by spinning threads from asbestos fibres, sometimes on brass wire, weaving this thread into a cloth and then impregnating it with a bonding material. This type of friction material can be subdivided into the laminated variety and the 'solid' woven variety. In the laminated variety the layers of cloth placed on top of each other and held together by the bonding material sometimes aided by stitching. In the 'solid' woven type the cloth is woven to the required thickness and this interlocked structure has much greater mechanised strength. Both the types generally incorporate metallic wire made of brass.

9.7.2 Moulded Friction Material

The moulded or composition type of lining is composed of asbestos fibres in their natural state mixed with a bonding material and then moulded in dies under pressure and high temperature. Metallic wire included in the structure to provide better wearing qualities.

9.8 BONDING MATERIALS

Basically the bonding materials are chemical compounds. There is a large variety of different bonding materials. These can be broadly classified as following:

(*a*) Compounds with Asphalt base (*b*) Vegetable gums (*c*) Rubber (*d*) Synthetic resins.

9.8.1 Compounds with Asphalt Base

Natural gums and oils are added to compounds with asphalt base. Their co-efficient of friction varies from 0.3 to 0.4 upto a temperature of 250 degree Celsius. The co-efficient tends to rise with rise in temperature. Excessive bonding material also helps to attain higher co-efficient of friction but as the excessive material may get removed at high temperature it is difficult to maintain high co-efficient of friction. The wearing properties of friction materials impregnated with this bonding material are good, particularly when the product is die pressed.

9.8.2 Vegetable Gums

The co-efficient of friction for this type of bonding material is slightly higher *i.e.*, between 0.35 and 0.45. This co-efficient of friction is maintained upto a temperature of 250 degree Celsius. This bonding material has got better wearing properties as compared to compounds with asphalt base.

9.8.3 Rubber

Rubber as bonding material has a co-efficient of friction of 0.6. It can be either in flexible form or in rigid form. Though it has high co-efficient of friction but the material tends to disintegrate under pressure.

9.8.4 Synthetic Resins

These can further be classified as alcohol soluble and oil soluble. Co-efficient of friction for alcohol soluble synthetic resins varies between 0.4 to 0.5 upto a temperature of 230 degree Celsius. At higher temperature the co-efficient of friction tends to reduce. The material is not affected by lubricating oil and can withstand high pressure.

The synthetic resins which are oil soluble have a co-efficient of friction 0.35–0.38 which can be maintained even at high temperature. Often these are used with vegetable gums and compounds with asphalt base and provide excellent wearing qualities.

Cotton is occasionally used instead of, or mixed with, asbestos and these fabrics can be made to give a co-efficient of friction of 0.6. But this high co-efficient of friction can not be maintained beyond a temperature of 150 degree Celsius. Cork is also used sometimes, but should be kept in oil. The cork has a co-efficient of friction of 0.3 and can withstand pressures upto 140 kN/m^2.

9.9 FLUID FLYWHEEL

It is a device for the transmission of motion from engine shaft to the gear box. It is a replacement of clutch in the transmission system.

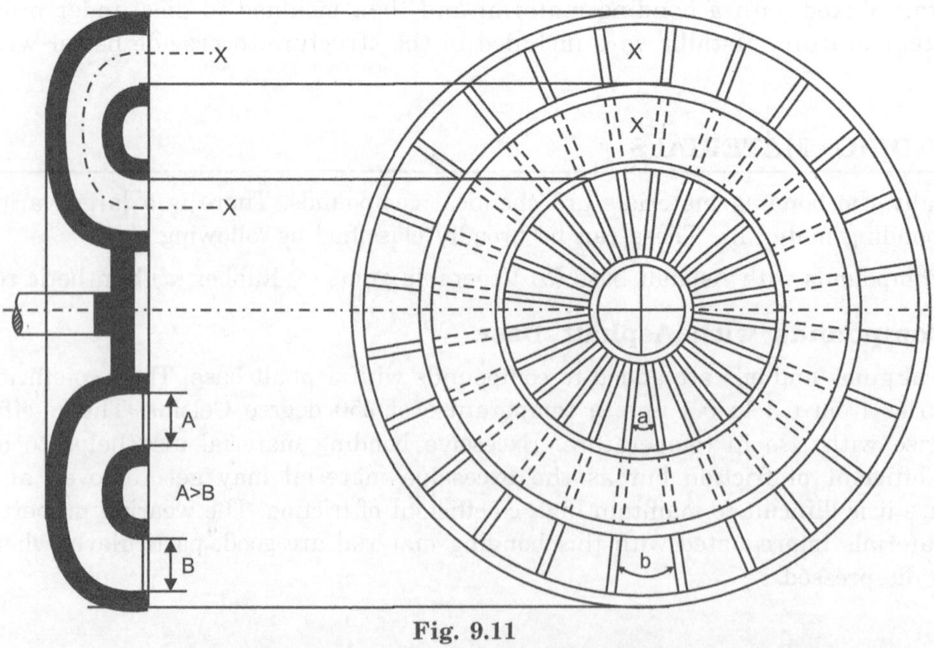

Fig. 9.11

It consists of two castings, also known as rotors, almost identical in form, one of which is fixed to the crankshaft of the engine and other to the gearbox shaft. These are roughly circular discs in which passages are formed. Since the areas of these passages perpendicular to their centre line XX in the sectional view, must be kept approximately constant and since the circumferential width of the opening '*a*' is less than that of '*b*' the radial size of the opening A is made greater than that of B (Fig. 9.11).

In a simplified form, the passages may be represented by tubes A and B with right angled corners as shown. These tubes are connected to the two shafts with axis X-X. The shaft on left is rotating at a speed of N rpm and that on right hand side is rotating with a speed of n rpm.

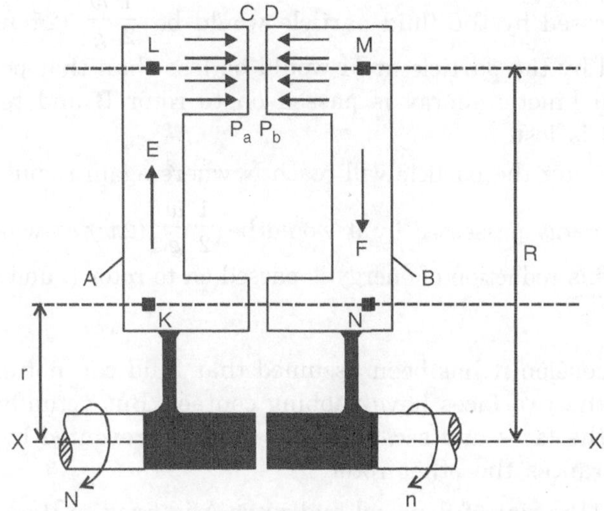

Fig. 9.12

Imagine these tubes to be full of fluid and their outer ends covered by diaphragm C and D as shown in Fig. 9.12.

The fluid would be exerting a pressure P_a on diaphragm C and it can be shown that $P_a \propto N^2$. Similarly, in tube B, the fluid would be exerting a pressure P_b on diaphragm, D. If $N > n$, $P_a > P_b$ and if the diaphragms were removed the fluid in A would commence to flow towards E and fluid in B towards F. Thus if $N \neq n$ the fluid in tube will flow round again and again if $N > n$ the flow will be as shown (*i.e.,* clockwise) and if $N < n$ the fluid will flow in counter clockwise direction. In case $N = n$ there will not be any flow. Since there will always be some resistance to flow the steady speed at which the fluid circulates will be proportional to $N^2 - n^2$. If the speeds of differ then tube A will not remain opposite to tube B. There will always be other tube, though in between replacement, the impinging of the fluid on tube walls (*i.e.,* on the webs of metal between the passage XX) may occur.

Having seen that a difference in the speeds of two rotors will cause a circulation of fluid from one rotor to other it can now be explained how the energy developed by the engine is transmitted to the gearbox, *i.e.,* from driving shaft to driven shaft.

Refer to Fig. 9.12, a particle of fluid at K is at *a* distance of r from axis XX. This particle has to rotate in a circle of radius r with angular speed of tube *i.e.,* N. The linear speed of particle in the circle is $2\pi r N$ and the kinetic energy possessed by particle is

$\frac{1}{2}\frac{w}{g}(2\pi r)^2 N^2$ where w is the weights of fluid particle. Now if the fluid is circulating as described above (N being assumed > n) fluid particle will arrive within short time at L rotating in a circle with radius R at speed N and kinetic energy possessed by the fluid particle would be 1/2 w/g $(2\pi R)^2 N^2$.

Since R > r, the kinetic energy possessed by fluid particle at L is greater than that possessed by it at K. The increase in kinetic energy as it moves from K to L is derived from the energy developed by the engine, the whole of which is utilised in increasing the kinetic energy of fluid as it flows from the centre to the outside of the tube A. Next, the fluid particle reaches at M. It will be rotation in a circle with radius R but at slower speed n. At this stage, the kinetic energy possessed by the fluid particle would be $\frac{1}{2}\frac{w}{g}$ $(2\pi r)^2 n^2$. As n < N, the kinetic energy possessed by the particle at M would be less than that possessed by it at L. Partly, this reduction in kinetic energy is passed on to rotor B and remaining would be converted into heat and is lost.

Finally a short time later the particle will reach N where again radius becomes r and the speed is n. The kinetic energy possessed by it would be $\frac{1}{2}\frac{w}{g}$ $(2\pi r)^2 n^2$ which is less than the energy possessed at M. This reduction of energy is passed on to rotor B and to the driven shaft.

Prevention of Leakage

In the previous discussion it has been assumed that fluid can not escape between the faces of the rotors and the two faces have rubbing contact. But actually there is a gap of about 1.5 mm between the faces and escape of the fluid is prevented by making one rotor with a cover and it entrances the other rotor.

The rotor B fixed to the rim of flywheel and rotor A is fixed to the gear shaft C. Fluid which escapes the rotors fills the space between outside the rotor and inside of the flywheel. Centrifugal force keeps this fluid in position and maintains a pressure at X. It also prevents the escape of fluid from the inside of rotor (Fig. 9.13).

9.9.1 Characteristics of the Fluid Flywheel

The difference in the speeds of driving and driven shaft is termed as 'slip'. Mathematically, slip = N − n. Its percentage with respect to the speed of driving shaft *i.e.*, N is percentage slip. Percentage slip = $\frac{(N - n)}{N} \times 100$ and is a measure of the difference in the speed of two shafts. If n is equal to N, percentage slip will be zero and if n = 0 then percentage slip would be 100. The graphical representation of variation of percentage slip with the speed of the driving shaft, for a particular engine, is shown in figure 9.14. It can be seen that at any speed less than about 600 rpm, slip is 100% that means clutch is disengaged. With the increase in speed from 600–1000 rpm slip comes down from 12% to 2%. The % slip at any engine speed, however depends upon the torque being transmitted, this curve being based on assumption that engine exerts full torque at every speed. Slip means direct energy loss and an increase in fuel consumption. Thus, it is an abuse of the fluid flywheel to allow the engine speed to remain between 1000–600 rpm, full throttle (torque) when slip is considerable. This is comparable to the slipping of ordinary clutch which also causes an increase in fuel consumption. The ordinary friction clutch would be damaged by prolonged slipping whereas the fluid flywheel will become very hot.

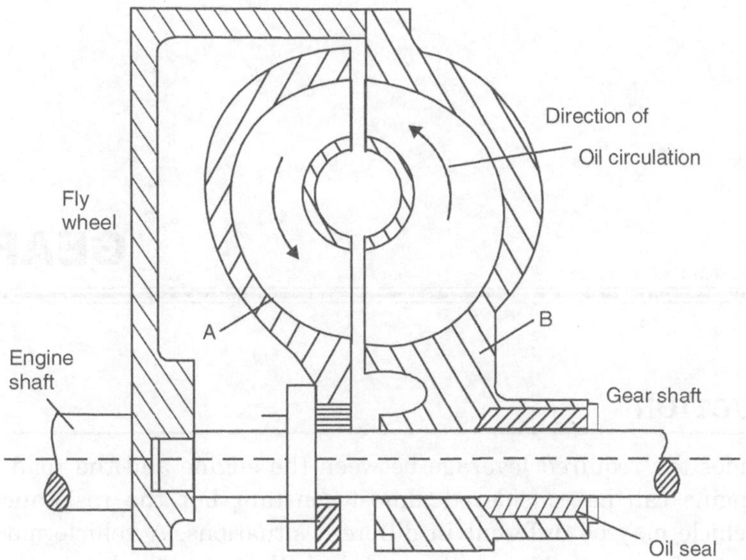

Fig. 9.13

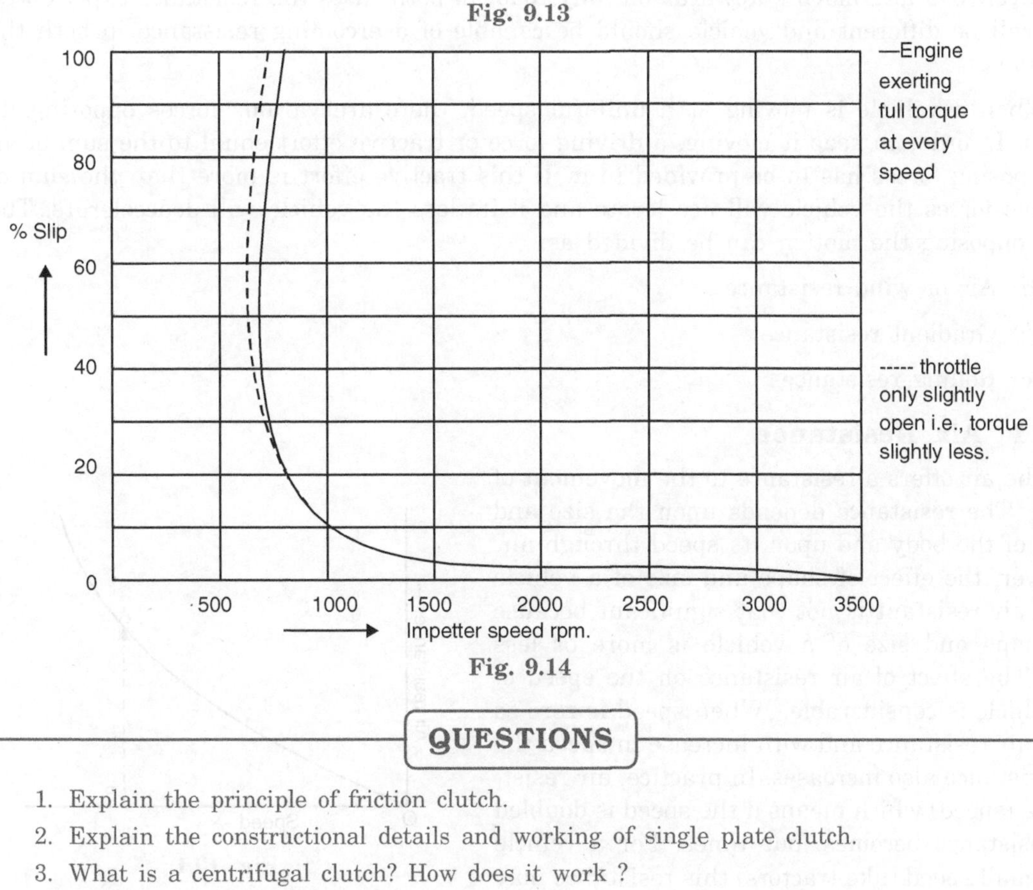

Fig. 9.14

QUESTIONS

1. Explain the principle of friction clutch.
2. Explain the constructional details and working of single plate clutch.
3. What is a centrifugal clutch? How does it work ?
4. Explain the working of electro-magnetic clutch.
5. Explain the principle of fluid flywheel.
6. Explain how the leakage is prevented in a fluid fly wheel.
7. What is slip ? Explain the effect of percentage slip on the speed of driving shaft.

10

GEAR BOX

10.1 INTRODUCTION

Gear box provides the required leverage between the engine and the road wheels. The output from the engine can be considered almost constant but the resistance to motion experienced by a vehicle may be different in different situations. A vehicle may be moving on a level road and moving upwards on hilly road, in both cases the resistance experienced by it will be different and vehicle should be capable of overcoming resistance in both the situations.

When a vehicle is moving at a uniform speed, there are various forces opposing its motion. In order to keep it moving, a driving force or tractive effort equal to the sum of all the opposing forces has to be provided to it. If this tractive effort is more than the sum of resistive forces the vehicle will accelerate and if its less the vehicle will deaccelerate. The forces opposing the motion can be divided as:

(*a*) Air or wind resistance

(*b*) Gradient resistance

(*c*) Rolling resistance.

10.1.1 Air Resistance

The air offers a resistance to the movement of bodies. The resistance depends upon the size and shape of the body and upon its speed through air. However, the effect of shape and size of a vehicle on its air resistant is not that significant because the shape and size of a vehicle is more or less fixed. The effect of air resistance on the speed of the vehicle is considerable. When speed is zero so is the air resistance and with increase in speed the air resistance also increases. In practice, air resistance $\propto$ (speed) which means if the speed is doubled the resistance becomes four times. For a vehicle with small speed (like tractors) this resistance may be negligibly small but with high speed vehicle (like racing cars) it is a very important parameter.

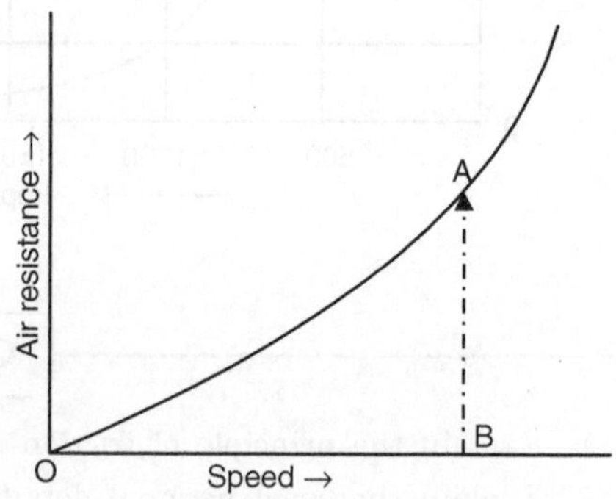

Fig. 10.1

10.1.2 Gradient Resistance

This is the resistance experienced by vehicle when moving upward on a gradient or hilly road. A simple analysis shows the magnitude of this resistance. As shown in Fig. 10.2, moving upward on gradient, vehicle has its weight, w acting vertically downwards. The weight, w can be resolved into horizontal (H) and verticle (V) components. To prevent the verticle to move backwards a force equal and opposite to H must be applied. H is the resistance due to gradient and the vehicle must overcome it to move upwards. It is dependant upon the steepness of gradient *i.e.*, angle θ and the weight of the vehicle and has nothing to do with the speed of the vehicle.

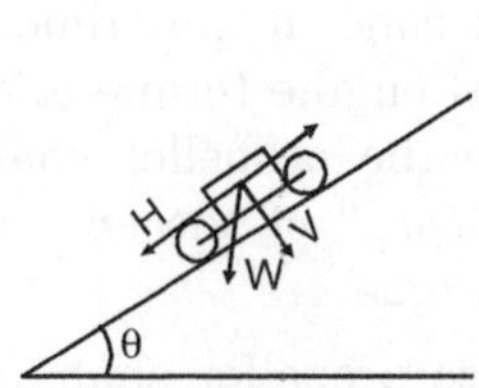

Fig. 10.2

10.1.3 Rolling Resistance

This includes all the remaining external resistances and also, the internal frictional resistances of the vehicle. Mainly the rolling resistance means the resistance (frictional) between tyres and energy dissipation due to impact of tyre and road surface. This depends upon the nature of road surface, nature of tyre surface, and total weight of the vehicle including load. The second part of rolling resistance depends upon the speed and up and down springing of vehicle. Due to springing of vehicle the resistance caused may be negligibly small and only a small part of the total rolling resistance. Due to limited information available, it is assumed that rolling resistance is independent of, speed.

10.1.4 Total Resistance

It is the sum of the three resistances discussed above. The rolling and gradient resistances are independent of speed whereas air resistance is dependent on speed.

Hence, as shown in Fig. 9.3, at a speed of OD km/hour, the total resistance AD is composed of air resistance AB, gradient resistance BC and rolling resistance CD.

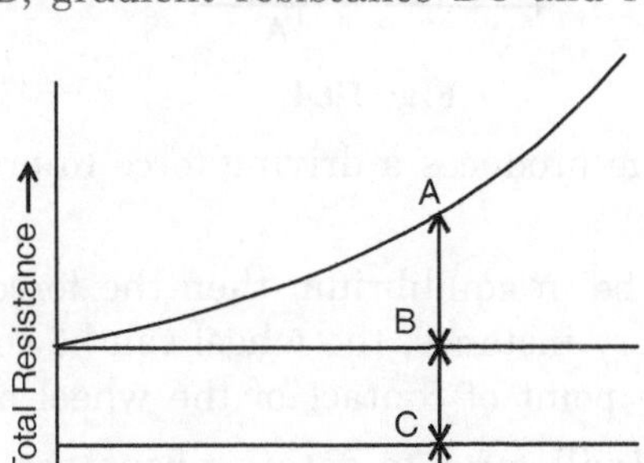

Fig. 10.3

10.2 TRACTIVE EFFORT

The source of tractive effort is engine. It turns the clutch shaft with torque T being transmitted to the gearbox. The gear box is located in between the clutch and the propeller shaft. The output of clutch, *i.e.*, rotation of clutch shaft, is the input to gear box and output of gear box is input to propeller shaft. Applying the principle of energy conservation, if frictional losses are neglected and if there is no storing of energy, the whole of the energy put into the gear box at the engine end must be given out at the propeller shaft end. Since the work done in unit time is measured by the product of torque and speed it means the product of engine torque and clutch shaft speed must be equal to the product of the torque acting on the propeller shaft and propeller shaft speed. If, therefore, the propeller shaft speed is $1/n_1^{th}$ of the engine speed then the propeller shaft torque must be n_1 times the engine torque.

Hence propeller shaft torque = $n_1 \times T_e$

n_1 — the gear box gear ratio between the engine (clutch) shaft and propeller shaft.

T_e — the engine torque, N–m.

The propeller shaft drives the road wheels through the final drive where another speed reduction occurs. If the road wheel speed is $1/n_2^{th}$ of the propeller shaft speed then the torque acting on the road wheel (the two wheels being considered to be single) will be n_2 times propeller shaft torque, again neglecting the frictional loss. The torque acting on the road wheel, $t_\omega = n_1 \times n_2 \times T_e$. n_1-the gear box ratio, n_2-the final drive ratio that is provided in the differential.

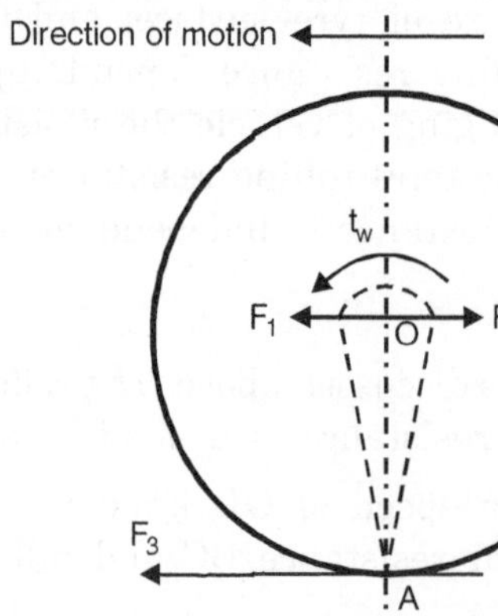

Fig. 10.4

The way in which this torque produces a driving force to propel the car along the road is as shown in Fig. 10.4.

If the wheel is regarded to be in equilibrium then the forces and couples acting upon it must be in equilibrium. At every instance, the wheel can be considered as a lever (shown by dotted line) fulcrumed at the point of contact of the wheel and the ground (A).

Under torque t_w the lever will tend to rotate about the point of contact with the ground (A) and the centre of the wheel (O) will tend to go forward and take the axle and the vehicle along with it. Let this force be F_1. The reaction F_2 of this force F_1 acts backwards on the wheel since the wheel is in equilibrium there must be an equal and opposite force F_3 acting at the point of contact.

F_2 and F_3 constitute a couple tending to turn the wheel in clockwise direction. Wheel being in equilibrium this must be equal to the torque t_w trying to rotate the wheel in counter clockwise direction. The magnitude of couple F_2, F_3 being $F_3 \times R$, where R is the radius of the wheel.

$$\therefore \qquad t_w = F_3.R \text{ or } F_3 = \frac{t_w}{r} = \frac{n_1\, n_2\, T_e}{r}$$ which is equal to tractive effort, P. or $P = \frac{n_1\, n_2\, T_e}{r}$

Now the values of n_2 and R are constant for any given vehicle and n_1 too is constant for a particular gear box. Hence $\frac{n_1 \times n_2}{r}$ is constant which can be taken as C.

Hence tractive effort, $P = C \times T_e$, where T_e is the engine torque and C is a constant depending upon wheel radius R, gear box ratio and final drive ratio.

10.2.1 Variation of Tractive Effort with Speed

The motion of the engine shaft is transferred to road wheels through suitable gearing in gear box. Therefore a particular vehicle speed corresponds to a particular engine speed. The tractive effort is proportional to engine torque therefore variation of tractive effort with engine speed will depend upon variation of engine torque with vehicle speed. The tractive effort also varies with the vehicle speed. Therefore the shape of the curve between tractive effort and vehicle speed is the same as that of engine torque vs. engine speed.

Thus the variation of engine torque with engine speed is similar to variation of tractive effort with vehicle speed. Thus the two curve are same as shown in figure 10.5.

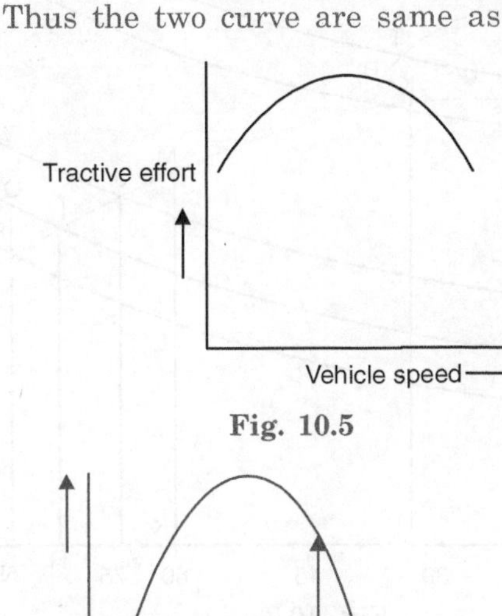

Fig. 10.5

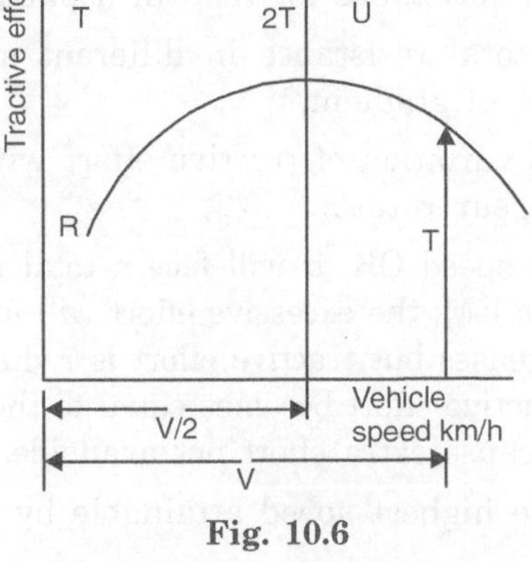

Fig. 10.6

Figure 10.6 represents variation of tractive effort with vehicle speed for two different gear box ratios. Curve between tractive effort and vehicle speed for a given gear box ratio, final drive ratio and road wheel radius is represented as RS. If the gear box ratio is altered so that the total gear ratio between engine and road wheels becomes double (than from previous case of curve RS) new curve TU will come into existence. It will be found that tractive effort T is available at vehicle speed V for curve RS. If the vehicle speed is reduced to V/2 tractive effort corresponding to curve TU will be 2T. This means that at a particular engine speed doubling the gear ratio will mean double tractive effort at half the vehicle speed.

10.3 PERFORMANCE CURVES

Figure 10.7 represents variation of total resistance with vehicle speed and variation of tractive effort with vehicle speed.

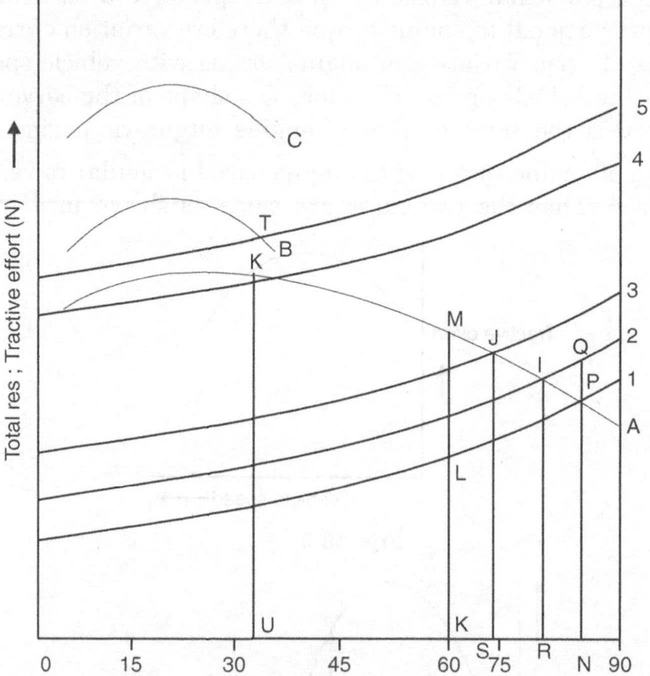

Fig. 10.7

Curve 1 is the curve of total resistance for road in a particular situation.

Curve 2, 3, 4, 5 represent total resistance in different situation. The resistance is increasing assumed to be because of gradient.

Curve A, B and C represent variation of tractive effort with vehicle speed. The three curves represent three different gear ratios.

Let vehicle be travelling with speed OK, it will face a total resistance of magnitude KL, while the tractive effort available is KM, the excessive effort will increase the vehicle speed. As the speed increases resistance increases but tractive effort is reduced until the vehicle attains a speed of ON at speed ON the tractive effort becomes equal to the resistance (point P). Hence the acceleration is not possible because extra effort not available.

Hence for curve 1, ON is the highest speed attainable by the vehicle.

The curve A represents the tractive effort when engine is running with throttle wide open. If speed beyond OK is not desired throttle will be closed thereby reducing the tractive effort. Once it becomes equal to KL the vehicle will run at constant speed of OK.

Now suppose the vehicle running with speed ON and it comes to a gradient where curve 2 is applicable. Here tractive effort available is NP but resistance is NQ. As resistance NQ is more the vehicle will slow down to speed of OR till the tractive effort and resistance become equal.

When the road becomes steeper as shown by curves 3, 4 and 5 the difference between resistance and tractive effort becomes more, more and more and speed that can be maintained becomes less, less and less (represented by points J, K etc). It can also be seen that for a gradient where total resistance is represented by curve C, the vehicle can not move as the tractive effort is always less than the resistance.

That means with a single gear ratio there can be a gradient like 5 where vehicle ceases to move. To overcome this gear ratio is changed and tractive effort is increased we get tractive effort curves B and C.

The vehicle can overcome higher resistance for second gear ratio which provides tractive effort curves B or C. The gear ratios in the gear box are generally four or five. For quick acceleration at low speed excessive tractive effort should be available. It can be observed (Fig. 10.7) that for vehicle moving at a speed of OK, excessive tractive effort, LM is available. Ideally the tracting effort should be just enough to overcome resistance. The vehicle while moves comes across infinite number of situation and each situation offers a particular resistance. Thus, ideally there should be infinite number of gear ratios each providing just enough tractive effort.

The above discussion assumes maximum speed to be the objective. If maximum economy is desired, the engine is required to be run at the highest torque possible for a given power output. Many recent developments like hydraulic and electrical mechanisms aim at providing higher number of gear ratios. Basically an engine size determines whether two, three, four or even more gear ratios be provided.

More than one gear ratio can be provided by having more than one pair of gears. This fact establishes the necessity of having more than one pair of gears or having a gear box in an automobile. These pair of gears are enclosed in a box *i.e.,* gear box. Another requirement, in case of automobiles is that these are required to more in backward direction also. Therefore, one pair of gears is needed to reverse the direction of motion. Gear box may be classified as *(i)* sliding-mesh type *(ii)* constant-mesh type.

There may be gear boxes which are combination of two types mentioned above spur gearing is used in all the gear boxes. The only difference is in the manner in which gears are brought into action. Sliding mesh type gear box is the simplest and the oldest.

10.4 SLIDING MESH GEAR BOX

The motion from the engine shaft, through clutch is transferred to the shaft, E which has pinion P at its end (Fig. 10.8). The shaft is supported in bearings. The pinion P meshes with spur wheel A. Wheel A is mounted on lay shaft L having its axis parallel to that of shaft E. This shaft is also supported in bearings at ends. Four spur wheels A, B, C and D are mounted on lay shaft L. Fifth wheel E continuously meshes with pinion F. The pinion F is free to revolve on a pin fixed in the casting. Third shaft K, known as main shaft, is arranged in line with shaft E. This shaft is also supported at one end in the casing and on other end in the spigot. The casing and the spigot are not shown in the figure for the sake of simplicity.

Splines are provided throughout the length of shaft K. Spur wheel G and a double wheel H_1 and H_2 are mounted on shaft K. These wheels can have axial movement but the shaft rotates along with these wheels. The selector forks make possible the axial movement of wheels so that these can mesh with wheels on lay shaft L. Again the selector forks have not been shown for the sake of simplicity. These forks are fixed on the rods that slide in the holes provided in the casing. The holes in the casing are provided with bushes for smooth movement.

Claw teeth are provided on spur wheel G and the pinion P. When these claw teeth come in contact with each other the motion from driving shaft is directly transferred to driven shaft. The contact is positive and therefore the speeds of driving shaft and driven shaft are equal.

The main shaft K is connected to the propeller shaft which in turn is connected to road wheels through differential and the axle. The lay shaft is supported in the casing with the help of thrust buttons. The left end of the main shaft K through thrust button is placed against the engine shaft. The right end of the shaft is supported in the wall of the casing. The engine shaft E is fixed at the right end with thrust button against the end of main shaft K. The gear box under consideration provides four forward and one reverse gear.

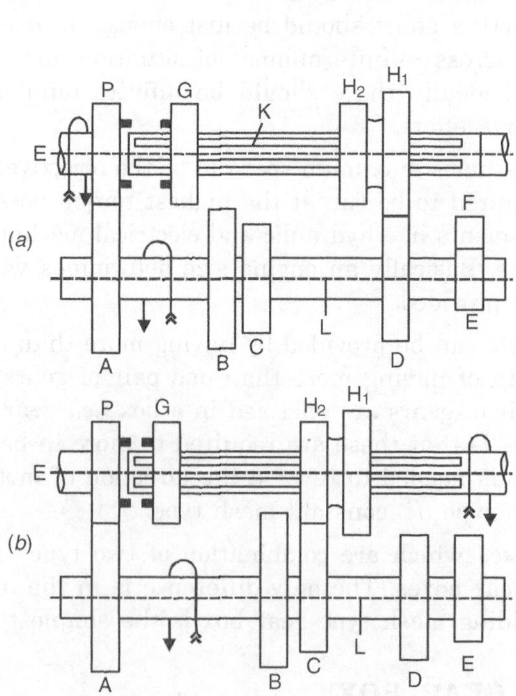

Fig. 10.8 (*a*), (*b*)

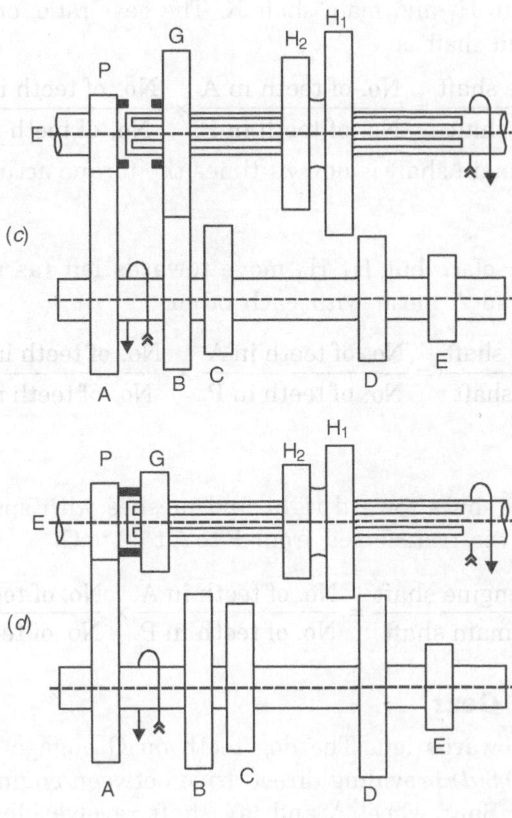

Fig. 10.8 (*c*), (*d*)

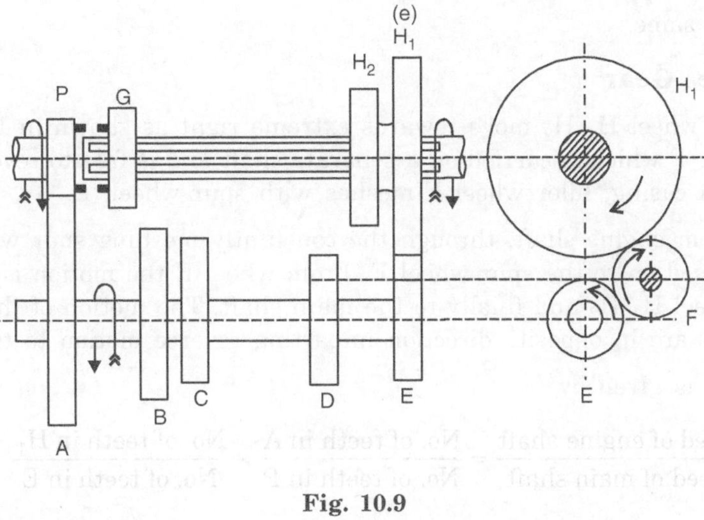

Fig. 10.9

10.4.1 First Gear

Gear H_1 H_2 slides along the main shaft K until H_1 meshes with D as shown in Fig. 10.8 (*a*). The motion comes from engine shaft through constantly meshing gears P and A to the lay shaft.

Through D it is transmitted to H_1 and main shaft K. The gear ratio, or the ratio between the speeds of the engine and main shaft is

$$\frac{\text{Speed of engine shaft}}{\text{Speed of main shaft}} = \frac{\text{No. of teeth in A}}{\text{No. of teeth in P}} \times \frac{\text{No. of teeth in } H_1}{\text{No. of teeth in D}} = n_1$$

The torque driving the main shaft is now n_1 times the torque acting on the engine shaft.

10.4.2 Second Gear

G remains at the same place but H_1 H_2 move towards left (as shown in Fig.10.8 (*b*)) and H_2 meshes with C. P and A mesh with each other.

$$\frac{\text{Speed of engine shaft}}{\text{Speed of main shaft}} = \frac{\text{No. of teeth in A}}{\text{No. of teeth in P}} \times \frac{\text{No. of teeth in } H_2}{\text{No. of teeth in C}} = n_2$$

10.4.3 Third Gear

Here, the spur wheel G shifts toward right and meshes with spur wheel B (as shown in Fig. 10.8 (*c*)). The motion is transferred from P to A to B to G.

$$\frac{\text{Speed of engine shaft}}{\text{Speed of main shaft}} = \frac{\text{No. of teeth in A}}{\text{No. of teeth in P}} \times \frac{\text{No. of teeth in G}}{\text{No. of teeth in B}}$$

10.4.4 Fourth or Top Gear

In this case G moves towards left. The dog teeth on G engage with similar teeth on pinion P as shown in Fig. 10.8(*d*) providing direct drive between engine and the main shaft. The gear ratio becomes 1:1. Spur wheel A and lay shaft revolve idly.

The direction of motion of layshaft is opposite to that of engine shaft and direction of motion of main shaft opposite to that of layshaft. Thus direction of motion of engine and mainshafts are the same.

10.4.5 Reverse Gear

Here the spur wheel H_1 H_2 move towards extreme right as shown in Fig. 10.9. Here it meshes with idler F which is carried on a separate shaft and is free to revolve. The shaft is fixed in gear box casing. Idler wheel F meshes with spur wheel, E.

The motion from engine shaft, through the constantly meshing spur wheels P and A and layshaft is passed on to the spur wheel E. From wheel E the motion is transferred to the idler F, the wheel H_1 H_2 and finally to the main shaft. The motion of the engine shaft and the main shaft are in opposite direction imparting reverse motion to the automobile.

The gear ratio is given by

$$\frac{\text{Speed of engine shaft}}{\text{Speed of main shaft}} = \frac{\text{No. of teeth in A}}{\text{No. of teeth in P}} \times \frac{\text{No. of teeth in } H_1}{\text{No. of teeth in E}}$$

Idler does not alter the ratio, it just reverse the direction of rotation.

There can be an alternate arrangement for reverse gear. Instead of simple reverse idler, there is a compound wheel with two spurs. This compound wheel is provided with its own shaft duly supported in the casing. One spur meshes with wheel on the lay shaft while the other spur meshes with the wheel on the main shaft. This helps to reduce the vertical distance between lay shaft and main shaft as the idler is mounted on independent shaft. Also, as the number of wheels on lay shaft is reduced the length of the shaft can be shortened. Both these factors help is reducing the overall size of the gear box.

10.5 SELECTOR MECHANISM

Working of sliding mesh gear box depends upon axial movement of spur wheel on the shaft. This movement is provided without disturbing the rotary movement of wheel. To make possible the movement of spur wheels, mechanism is employed consisting of forks (Fig. 10.10). The number of forks depend upon the number of spur wheels. The movement of each wheel is handled independently. Grooves are provided in the boss of wheels where fork fits into. These forks may slide on the rods that are fixed in the gear box casing. Alternatively, the forks are fixed on the rods and rods slide. The axis of rods is parallel to the axis of the shaft. The mechanism includes gear lever which is operated by the driver of the vehicle. The 'push or 'pull' exerted by the driver on the gear lever cause the movement of forks through various components of control mechanism.

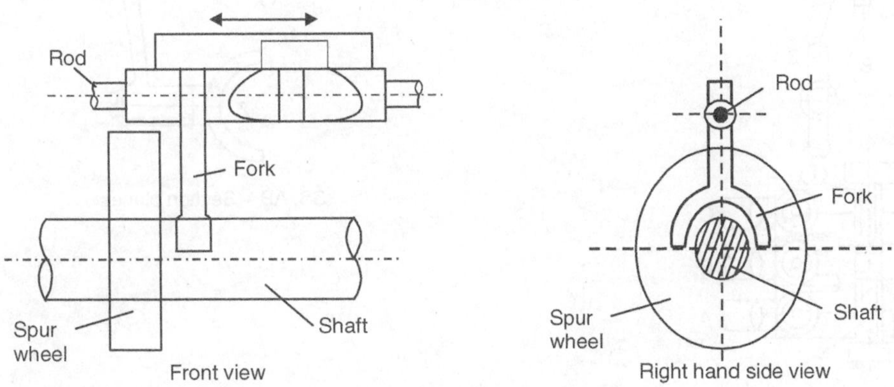

Fig. 10.10

10.5.1 Sliding Type Selector Mechanism

Figure 10.11 represents the three sectioned views of the sliding type selector mechanism. There are three selector forks X, Y and Z for three spur wheels in the gear box. Forks X and Y slid on rods A and B fixed in the casing. Z is carried by pivoted lever Q. Q is actuated by member that slides on the rod C, the forks are moved by a fore and aft motion of the gear lever D. This motion is taken up by a shaft F pivoted in the casing. Its inner end is secured to the striking lever E. The particular fork requiring motion is selected by the sideways sliding motion of member D, F and E. The forks are kept in their positions by springs. The grooves are cut on rods A, B and C where springs get fixed.

The simultaneous movement of two forks is prevented with the help of a locking piece. This slides on cross rod N that is secured in casing. G is provided with horns I and J and between these horns is situated the end of the striking lever K, so that the sideways movement of K causes the member G to slide on its rod. The gap between I and J is only slightly wider than each of the sliding members. This makes possible movement of one sliding member at a time.

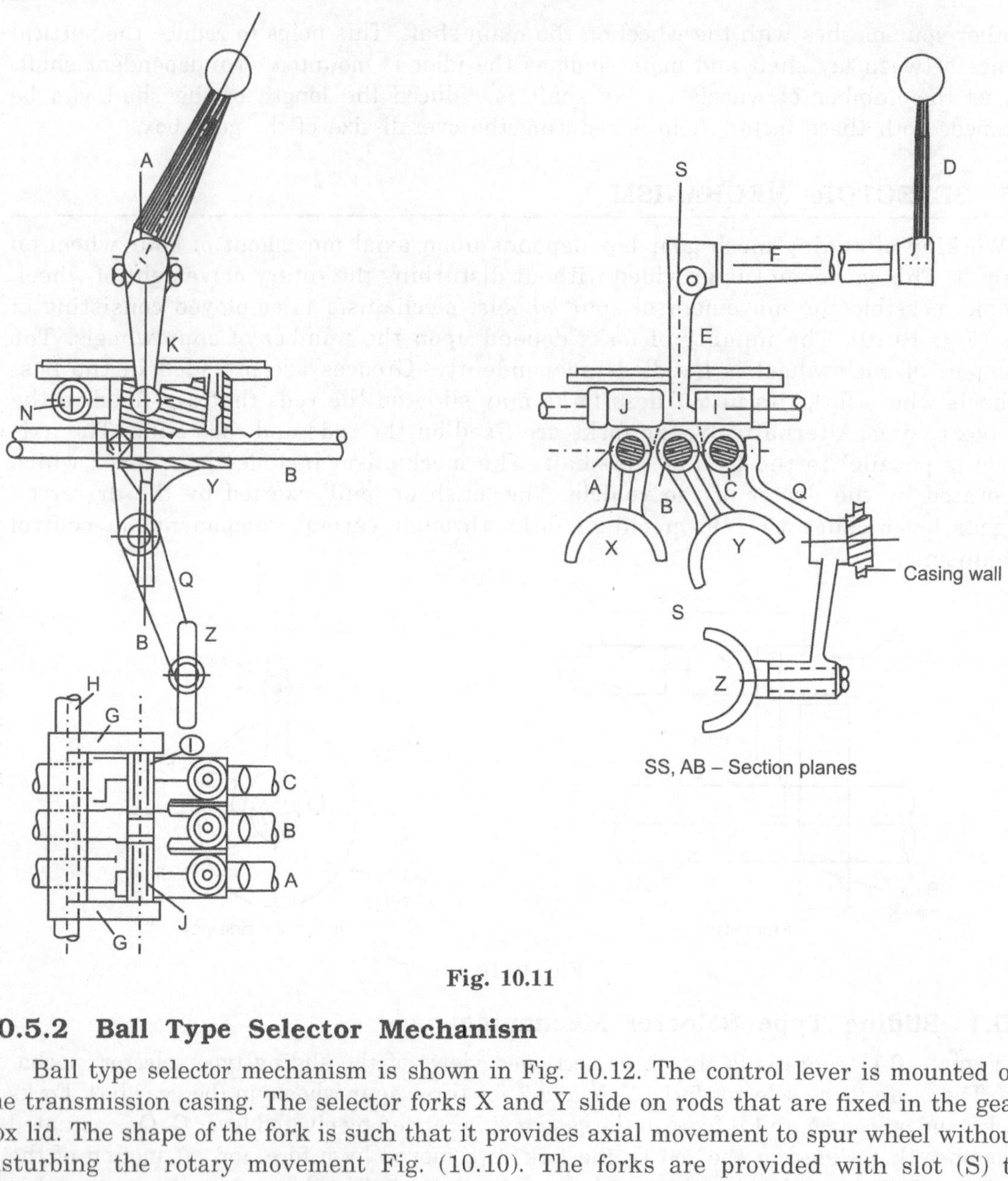

Fig. 10.11

10.5.2 Ball Type Selector Mechanism

Ball type selector mechanism is shown in Fig. 10.12. The control lever is mounted on the transmission casing. The selector forks X and Y slide on rods that are fixed in the gear box lid. The shape of the fork is such that it provides axial movement to spur wheel without disturbing the rotary movement Fig. (10.10). The forks are provided with slot (S) to accommodate the end (E) of the arm. E is the lower end of the gear lever L. These is ball joint J. By pushing/pulling the lever, the end E engages either of the selector forks. Sliding motion is provided to forks along the rod on which these are mounted.

Two small plungers P_1 and P_2 are provided to prevent sudden and simultaneous movement of forks. In neutral position, slots S are opposite to each other. Holes are provided in forks and small springs force the plungers into these holes. To move the fork, the plunger locking it, must be pressed. Sideways motion of gear lever presses the plunger and unlocks the fork. When one plunger is pressed other keeps the other fork locked. After the gear is engaged properly, the forks are locked by springing into recess R in the forks.

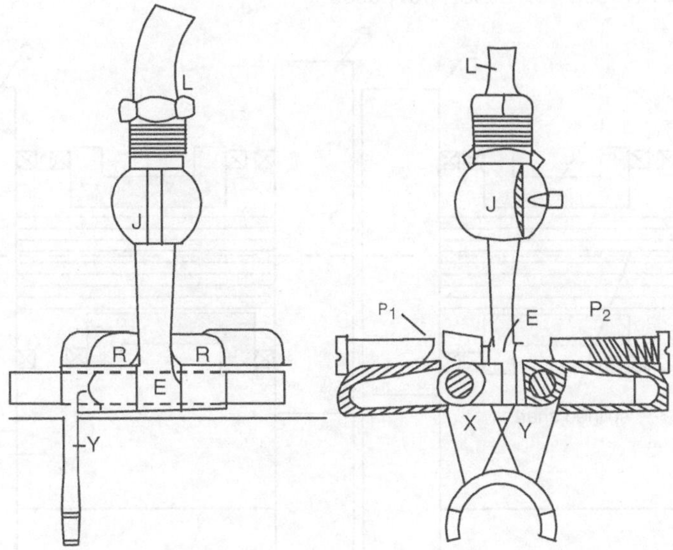

Fig. 10.12

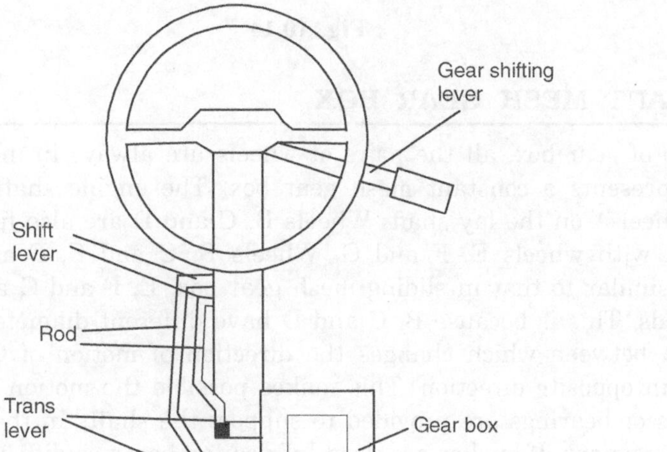

Fig. 10.13

In old cars, the gear shifting mechanism is accommodated in steering column. The gear shifting lever is provided in the steering wheel (Fig. 10.13). This provides extra space on the floor which can be utilised for other purposes.

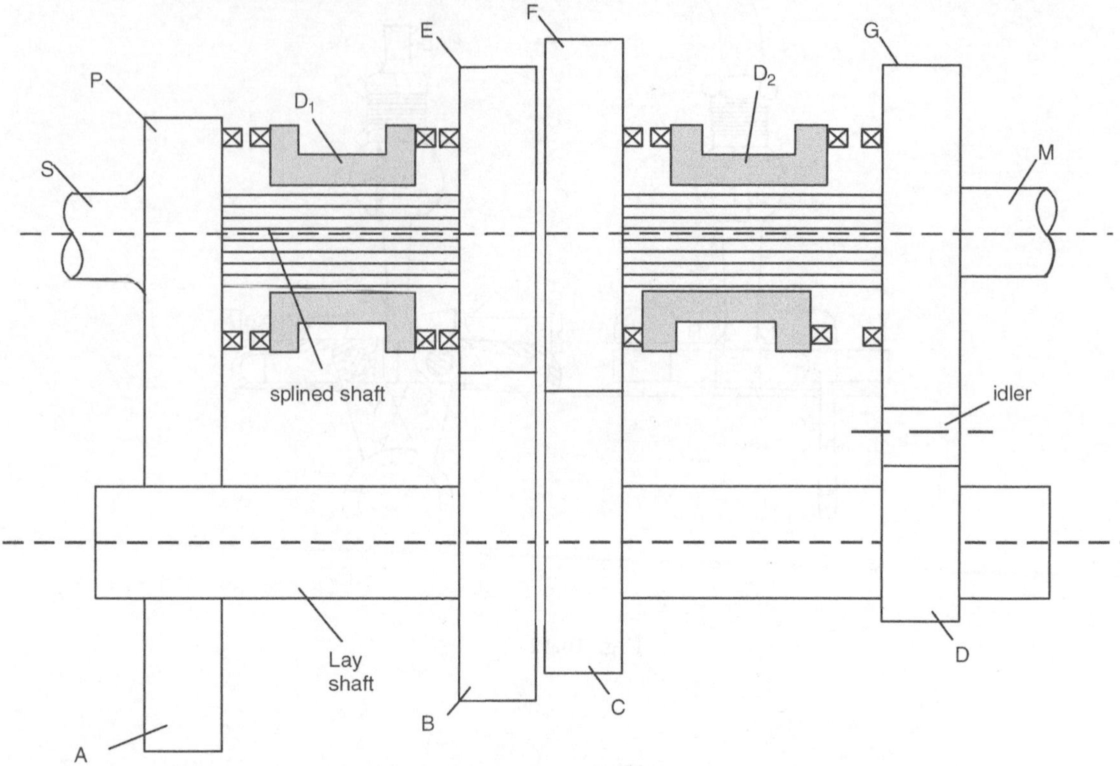

Fig. 10.14

10.6 CONSTANT MESH GEAR BOX

In this type of gear box, all the pairs of wheels are always in mesh with each other. Figure 10.14 represents a constant mesh gear box. The engine shaft, S has a pinion P, meshing with wheel A on the lay shaft. Wheels B, C and D are also fixed on lay shaft and constantly mesh with wheels E, F and G. Wheels B, C and E, F are twin wheels. The arrangement is similar to that in sliding mesh gear box. E, F and G are constantly driven at different speeds. This is because B, C and D have different diameters. Wheels G and D have an idler in between which changes the direction of motion of G. The wheel G and pinion P rotate in opposite direction. This makes possible the motion of vehicle in reverse direction. Bushes or bearings are provided to support the shafts in the gear box casing for smooth rotary movement. If bushes are used lubrication becomes difficult. This causes wear and noise. Roller bearings, when used, can ease the lubrication but make the gear box bulky. Needle-roller bearings are most beneficial as these allow the easy lubrication and do not make the gear box bulky. These have not been shown in the figure for the sake of simplicity. Between wheels E, F and pinion P a dog clutch, D_1 is provided. Similarly, between E, F and wheel G another dog clutch, D_2 is provided. These dog clutches are required to have free movement axially but there should be no relative rotary motion between the shaft and the clutches. To achieve this, the shaft is splined as shown.

Alternatively, the shaft can be provided with square cross-section. The dog clutches are provided with the teeth and similar teeth are provided on the pinion, P and wheels E, F and G. If dog clutch D_2 slide towards left it's teeth mesh with teeth on wheel F. The motion would be transferred from pinion, P to wheel A to wheel C to wheel F and to the main shaft M. The dog clutch D_1 remains in neutral position. Next dog clutch D_1 move towards right and its teeth mesh with teeth on wheel E. Other dog clutch attains neutral position. The motion, in this case, is transferred from pinion, P to wheel A, to wheel B, to wheel E and to the main shaft M. When dog clutch D_1 moves towards left its teeth directly mesh with the teeth on pinion P and motion from engine shaft, E is transferred directly to the main shaft M. The right ward movement of dog clutch D_2 causes the meshing of teeth of wheel F and G. In this case, the motion is transferred through idler and therefore wheel G moves in opposite direction causing motion of the vehicle in reverse direction.

In constant mesh gear box, helical or double helical gear wheels can be used. These operate quietly and therefore the working of the gear box is less noisy. The teeth on the dog clutch are easier to engage. These are less liable to damage. In this type of gear box, synchromesh device can be readily fixed which is an additional advantage.

10.7 SYNCHROMESH DEVICE

Shifting of gears can be inconvenient and if not done properly it can cause jerks and even damage the gear wheels. For the convenience of the driver, the shifting of gears should be as smooth as possible. Synchromesh device is fixed in the gear box and makes possible the smooth shifting of gears. The wheels to be engaged are initially brought into frictional contact and when friction has almost equalized their speeds, positive contact is made. The device can be readily employed in constant mesh gear box and with some modifications in sliding mesh gear box.

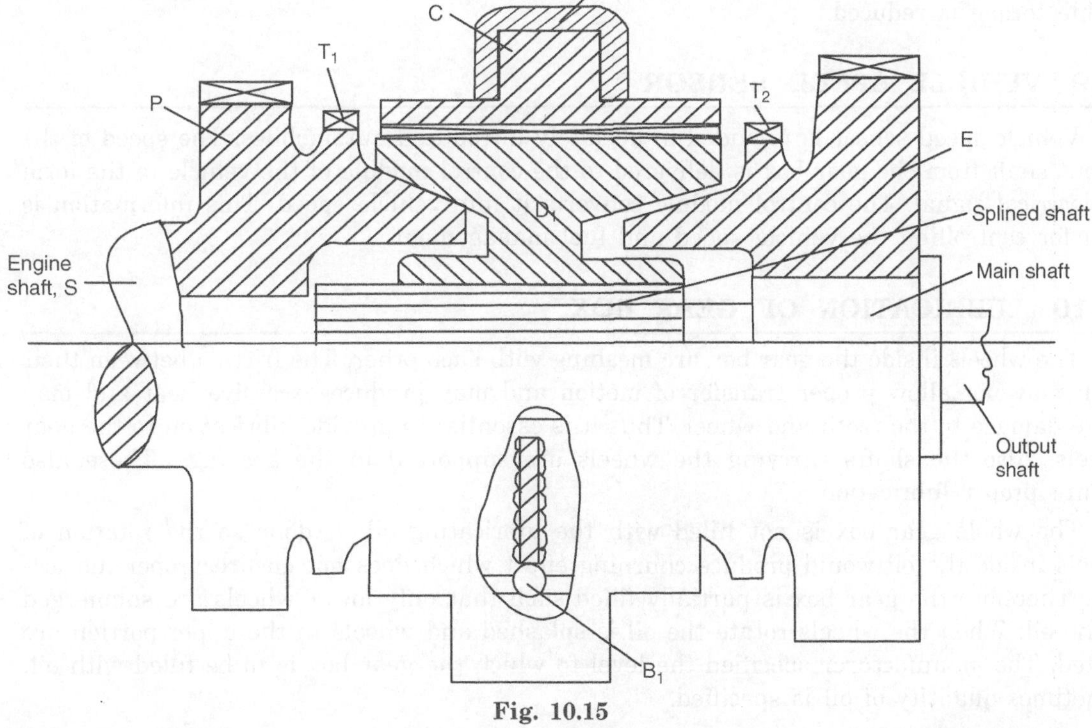

Fig. 10.15

Fig. 10.15 represents the synchromesh device applied to the direct and the lower gear of a constant mesh gear box. S is the engine shaft and the integral pinion P meshes with wheel fixed on the lay shaft that has not been shown. The wheel E is free to rotate on the main shaft and permanently meshes with another wheel fixed on the lay shaft. Wheels P and E have dog tooth portions T_1 and T_2 and conical portion with slanting surfaces. D_1 is free to slide on splines on the main shaft. D_1 has conical portion corresponding to conical portion on P and E. As D_1 slides towards left contact between conical surfaces occurs. Due to friction between conical surfaces motion is transferred. D_1 has teeth exactly similar to those on T_1 and T_2. Another member C can slide. Its sliding towards left causes positive contact between P and D_1 and transfer of motion from pinion to splined shaft. Before positive contact is made it has to overcome the resistance due to six spring loaded balls (only B_1 shown in the figure). The speeds of the P and D_1 being almost equal a little effort is required to overcome this resistance. These balls do not allow the accidental removal of the contact. The movement of C is caused by selector fork K operated by the driver through gear lever. In between shifting of gears the lever is brought to neutral position. Because constant load is applied during shifting of gears this type of synchromesh device is classified as constant load synchromesh.

10.8 OVERDRIVE

In a gear box, for lower gears, smaller gear wheels drive larger gear wheels. Thus, gear ratio less than one is provided. This also means that driven shaft rotate at lesser speed than driving shafts. Under this condition, at high cruising speed the engine speed may increase and may go beyond the permissible economical limit. This means higher consumption of fuel. To avoid this situation the large gear wheel is made to drive small gear wheel at top gear. This provides high speed of driven shaft while driving shaft is at low speed. This makes possible high cruising speed even at low engine speed which is within the permissible economic limit. Though this provides economical running of automobile at high speed but engine torque is reduced.

10.9 VEHICLE SPEED SENSOR

Vehicle speed sensor is provided in all lately introduced automobiles. The speed of the output shaft from the gear box is delivered to the control module of the vehicle in the form of electrical signal. The control module converts it into vehicle speed. This information is vital for controlling the vehicle speed and fuel management.

10.10 LUBRICATION OF GEAR BOX

The wheels inside the gear box are meshing with each other. The friction between their teeth may not allow proper transfer of motion and may produce excessive heat and may cause damage to the teeth and wheel. Thus it is essential to provide lubrication to the gear wheels. Also the shafts carrying the wheels are supported in the bearings. These also require proper lubrication.

The whole gear box is not filled with the lubricating oil. If done so the rotation of wheels inside the oil would produce churning effect which does not ensure proper lubrication. Therefore the gear box is partially filled such that only lower wheels are submerged in the oil. When the wheels rotate the oil is splashed and wheels in the upper portion are wetted. The manufacturer specified the level to which the gear box is to be filled with oil. Sometimes quantity of oil is specified.

In a gear box, the oil is pressed in between the teeth of the wheels. Thus it should be capable of taking up high pressure. The temperature is not that high in the gear box. Therefore, alteration in the viscosity of oil at high temperatures does not require consideration.

Leakage of oil may occur as the air inside the gear box expands. To avoid this, gear box is provided with a vent so that the expanded gas comes out. The ball bearings supporting the shaft are provided with large washers to keep any particle of grit or chip outside the bearings.

The external parts, such as selector mechanism parts are lubricated with thin machine oil or engine oil. Every manufacturer provides the instructions to the user that are vital and need to be followed for proper working of gear box.

10.11 TORQUE CONVERTER

Torque converter is a device that transfers motion from driving shaft to driven shaft in a particular ratio. Its function are similar to that of gear box, the difference being that speed can be changed with a continuous variation of ratio from the lowest to the highest.

A torque converter consists of the following components: (*a*) Driving element or impeller connected to the engine shaft (*b*) the driven member or rotor connected to the driven shaft *i.e.*, propeller shaft and (*c*) the 'fixed' element or reaction member fixed to the frame. The fixed element makes it possible to obtain change of torque between input and output shafts.

Torque converter is represented in Fig. 10.16 *a*, *b* and *c*. This is a three stage converter. In three stage converters, output torque is available at three different stages. The torque converter consists of impeller connected to driving shaft, rotor connected to the driven shaft and fixed casing which also acts as enclosure for impeller and rotor. Figure 10.16(*a*) represents elevation, and side view of the impeller which have been suitably sectioned to show the inner details. Fig. 10.16(*b*) represents the rotor. The two sectioned views shown are elevation and side view. Figure 10.16(*c*) represents the assembled view, upper half sectioned to show the inner details.

Impeller, Fig. 10.16(*a*), is provided with set of blades I_1, at the outer edge of disc. These blades are covered by a ring shaped disc I_C. The impeller is at the end of driving shaft and it rotates alongwith it.

Rotor, Fig. 10.16(*b*), is provided with three set of blades namely R_1, R_2 and R_3. As shown in the side view sets of blades R_1 and R_2 are on one side of tubular ring and set of blades R_3 is on the other side of tubular ring. Set of blade R_2 are part of disc connected to driven shaft. Sets of blades R_1 and R_3 are provided with a cover RC_1 and RC_3 in the form of ring shaped disc. Elevation shows the three set of blades. The view has been suitably sectioned to show the inner details.

Figure 10.16(*c*) represents the assembled view of torque converter. Third component is the casing which acts as enclosure for the impeller and rotor and is provided with two sets of blades. One set of blades is F_1 located at the top and another set of blades is F_2 located on right hand side. This component provides the reaction component to the fluid flow and is also referred to as reaction member.

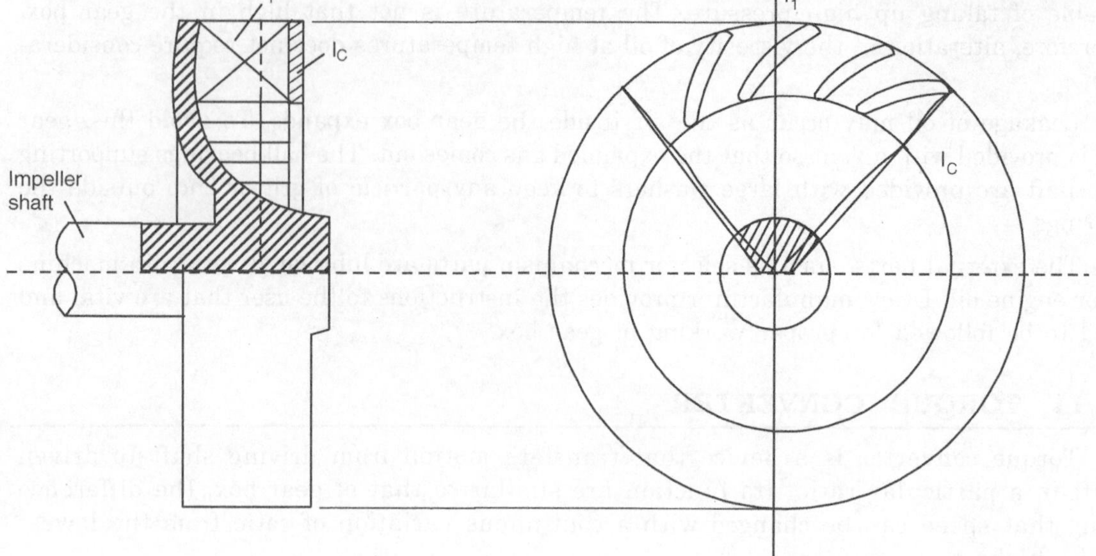

Fig. 10.16(*a*)

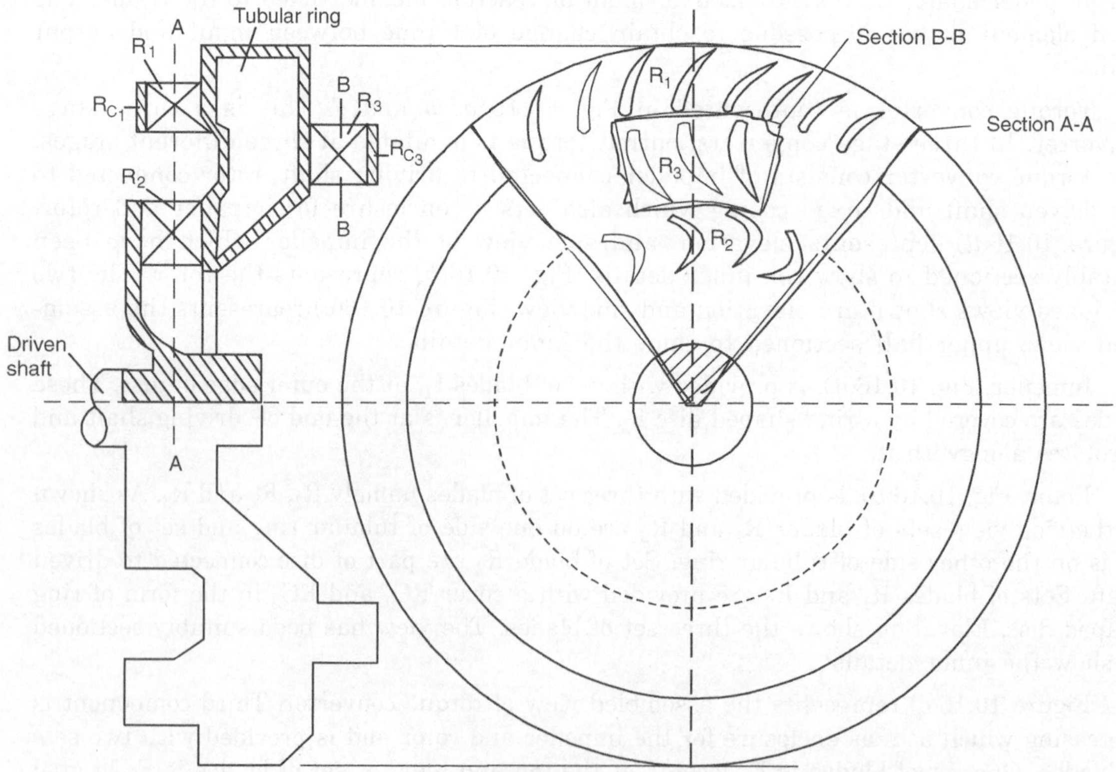

Fig. 10.16(*b*)

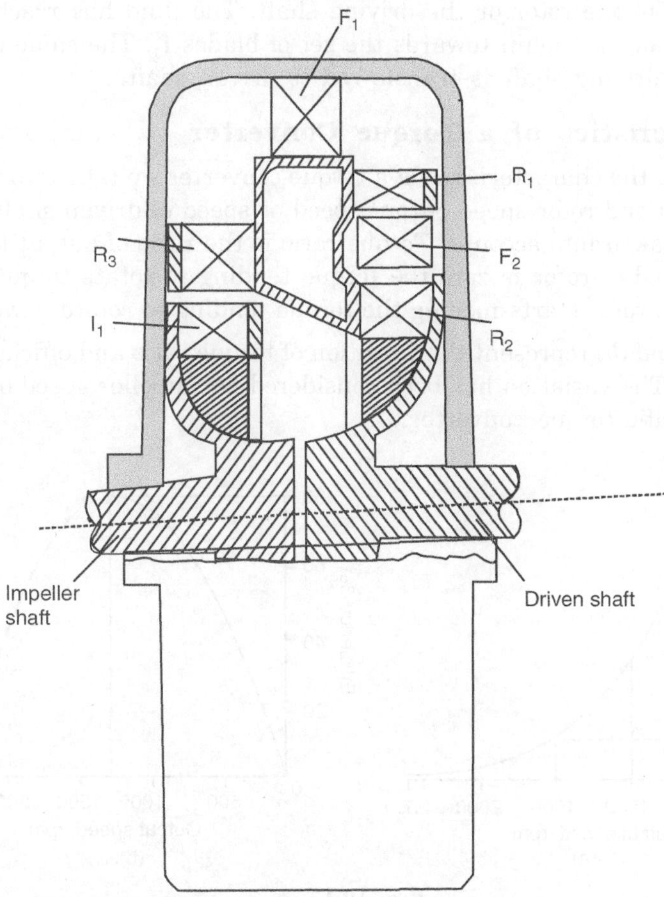

Fig. 10.16(*c*)

10.11.1 Working of the Torque Converter

Considering the flow of fluid from the lower portion *i.e.,* eye of the impeller, the fluid passes through the set of blades I_1, part of impeller. The fluid files off the blades I_1 and strikes against the set of blades, R_3 and is deflected. The fluid comes out with certain velocity. Its tangential component is lost and as the fluid leaves the blades R_3 it has only radial component. The particle loses momentum which is gained by set of blades R_3 (or rotor). This is the first stage when energy is transferred from impeller *i.e.,* driving shaft to rotor *i.e.,* driven shaft. The flow of fluid is now upwards in the casing. The fluid enters the set of blades F_1 located at the top of the casing. When the fluid flows through these blades, the axial velocity of fluid particle is changed into velocity having significant tangential component. When fluid is deflected in blades F_1, the casing receives a backward thrust and would have rotated back, had it not been fixed. As a reaction to it the fluid particle gets forward thrust. The fluid now flows downwards and passes through the set of blades R_1. Here again the tangential component of velocity is removed which causes loss of momentum. This loss is the gain of blades R_1. This is second stage when energy is transferred to the driving or to the driven shaft. The fluid next enters the set of blades F_2 which are fixed to the casing. While passing through blades F_2, the tangential component of fluid particle velocity is restored which is removed as the fluid flows through R_2, the set of blades with rotor. This means loss of momentum by fluid and gain of momentum by rotor. This is third stage where energy is transferred to the rotor or the driving shaft. The fluid has reached in the lower most part and starts moving again towards the set of blades I_1. The same cycle is repeated and energy from the driving shaft is transferred to driven shaft.

10.11.2 Characteristics of a Torque Converter

While considering the characteristics of a torque converter we take into account impeller speed *i.e.,* input speed and rotor speed output speed or speed of driven shaft. Also efficiency and torque ratio are taken into account. Torque ratio is the ratio of output torque and input torque. When the speed of rotor is zero the torque tending to rotate torque ratio would be the highest. When the rotor starts moving the torque tending to rotate it would be reduced.

Figure 10.17(*a*) and (*b*) represent the variation of torque ratio and efficiency with respect to speed of the rotor. The variation has been considered for impeller speed of 2500 rpm. The values are for a specific torque converter.

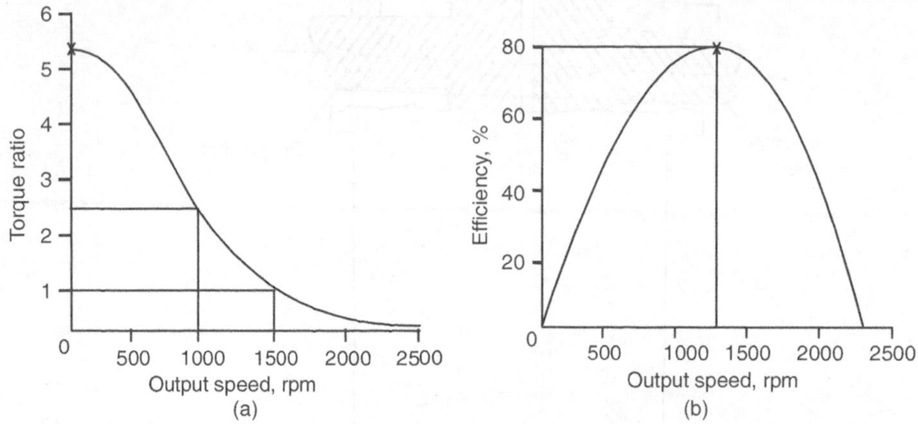

Fig. 10.17

The graph (*a*) shows the variation of torque with respect to rotor speed (or output shaft speed) when impeller speed (*i.e.,* input) is constant. When the rotor speed is zero the torque tending to rotate it is 5.25 times the torque developed by the engine at 2500 rpm. As rotor starts moving and gathers speed (at about 1000 rpm), the ratio driving torque (output) to engine torque (input) becomes 2.5 and at about 1600 rpm, the ratio becomes equal to one becoming one. At about 2500 rpm the ratio is (*i.e.,* driving torque) becomes almost equal to zero.

As shown in (*b*) efficiency is zero when rotor speed is zero as no work is available as output even when input is being given. With increase in rotor speed the efficiency increases becoming 80% at rotor speed of about 1200 rpm. Further the driving torque decreases thereby reducing the efficiency which ultimately becomes equal to zero.

Thus only in a narrow range of rotor speed the efficiency is reasonably good. 80% efficiency means 20% of the engine power being wasted and being converted into heat there by raising the temp of converter fluid. The fall off of efficiency at low speed end is acceptable as that occurs while starting and moving over gradients but fall off at high speeds is undesirable and is controlled by substituting a direct drive for the torque converter at high speeds.

10.11.3 Direct Drive Torque Converter

As has been seen that efficiency of the torque converter falls with increase in output speed (Fig. 10.17). To avoid this situation the torque converter is by passed at high speed and motion from engine shaft is directly transferred to the output shaft with the help of clutch plate as shown in Fig. 10.18. Clutch plate, C can move either way to left or right. Initially it is kept towards left in contact with impeller plate, I and motion is transferred to the output shaft through torque converter with the help of drive, D. When the output speed reaches a value where maximum efficiency is available. Plate, C is moved towards right and comes in contact with plate, P (through friction surfaces–as shown) and motion is transferred to shaft, S and finally to driven shaft. This shifting of clutch plate, C is done with the help of mechanism. The friction between the surfaces causes the transfer of motion which is almost 100% and therefore the efficiency of the torque converter is also almost 100%. The rotor of the torque converter comes to rest after a while. Figure 10.19 shows graphical representation of efficiency with output speed.

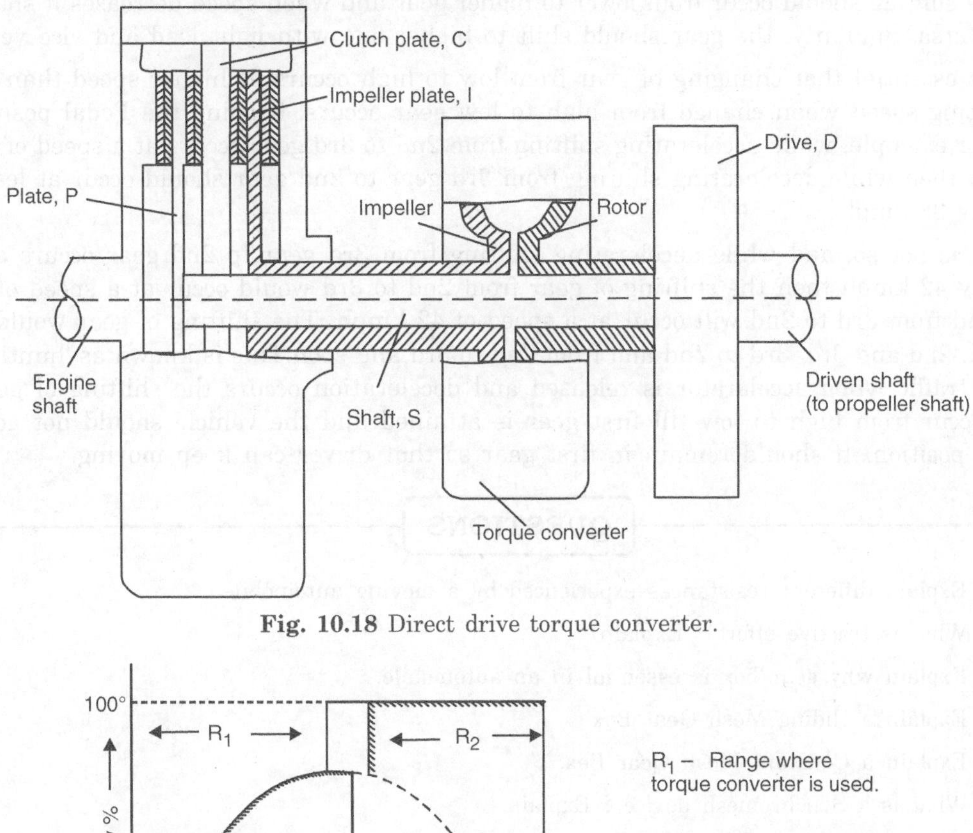

Fig. 10.18 Direct drive torque converter.

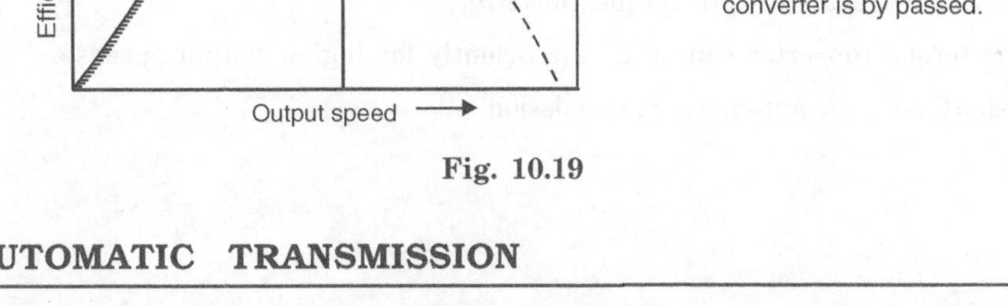

Fig. 10.19

10.12 AUTOMATIC TRANSMISSION

Automatic transmission is an attempt to provide convenience to the driver of the automobile. Driving vehicles with manual transmission is quite cumbersome. Clutch pedal has to be pressed and shifting of gears is required to be done repeatedly. Both these operations make driving quite inconvenient.

There are a number of automobiles available with automatic transmission. In most of them, control levers or buttons are provided. By operating these levers the vehicle is shifted to gear from neutral or to move the vehicle in reverse direction. These levers also enable the shifting of gears in a 'range' for example when you are moving on a plain road you have to drive the vehicle in a particular 'range'. Similarly, while moving upwards on hilly road you have to opt for another 'range'. Within a particular 'range' the shifting of gears is automatic.

There are two parameters to be considered for shifting of gears from lower one to higher one. One is the speed of the vehicle and another is load on the engine. When the speed increases shifting should occur from lower to higher gear and when speed decreases it should be vice versa. Similarly, the gear should shift to higher one with light load and vice versa.

It is essential that changing of gear from low to high occurs at higher speed than the decelerating speed when change from high to low gear occurs, keeping the Pedal position same. For example, while accelerating shifting from 2nd to 3rd gear occurs at a speed of say 40 kmph then while decelerating shifting from 3rd gear to 2nd gear should occur at lesser speed say 38 kmph.

If it is not so, and while decelerating shifting from 3rd gear to 2nd gear occurs at a speed say 42 kmph then the shifting of gear from 2nd to 3rd would occur at a speed of 40 kmph and from 3rd to 2nd will occur at a speed of 42 kmph. The shifting of gear would be cyclic *i.e.*, 2nd and 3rd, 3rd to 2nd and from 2nd to 3rd and soon.This is known as 'hunting'. Also, in traffic when accelerator is released and deceleration occurs the shifting of gears should occur from high to low till first gear is attained and the vehicle should not go to 'neutral' position. It should remain in first gear so that driver can keep moving.

QUESTIONS

1. Explain different resistances experienced by a moving automobile.
2. What is tractive effort ? Explain.
3. Explain why gear box is essential in an automobile.
4. Explain a sliding Mesh Gear Box.
5. Explain a Constant Mesh Gear Box.
6. What is a Synchromesh device ? Explain.
7. Explain a Torque Converter.
8. Explain the characteristics of torque converter.
9. How the torque converter can be used efficiently for higher output speeds.
10. Write short note on automatic transmission.

11

PROPELLER SHAFT, DIFFERENTIAL AND AXLE

11.1 PROPELLER SHAFT

The motion from gear box is transferred to differential which is the next component of the transmission system. The differential is located in the middle of the rear axle and the vehicle is rear wheel drive. Alternatively, the vehicle may be front wheel drive or in some cases, motion from the engine may be transferred to both-front and rear axles. In case of rear wheel drive vehicle the distance between the gear box and differential may be considerable and the propeller shaft connecting the two has considerable length. In case of front wheel drive the distance between the two may be small and so small propeller shaft would be required. To study the propeller shaft, it is proper to consider vehicle with rear wheel drive.

The wheels are mounted on the axle at its two ends. In between the wheels and the axle is suspension system. The suspension system has leaf springs providing up and down movement to the axle. Thus differential also has swinging up and down movement. The gear box is located in the front portion of the vehicle is fixed. Therefore motion is being transferred through propeller shaft from one fixed end to other end which is swinging. Due to this the angle between the gear box end and the differential does not remain same and varies as depicted in figure 11.1. Also the distance between the two ends does not remain same and varies which has also been depicted in the figure. It can also be noticed from the figure that the gear box and the differential are located at different heights. To cope up with these conditions the propeller shaft is provided with universal joints at the two ends and a slip joint. The universal joint takes up the variation in the angle and slip joint takes up the variation in the distance between two connected points.

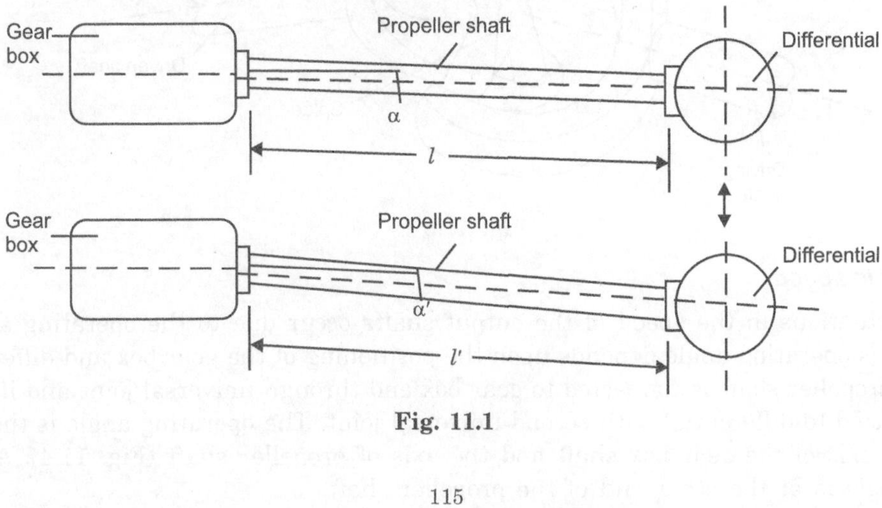

Fig. 11.1

11.2 UNIVERSAL JOINT

When motion is to be transferred from one shaft to another shaft and the axis of the two shafts are not aligned and are making some angle universal joint is used. This is also known as constant velocity joint. Universal joint consists of two yokes and a cross shaped member known as spider (Fig. 11.2 a, b). These parts are joined together as shown in Fig. 11.3. One yoke is connected to the driving shaft and the other is connected to driven shaft. The four arms of the spider are assembled in the needle bearings in the two yokes. The driving shaft provides motion to the spider that rotates. This motion to yoke connected with driven shaft through the spider. The needle bearings provide the swinging movement to yoke around trunnions. This joint has some limitations such as the speed of the driving shaft and the driven shaft may not be same. The transfer of motion may be accompanied by some vibrations. The Fig. 11.3 represents a simple universal joint.

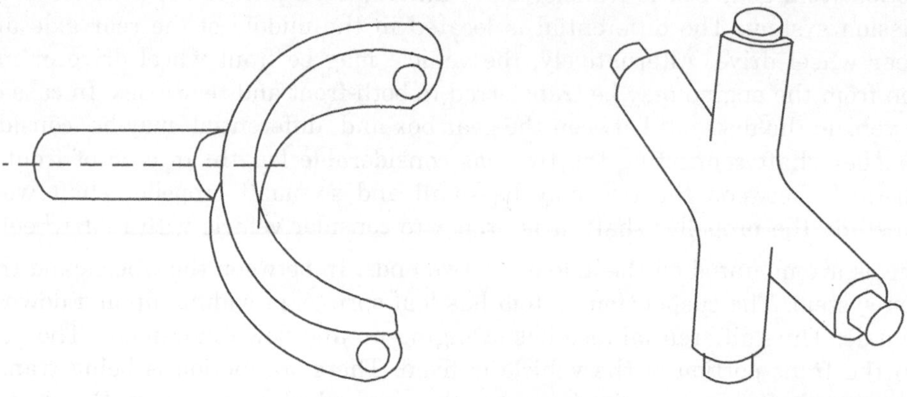

(a) Yoke (b) Spider

Fig. 11.2

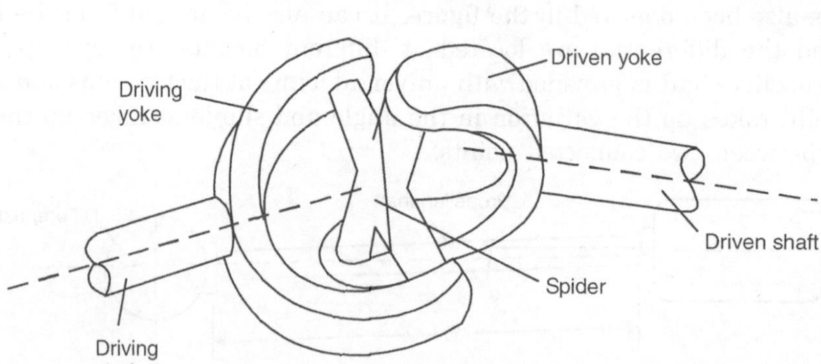

Fig. 11.3

Fluctuations in speed

The fluctuations in the speed of the output shafts occur due to the operating angle of the shaft. This operating angle depends upon the positioning of the gear box and differential. One end of propeller shaft is connected to gear box end through universal joint and its other end is connected to differential with second universal joint. The operating angle is the angle between the axis of the gear box shaft and the axis of propeller shaft (Fig. 11.4). Another operating angle is at the other end of the propeller shaft.

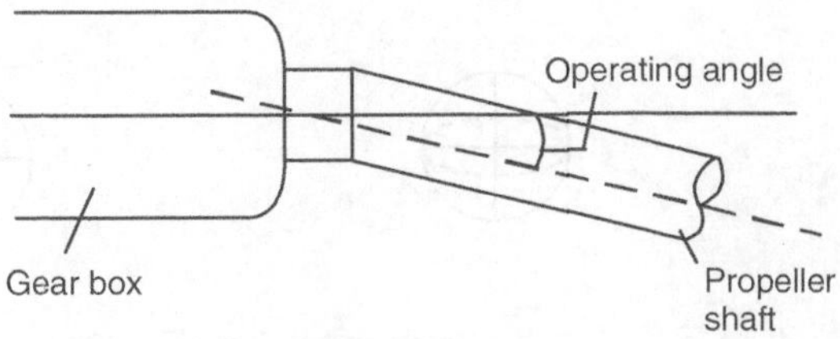

Fig. 11.4

The speed of the yoke connected to the gear box shaft is constant and it has circular motion. If seen in the same view, the yoke of the propeller shaft does not have circular motion; rather it has elliptical motion, as the axes of the two shafts are not aligned (Fig. 11.5). The shape of the ellipse is such that the distance travelled near the major axis end is smaller than the distance travelled near the minor axis ends. Considering the movement of driven yoke through an angle of θ degrees in the vicinity of 0 degree (*i.e.*, in the vicinity of major axis) and through an angle of θ degrees in the vicinity of 90 degree (*i.e.*, in the vicinity of minor axis). It would be clear from the Fig. 11.6 that the distance travelled in first case is more than the distance travelled in the second case. Hence the speed of yoke would be more when it is in the vicinity of 0 degree because in the same time interval it has to travel more distance. In second case the speed would be less because in the same time interval it has to travel less distance.

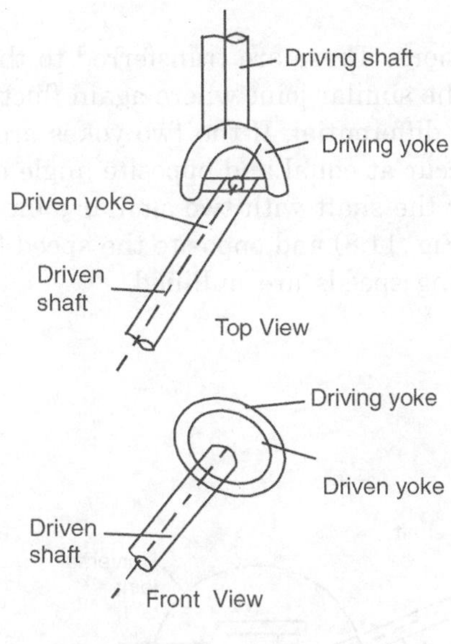

Fig. 11.5

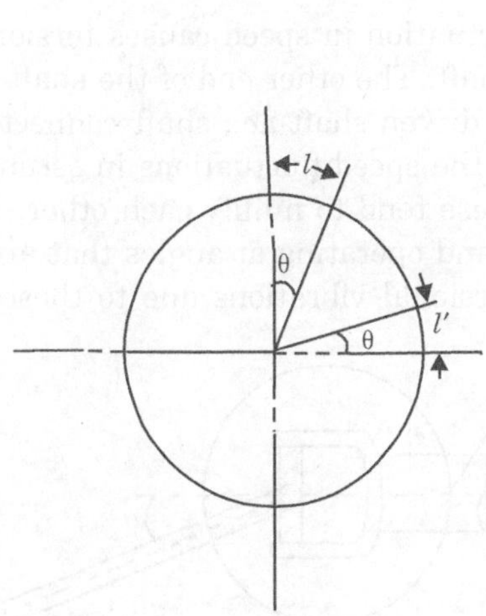

Fig. 11.6

Now, consider the rotation of yoke through 90 degrees. While the speed of the driving yoke remains constant the speed of the driven yoke would be less initially, it will increase gradually, and would be maximum when it has travelled through 90 degrees. Its average speed would be same as the speed of the driving yoke. The variation in the next 90 degrees would from maximum to minimum as shown in the graph (Fig. 11.7). This variation in the speed also depends upon the angle between the axes of the shaft. If the angle between the axes of the shafts is small the variation would also be less.

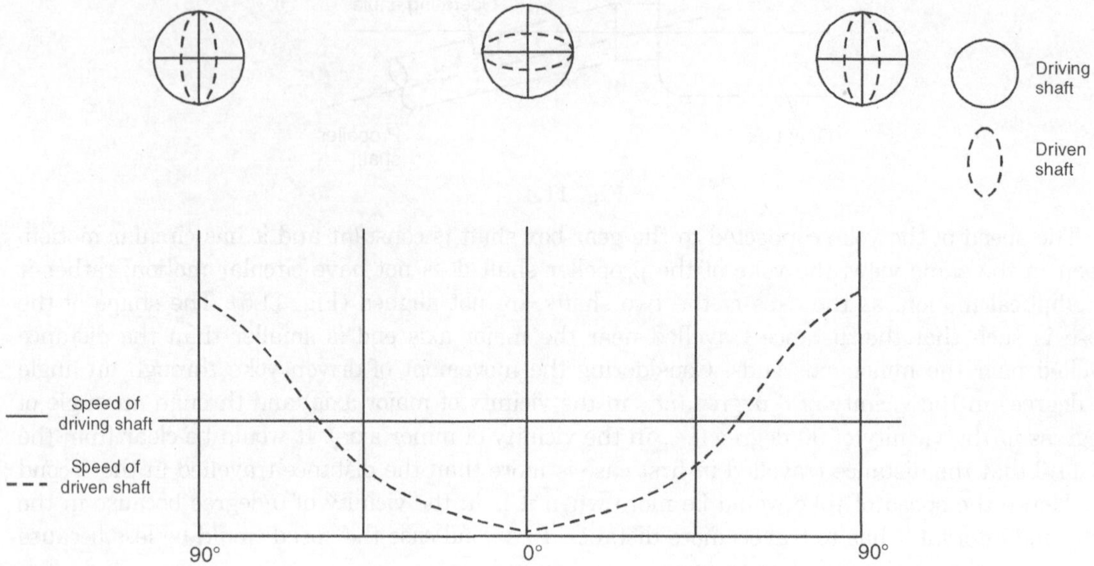

Fig. 11.7

The fluctuation in speed causes torsional vibrations. These are transferred to the other end of the shaft. The other end of the shaft has got the similar joint where again fluctuations occur in the driven shaft *i.e.*, shaft connected to the differential. If the two yokes are in the same plane, the speed fluctuations in second joint occur at equal and opposite angle α to the first joint these tend to nullify each other. Hence for the shaft with two similar joints in the same plane and operating at angles that are equal (Fig. 11.8) and opposite the speed fluctuations and torsional vibrations due to these fluctuating speeds are nullified.

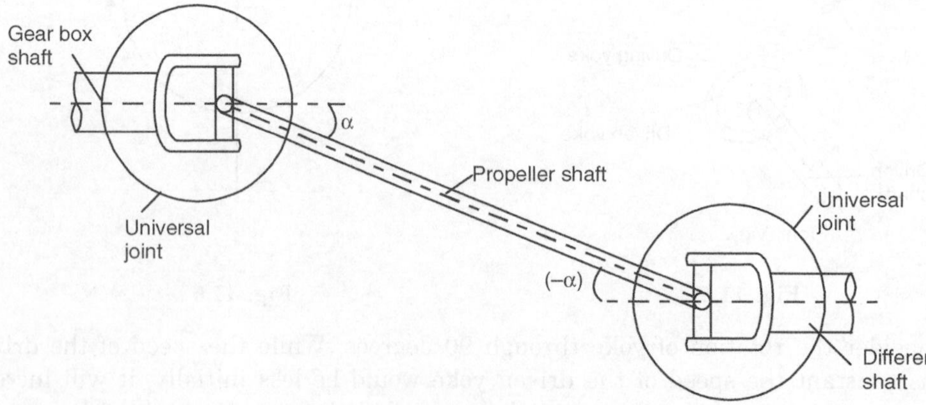

Fig. 11.8

Rzeppa joint

One of the most commonly used constant velocity joint is Rzeppa joint (Fig. 11.9). The speed fluctuations are nullified even when the angle between the axes of the two shafts is high. It has got six steel balls. These balls move in grooves in inner and outer race provided in driving and driven shaft. A cage keeps the balls in position. The torque is transmitted through the balls from inner race to the outer race. The balls always remain in a plane which is bisecting angle θ between the planes perpendicular to the axes of the shafts as shown in the Fig. 11.9.

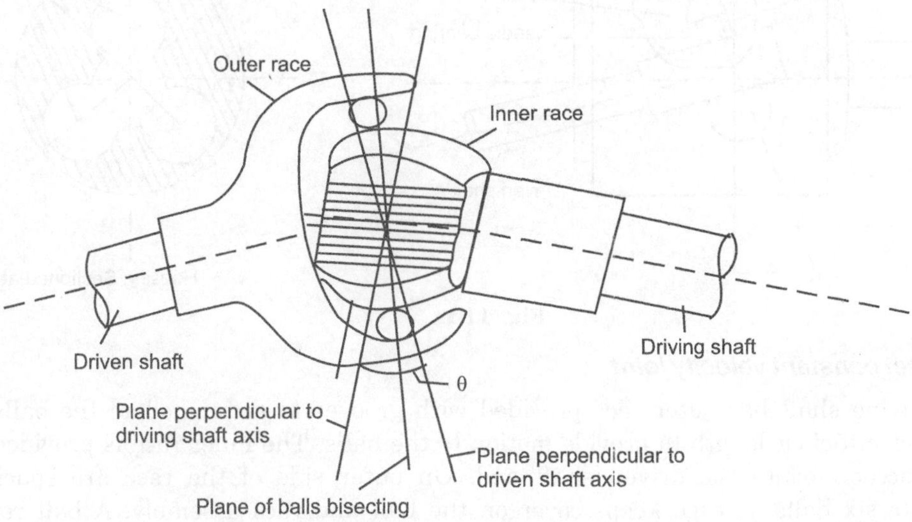

Fig. 11.9

In some automobile design the up and down movement of wheel may be controlled with the help of constant velocity joint provided in the axle half shaft. The half shaft at differential end is provided with the joint. The ball joint moves in the slot provided in the half shaft (as shown in Fig. 11.10). The sliding movement helps the wheel to move up and down.

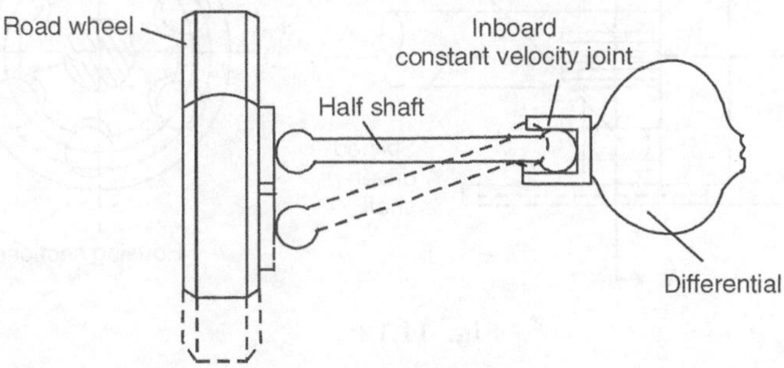

Fig. 11.10

Plunging tripod constant velocity joint

The driving shaft has one end connected to differential and the other end connected to the joint. Three balls are mounted on the needle bearings which fits on the spider with the help of trunnion. The spider is provided with splines on the inner side. Similar splines are provided on the outer side of driven shaft which is half shaft here. There is housing that accommodates the driven shaft end. It is provided with grooves to accommodate the three balls. Up and down movement of the wheel causes the to and fro motion of the balls in the grooves. The assembled view of the joint is shown in the figure 11.11.

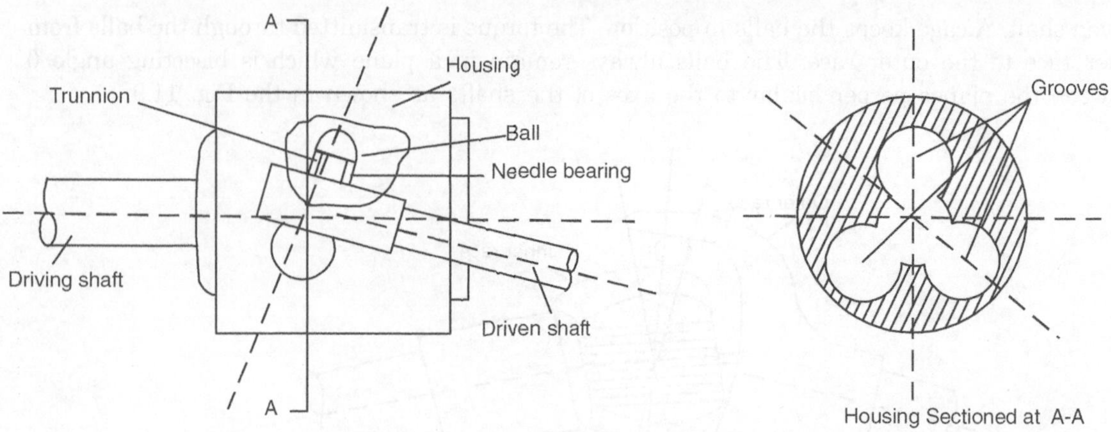

Fig. 11.11

Double offset constant velocity joint

The driving shaft has outer race provided with grooves to accommodate the balls. The grooves have sufficient length to provide motion to the balls. The inner race is provided with splines to accommodate the driven shaft end. On outer side of the race are spacing to accommodate six balls. A cage keeps cover on the ball and race assembly. A ball retainer is provided that keeps the balls in place. Figure 11.12 represents a double offset constant velocity joint.

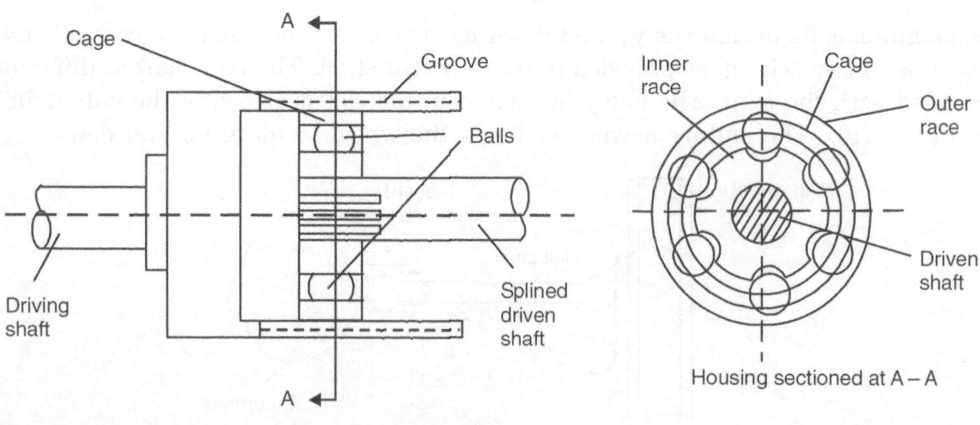

Fig. 11.12

Drive for automobiles with rear-mounted engine

In some designs a car may be rear wheel driven with engine mounted on the rear side. The half shafts are provided with plunging constant velocity joints at the ends of both the half shafts.

Drives for automobile with four wheel drive

In a four wheel drive motion is transferred to all the four wheels of the automobile. The motion is transferred on the rear side through propeller shaft with two constant velocity joints one on the gear box side and other on the differential side. To transfer motion on the front side universal joint is used similar to independent rear wheel suspension system. In this system, wheel on side can move up and down without affecting the wheel on the other side. The wheels are driven by half-shafts having universal joints at both the ends.

11.3 DIFFERENTIAL

The differential is yet another component of transmission system. The differential performs the following two functions:

(*a*) The live axle is at 90° to the propeller shaft. Axle where motion is transferred is known as live axle. The motion which is available at the propeller shaft end is to be rotated through 90°. This is performed with the help of a pinion and wheel arrangement. The axes of the pinion and the wheel are at 90° to each other.

(*b*) When the automobile is moving on a curved path its inner wheels are traversing a circle of smaller radius and outer wheels are traversing a circle of bigger radius (as shown in Fig. 11.13). Thus the wheels on the outer side traverse a longer distance than the wheels on the inner side. As all the four wheels are part of the automobile they have to move together. They are required to cover different distances in the same time period. Therefore it is required that speed of the outer wheels should increase and speed of inner wheels should decrease. This variation in the speeds of the wheels on inner and outer side is done by the differential.

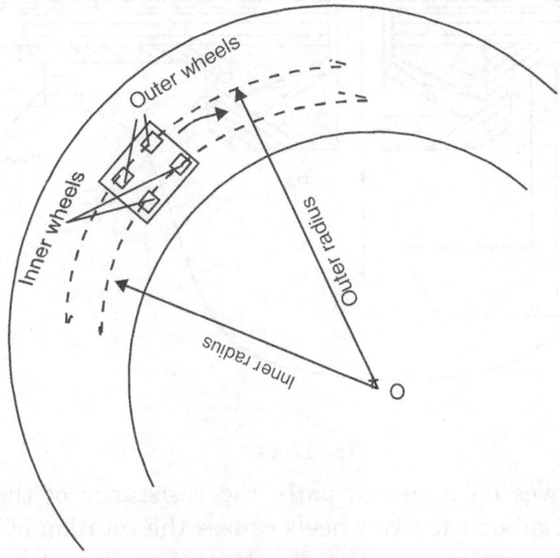

Fig. 11.13

Figure 11.14 represents a differential. The view has been sectioned to provide the inner details. The motion from the constant velocity joint of the propeller shaft is transferred to the pinion P_1. The crown wheel, CW is meshing with pinion P_1 and motion from propeller shaft through pinion, P_1 is transferred to crown wheel. As the axes crown wheel and pinion are at right angle to each other the first requirement of the differential is fulfilled *i.e.*, motion is rotated through 90°. The crown wheel, CW is fixed on a cage C. The cage consists of a set of bevel pinions B_1 and B_2 and pair of pinions P_2 and P_3. The two have their axes at right angle to each other. The size of bevel pinions and the pinions is same. Bevel pinions, B_1 and B_2 are fixed on the splined portion of half shafts H_1 and H_2. The half shafts support the cage and crown wheel. The half shafts are supported in bearings on both the sides to provide smooth rotary motion and support. The two half shafts have road wheels at their ends. These half shafts are the 'live' axle which is generally located on the rear side of the vehicle.

When the vehicle is moving on straight path the motion gets transferred to crown wheel CW and to two half shafts H_1 and H_2. The cage, in this situation, moves as a single unit. There is no relative movement between the bevel pinions and the pinion. Therefore the two half shafts rotate at the same speed.

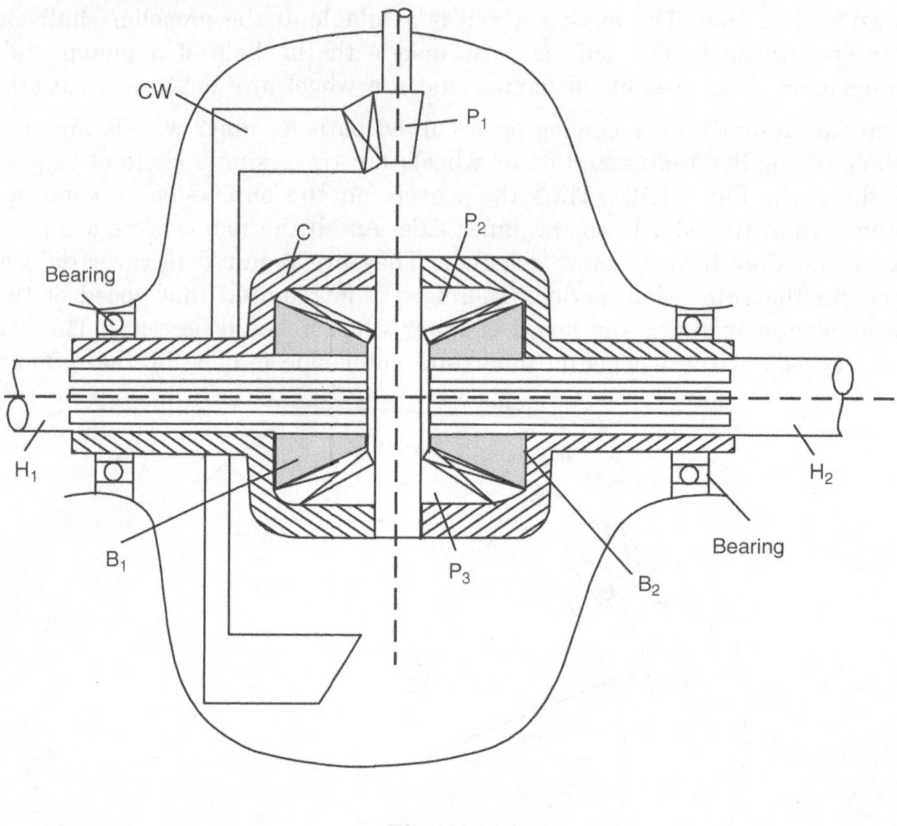

Fig. 11.14

When the vehicle moves on a curved path, the resistance of the inner wheel starts increasing. The unequal load on the two wheels causes the rotation of bevel pinions B_1 and B_2. The pinions P_2 and P_3, just reverse the direction of motion of bevel pinions. If there is an increase in the speed of the bevel pinion B_1 the speed of the B_2 will decrease. The increase on one side and the decrease in speed on the other side would be equal as the sizes of the wheels are equal. This fulfills the second requirement of the differential that on a curved path the inner wheels rotate at decreased speed and the outer wheels rotate at increased speed.

11.3.1 Differential with Lock

The differential divides the torque evenly between the two half shafts. As long as the wheel tyre has a grip on the road, the necessary frictional resistance is there which causes the forward motion. Now, if one wheel tyre is on the slippery ground, due to lack of frictional resistance, the wheel would not move forward and instead would continue to spin. As the resistance has been reduced the torque delivered to both the wheels changes. The vehicle can not move and the wheel on slippery ground continues to spin.

This situation does not allow the movement of the vehicle. The speed of the differential gears increases causing a generation of heat which may burn the lubricant film and cause damage to wheels. Even axle can break if the wheel continues to rotate for a long time.

This situation can be prevented by using differential with lock or limited slip differential. These differentials can be classified as (*a*) Clutch Packs type and (*b*) Brake Cones type.

Clutch Pack Type Differential

(*a*) In this type of differential, there are two sets of clutch plates as shown in the Fig. 11.15. The clutch plates have friction coatings on both the sides. These are splined internally. The hub is provided with external splines. The plates can slide on the shaft but rotate along with the shaft. The plates are placed between the bevel pinions and the differential casing. The springs are provided that keep a pressure on the clutch plates. These are generally coil springs and are retained by retainers. The clutch plates keep a grip and bevel pinions are locked to the casing. The casing and the half shafts rotate at the same speed and do not allow the slippage of wheel on slippery surface. The differential is so designed that when the vehicle is taking a turn the clutch plates would permit slippage and differential action would start at a particular value of torque.

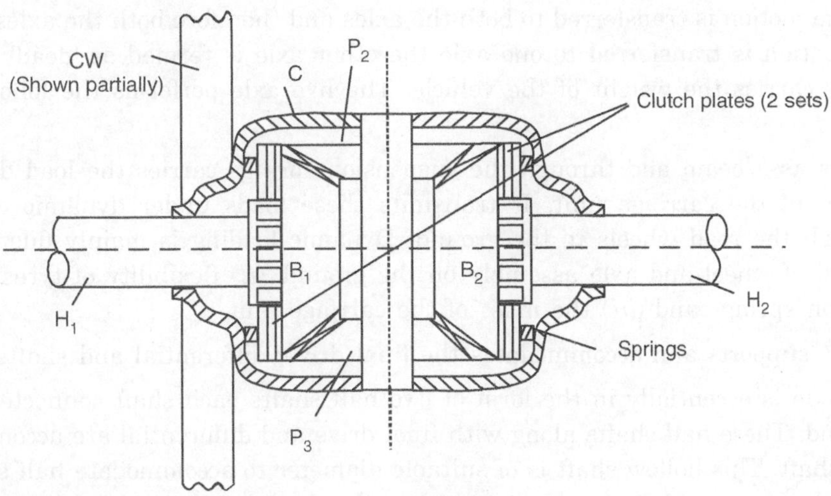

Fig. 11.15

(*b*) In second type of differential, brake cone is used instead of clutch plates. The cones are located between the bevel pinions and the differential casing (Fig. 11.16). Friction material is coated on the external surface of the cone. It provides friction contact between the cone surface and inner surface of the casing. A passage for lubricant is provided on the friction surface in the form of spiral thread. Coil springs are provided to keep initial contact. These can be four or six in number. The cones are enclosed between the bevel pinions and the casing and rotate along with the casing and keep both half shafts locked. When vehicle moves on a curved path, designed torque is needed to overcome the friction after which the differential action begins.

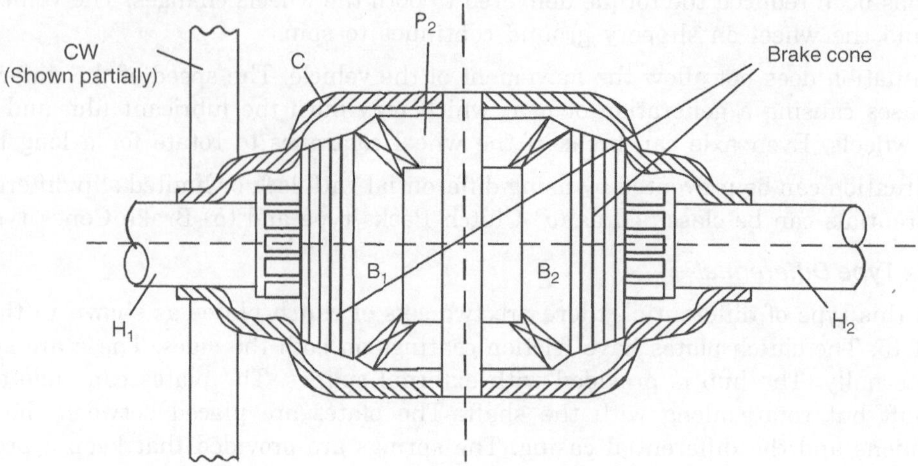

Fig. 11.16

11.4 LIVE AXLE

It is the axle where the motion from the propeller shaft is transferred. In case of four wheel drive, the motion is transferred to both the axles and therefore both the axles are 'live' axles. When motion is transferred to one axle the other axle is termed as 'dead' axle. The dead axle only carries the weight of the vehicle. The live axle performs the following two functions:

(*a*) It acts as a beam and through the suspension springs carries the load due to the weight of the carriage unit. It transmits these loads under dynamic conditions through the road wheels to the ground. Dynamic loading is mainly due to (*i*) the motion of wheel and axle assembly on the ground, (*ii*) flexibility of tyres and suspension springs and (*iii*) the mass of the carriage unit.

(*b*) It also supports and accommodates the final drive, differential and shafts.

The live axle is essentially in the form of two half shafts each shaft connected to road wheel at the end. These half shafts along with final drive and differential are accommodated in the hollow shaft. This hollow shaft is of suitable diameter to accommodate half shafts. Its middle portion has spherical shape to accommodate the final drive and the differential. It is strong enough to take up the load of carriage unit and suspension springs.

The final drive

Sometimes while transferring the motion from propeller shaft to the differential it is essential to reduce its speed. By reducing the speed the torque can be increased. This is done with the help of gears and the gear drive is known as final drive. If the speed is to be reduced up to about 7:1 a single pair of gears is sufficient. This is also known as single stage gearing. If the reduction of speed is needed in the higher ratio than this the reduction may not be possible in single stage. Then this is carried in two or even three stages. Worm wheels are used for the gear drive as these are silent. Bevel and hypoid bevel drives are also quite commonly used being less costly. The disadvantage being that sliding action of the worm teeth produces heat. This could be tackled by providing sufficient amount of lubricant.

A single reduction axle has been shown in the Fig. 11.17. It has a hollow casing C. The ends of the casing have road wheels. There are suspension springs and through these springs the weight of the carriage unit is supported by the casing. The suspension springs are located in between the casing and the wheels. In turn the casing is supported by the road wheels. The casing acts as a beam, supported at the ends and loaded in the middle portion. The casing has to be stiff enough to take up the load of the carriage unit and other components.

The motion from the propeller shaft is transferred through the shaft S which is integral with pinion P_1. The shaft is duly supported in the bearing provided in the casing. Inside the axle the pinion P_1 meshes with crown wheel. The crown has differential cage as explained in differential. The half shafts are supported in bearings B_1 B_2 in the casing. At the end the half shafts are bolted to the hub of the road wheels. Road wheels are kept in place with the help of shoulders that are provided in the casing and the nut, N.

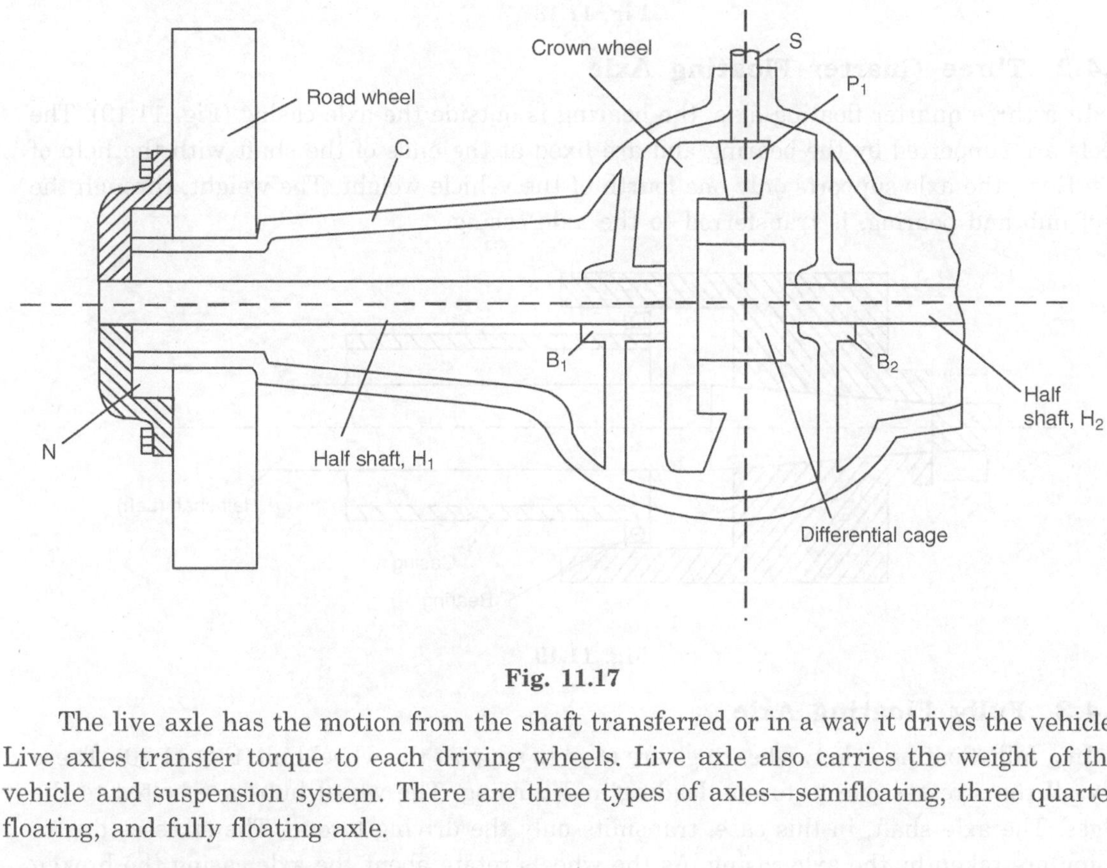

Fig. 11.17

The live axle has the motion from the shaft transferred or in a way it drives the vehicle. Live axles transfer torque to each driving wheels. Live axle also carries the weight of the vehicle and suspension system. There are three types of axles—semifloating, three quarter floating, and fully floating axle.

11.4.1 Semi-floating Axle

Semi-floating axle support the weight of the vehicle. It is supported at the ends by bearings (Fig. 11.18). These bearings are located in the axle casing and reduce the rotational friction. The axle transmits only driving torque. That's why it is termed as floating axle. The driving wheels are bolted on the outer end of the axles. The bearings support the axle. The axle is held in place with the help of bearing retainer or by a C shape washer. The bearing retainer is belted to the flange. The axle shaft pushes on the bearings as they rotate. The axle casing, suspension springs and chassis experience force. The axle experiences a bending stress as the vehicle moves on a curved path. The disadvantage of C shape washer as retainer is that the driving wheel comes out in case the axle breaks.

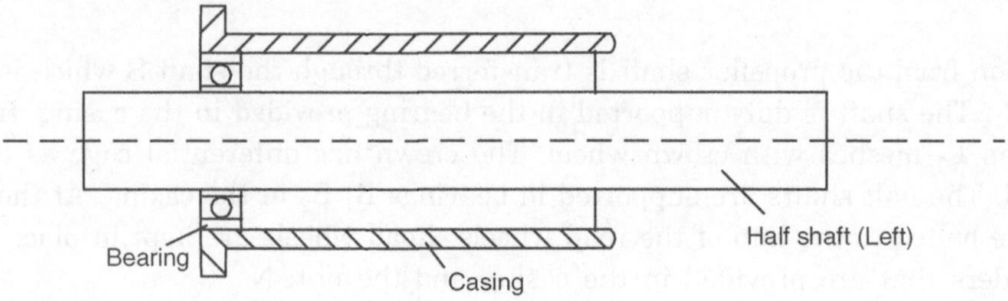

Fig. 11.18

11.4.2 Three Quarter Floating Axle

In a three quarter floating axle, the bearing is outside the axle casing (Fig. 11.19). The wheels are supported by the bearing and are fixed at the ends of the shaft with the help of bolts. Here, the axle supports only one fourth of the vehicle weight. The weight, through the wheel hub and bearing, is transferred to the axle casing.

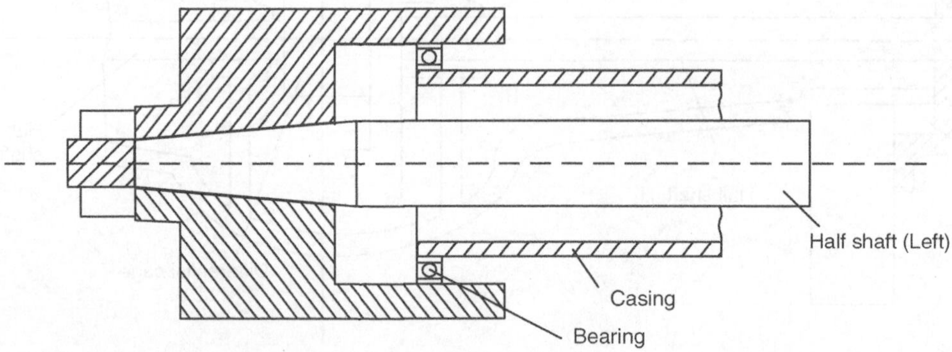

Fig. 11.19

11.4.3 Fully Floating Axle

In a fully floating axle two bearings are used to support the wheel hub (Fig. 11.20). These carry all the stresses due to torque loading and turning. The wheel hubs are bolted to the flanges. The axle shaft, in this case, transmits only the driving torque. The stresses due to turning are taken by the axle casing. As the wheels rotate about the axle casing the broken axle can be removed and the vehicle can be towed.

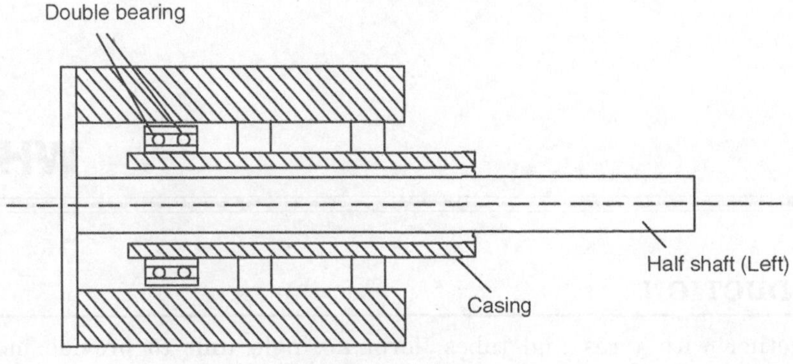

Fig. 11.20

QUESTIONS

1. What are the functions of a propeller shaft? Explain.
2. What is a constant velocity joint? What are the shortcomings of the constant velocity joints?
3. What are requirements as the vehicle moves on a curved path?
4. Explain the constructional details and working of differential.
5. Explain the function of clutch plates provided in the differential.
6. What are the functions of live axle?
7. Explain semi-floating, three quarter and fully floating axles.

12

WHEELS

12.1 INTRODUCTION

Wheels, together with tyres and tubes, form a single unit to provide motion to an automobile. These support the automobile on road and take up its load. Generally there are four wheels but often to enhance the load carrying capacity the number of wheels may be more than four. Tyres and tubes are made of soft rubber. The tubes are inflated and provide flexible support to the automobile. The wheels should be:

(*a*) capable of providing flexible support and cushioning effect. The tyres should be flexible but at the same time should be able to maintain their shape and provide support.

(*b*) strong enough to take up the load.

(*c*) balanced statically as well as dynamically for the stability of automobile. It is essential that the wheels are balanced so that while moving on the road the automobile is stable.

(*d*) as light as possible and

(*e*) easy to remove from axle and easy to mount on the axle. This is an important requirement. The tube may lose the air inside it if it is pricked by some sharp object say, a nail. In that case the tyre becomes flat and the wheel loses its circular shape and it is impossible for the automobile to move. The wheel has to be replaced by another wheel. Hence, it is essential that flat wheel can be removed and other wheel can be mounted easily.

12.2 TYPES OF WHEEL

The wheels may be classified, on the basis of their construction, as steel disc wheel, wire wheel.

Disc wheel

These wheels are made of steel and are simple and robust in construction as shown in Fig. 12.1. These are mounted on the axle with the help of tapered mounting nuts. These are known as lug nuts. Generally there are four nuts but in case of large wheel their number may be more. Holes are provided in the disc for better heat dissipation. As the holes tend to weaken the disc, these are swaged. Through swaging, the disc plate is turned inwards smoothly around each hole. This compensates the loss of strength due to holes. The outer part of the wheel is a steel rim which is well based so that tubes and tyres can be properly accommodated. The rim and disc may be a single unit, as in case of small wheels or in two parts as in case of big wheels.

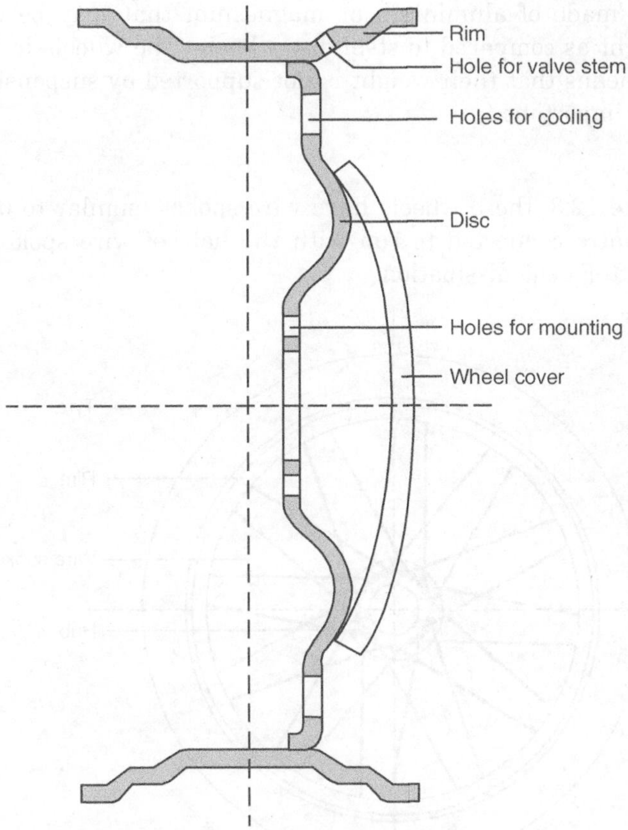

Fig. 12.1

The wheels may be inset, zero set or out set as shown in Fig. 12.2. This depends upon the position of rim in relation to attachment face of the disc. In some wheels, the rim can be mounted on either side of the disc to provide inset or outset. This increases or decreases the wheel track as per requirement of the automobiles. These are known as reversible wheels. For heavy vehicles, instead of well type rims flat based rims are used which provide better support.

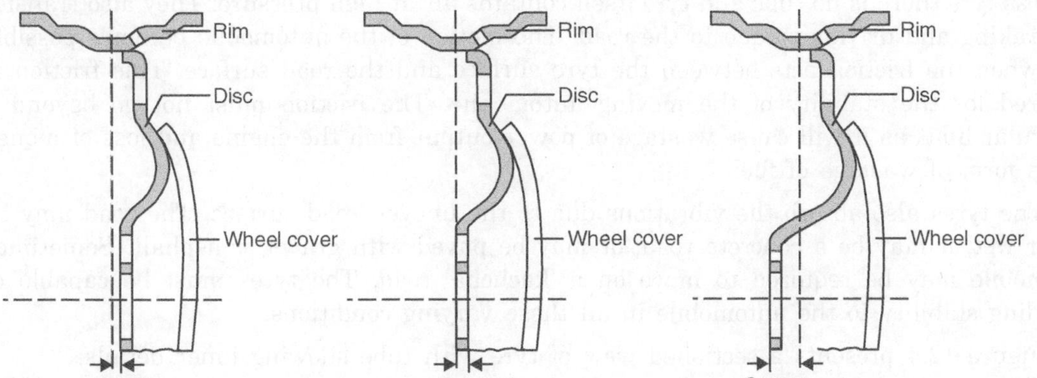

Fig. 12.2

The rims may be made of aluminium or magnesium that may be die-cast or forged. These are light in weight as compared to steel disc wheels. The wheels in an automobile are unsprung parts. This means that their weight is not supported by suspension springs. Hence reduction in weight is important.

Wire wheels

As shown in Figure 12.3, these wheels have wire spokes (similar to bicycle wheel). The wheel hub is in the centre connected to rim with the help of wire spokes. These are light in weight and have better heat dissipation.

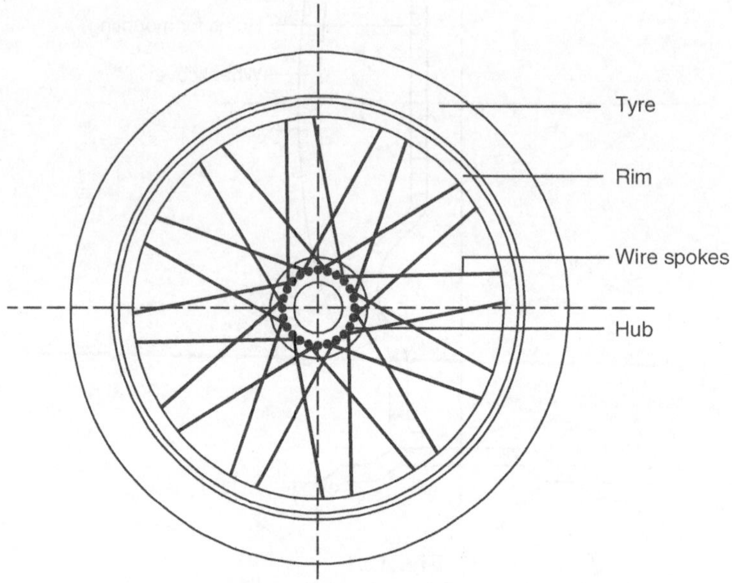

Fig. 12.3

12.3 TYRE

The tyres are required to carry the load of the automobile. The tyres may be with tube or tubeless. In the former, the tube is inside the tyre and contains air at high pressure. In tubeless tyre there is no tube and tyre itself contains air at high pressure. They also transfer the braking and driving torque to the road. The motion of the automobile becomes possible only when the friction acts between the tyre surface and the road surface. This friction is required for the stability of the moving automobile. The friction must not go beyond a particular limit as it will cause wastage of power output from the engine and loss of money in the form of wastage of fuel.

The tyres also absorb the vibrations due to the uneven road surface. The road may be dry or wet, it may be a concrete road, or may be paved with gravel or asphalt. Sometimes automobile may be required to move on a 'kachcha' road. The tyres must be capable of providing stability to the automobile in all these varying conditions.

Figure 12.4 presents a sectioned view of tyre with tube showing inner details.

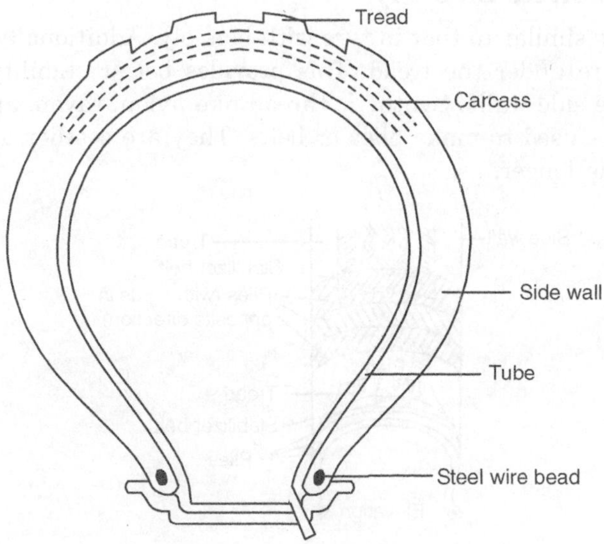

Fig. 12.4

Inner most part is the tube. The tyre is griped in the rim as shown. The tyre is provided with a steel wire (visible as bead). It is circular and provides shape to the tyre. If the wire breaks the shape of the tyre may not remain circular. The side wall of the tyre is needed to be flexible. The tread is the part which is in contact with the road surface. It is provided with grooves and this groove design provides stability to the automobile when it is moving on the wet surface. Carcass is the inner part of the tyre. It consists of a number of layers. The number of layers depends upon the load to be carried by the tyre. Each layer comprises of threads set in pattern and rubber impregnated upon them. The threads may be made of cotton, synthetic material or even metal in the form of metallic wires. On the basis of carcass the tyres are classified as bias ply, belted bias ply type or radial ply type.

12.3.1 Tyre with Bias Ply

This is the oldest design in tyres. Cotton thread is placed in such a manner that it forms a crisscross design. The angle of thread is kept approximately 35° with the centre line of the tyre. This angle may vary according to design. Lower angle gives stability at high speed but the ride becomes harsh. The number of plies depends upon the load to be carried by the tyre. Generally, these are between 2 and 4.

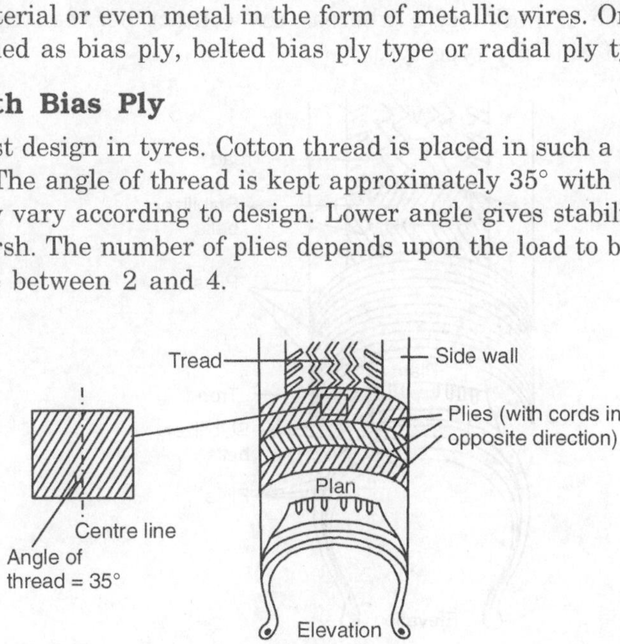

Fig. 12.5

12.3.2 Tyre with Belted Bias Ply

The pattern of ply is similar to that in tyre with bias ply. Additionally, two or even more belts run around the tyre under the tread. This provides better stability to the tread and enhanced strength to the side walls. Synthetic thread like nylon, rayon, and even steel wires or their combinations are used to make plies or belts. They are costlier as compared to bias ply tyres but their life is longer.

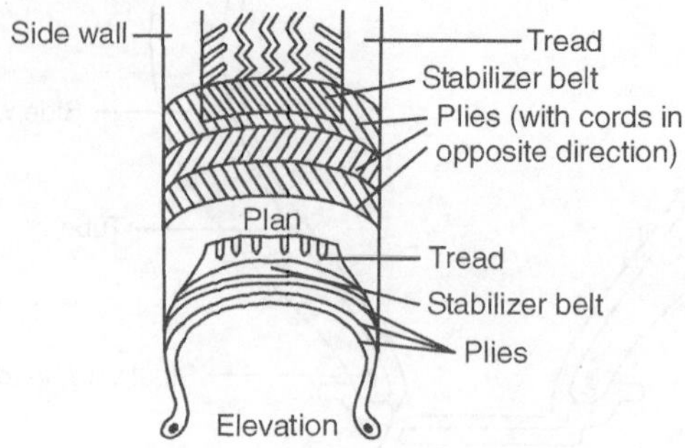

Fig. 12.6

12.3.3 Tyre with Radial Ply

Here the cords extend between the two steel wire beads at an angle of 90° to the centerline of the tyre. The tyre is provided with two or more layers of inflexible belts under the tread. The material for thread varies from synthetic material like rayon or nylon to fibre glass and even steel wires. This gives better strength to the tread area and flexibility to the side wall. Radial tyres offer better anti skid property and provide stable ride to the occupants of the automobile. Also while cornering, the radial tyres provide better stability as compared to bias ply tyres. As shown in the Fig. 12.8, there is no loss of contact between the tyre tread and road surface in case of radial tyre whereas the contact between the tread and road surface is lost in case of bias ply type tyre.

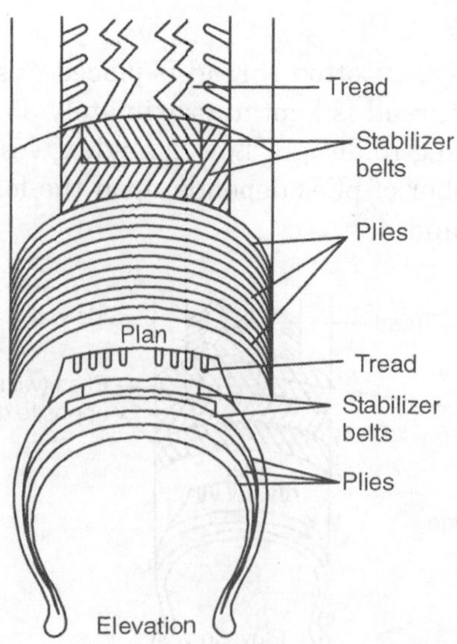

Fig. 12.7

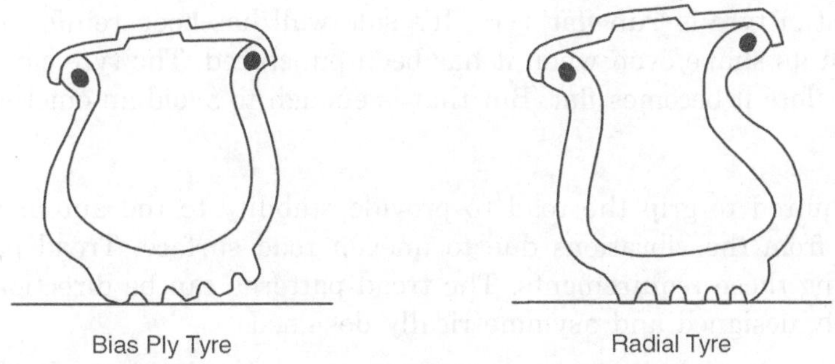

Fig. 12.8

12.3.4 Tubeless Tyre

These tyres do not have tube inside. The air at high pressure is filled in the tyre itself. It has got a soft inner lining that prevents the air from leaking. The lining in a puncture free tyre forms a seal around the object (such as nail) that has entered in the tyre and prevents the puncturing of tyre. The sealing material can fill the hole formed when the object (nail) has been removed. This way no immediate attention is needed to repair the tyre. The tyre is provided with one way valve that allows the inward entry of the air. A rubber cap above the valve also prevents the leakage of air. The rim is specially designed so that leakage of air does not occur.

Figure 12.9 presents the cut section of a tubeless tyre. The carcass consists of layers or plies of rubber impregnated cords. The number of plies may be 2, 4 or even high depending upon the load on the tyre. The tread is the uppermost portion of the tyre. It remains in contact with road surface. The grooves provided on the tread provide stability to the automobile particularly when the road surface becomes wet or slippery. Some small cuts are provided into the ribs of the tread. These open as the tyre flexes on the road and provide stability to the automobile during motion. The sidewalls have thinner cross-sectional area and provide flexibility to tyre. The steel wire bead provides grip between rim and the tyre and also gives circular shape to the tyre. Cords used in the plies are made of rayon, nylon, fiber glass and even steel wires. Rayon cords have low cost and give comfortable ride but are not very strong. Nylon cords are tough. Fiber glass cords are much tougher and give a smooth ride. Steel wire used as cords is tough but gives slightly rough ride.

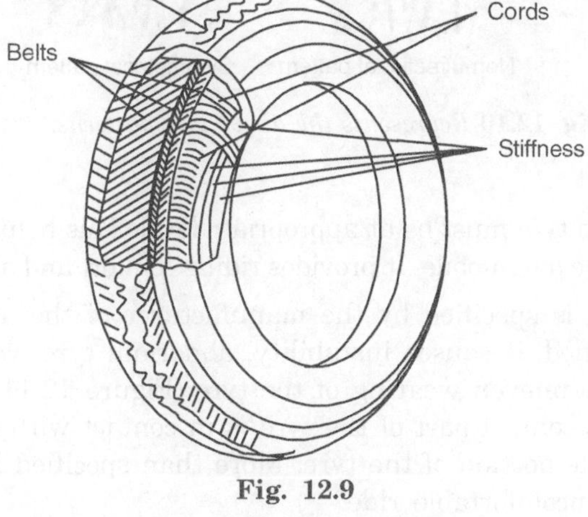

Fig. 12.9

Another variant of tyre is 'run flat' tyre. It's side wall has been reinforced. This allows the tyre to maintain its shape even when it has been punctured. The tyre can move through a limited distance before it becomes flat. But that is enough to avoid an emergency stoppage.

Tread pattern

The tyre is required to grip the road to provide stability to the automobile. It should provide cushioning from the vibrations due to uneven road surface. Tread pattern plays a vital road in fulfilling these requirements. The tread patterns can be directional, non directional, symmetrically designed and asymmetrically designed.

Directional tyre: It is designed to rotate in a particular direction. It offers good performance only when it rotates in the direction for which it has been designed. If made to rotate in other direction its performance may not be up to the mark. Rear wheels of tractor have directional tyres.

Non-directional tyre: It can be made to rotate in any direction. Its performance is not affected with direction of rotation.

Tyre with symmetrically designed pattern: It has symmetric pattern of tread about the centre line of the tyre.

Tyre with asymmetrically designed pattern: The pattern of the tread in not symmetrical about the centre line. It is a directional tyre and provides better grip when the automobile is moving. The inside half of the pattern help to achieve better grip during motion on straight path and outside half of the pattern help to achieve the same during motion on the curved path.

On a wet road surface, the water is required to be moved away from the tyres. Channels in treads of the tyre perform this function. Though due to these channels, the area of contact between the tyre and the road surface decreases but they provide stability to the automobile during rainy weather conditions.

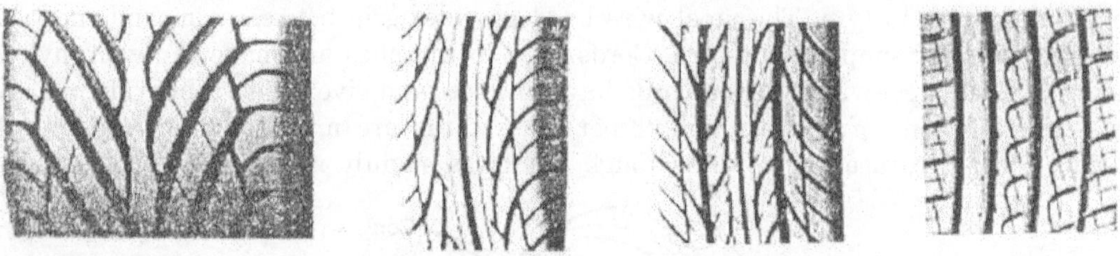

Directional pattern | Non-directional pattern | Symmetrical pattern | Asymmetrical pattern

Fig. 12.10 *Represents the above tyre patterns.*

Inflation pressure

Inflation pressure in the tyre must be at appropriate level. This helps to achieve longer tyre life. It provides stability to the automobile. It provides riding comfort and also better fuel economy.

The inflation pressure is specified by the manufacturer of the automobile. If inflation pressure is less than specified, it causes instability, abnormal tyre wear and enhanced fuel consumption. It also causes uneven wearing of the tyre (Figure 12.11). If inflation pressure is more than specified. The central part of the tyre is in contact with the road surface. This will cause wearing of middle portion of the tyre. More than specified inflation pressure also causes tyre bruising and uncomfortable ride.

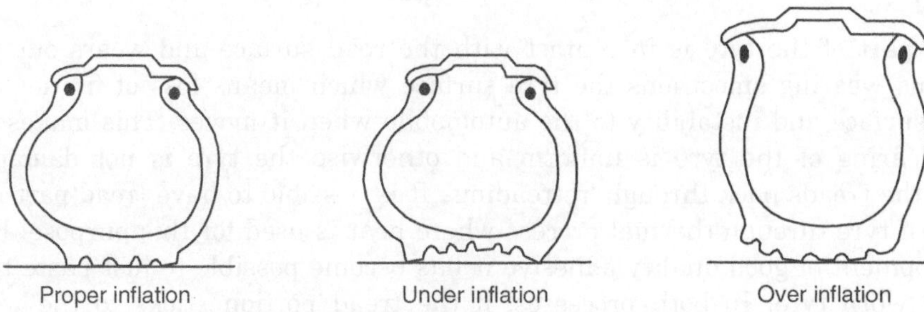

Fig. 12.11

Due to the properties of the tube material, the tubes are able to retain the air for a long time. That is why; now-a-days it is not essential to check the inflation pressure frequently. Some automobiles are provided with sensors strapped in centre of the wheel that senses the tyre pressure. When tyre pressure is reduced the sensor causes illumination of warning light in the control panel of the automobile.

Tyre rotation

The front and rear tyres perform differently. This is mainly due to different amount of load being taken up by them. Generally, the rear tyres take up more loads as compared to front tyres. Due to this, the tyres wear out differently on front and rear side. This unequal wearing of the tyres causes instability to the automobile. The tyres are required to be rotated so that these wear equally. Tyre rotation means that wheel and tyre assemblies are shifted from one end of the axle to other or from one axle to other. Tyre rotation can be done for a rear wheel drive without spare tyre as shown in Fig. 12.12(*a*) and (*b*). Tyre rotation for front wheel drive without spare tyre has been shown in Fig. 12.12(*c*). Figure 12.13 shows tyre rotation with spare tyre. It is recommended that spare tyre is not used as regular tyre.

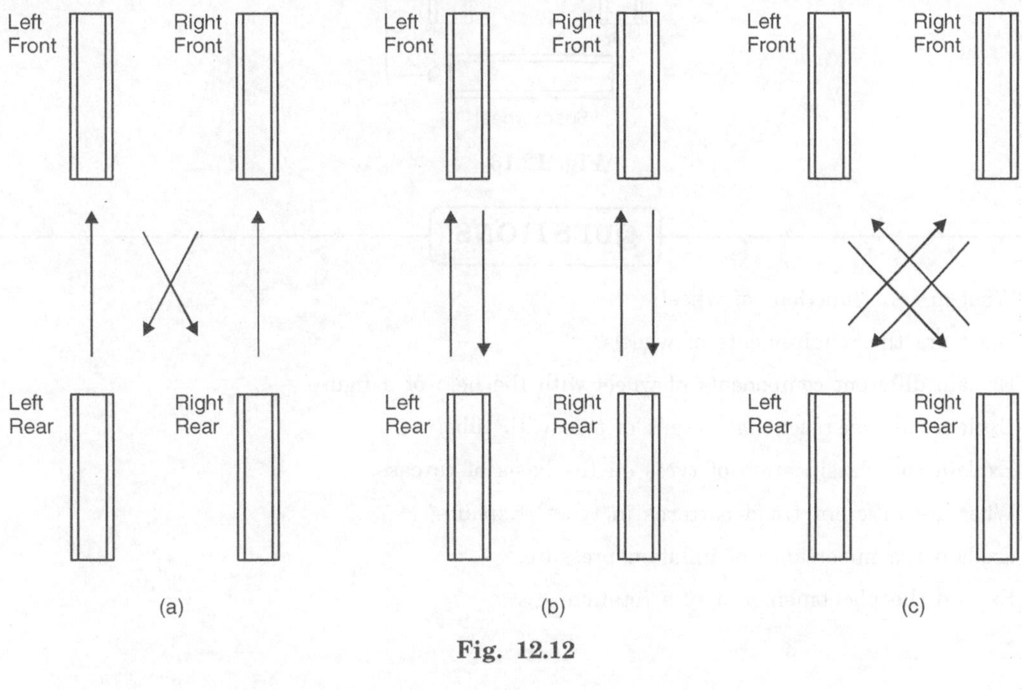

Fig. 12.12

Tyre retreading

The tread part of the tyre is in contact with the road surface and wears out with the use of tyre. This wearing smoothens the tyre surface which means loss of friction between tyre and road surface and instability to the automobile when it moves. This makes the tyre unusable. If wearing of the tyre is uniform and otherwise the tyre is not damaged it is possible to get the treads back through 'retreading'. It is possible to have tread part mounted around the worn tyre through thermal process where heat is used for this purpose. Recently, with the development of good quality adhesive it has become possible to just paste the tread portion on the worn tyre. In both processes, if the tread portion sticks to the worn tyre properly, the tyre can be used again till the new tread portion wears out.

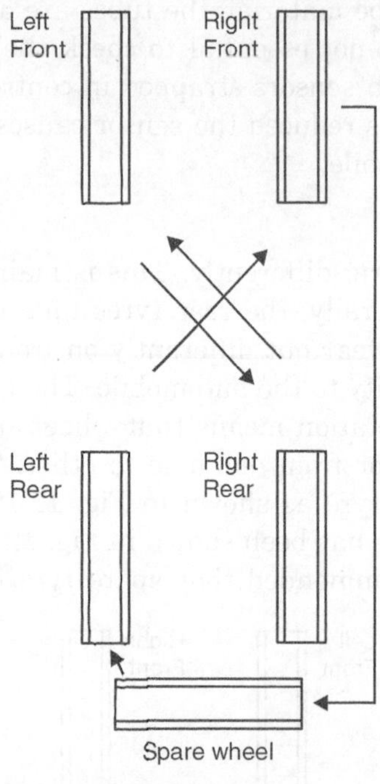

Fig. 12.13

QUESTIONS

1. What are the functions of wheel?
2. What are the requirements of wheels?
3. Explain different components of wheel with the help of a figure.
4. Explain the constructional details of tyre with tube.
5. Explain the classification of tyres on the basis of carcass.
6. What are different tread patterns in tyres? Explain.
7. Explain the importance of inflation pressure.
8. Explain the phenomenon of tyre rotation.

13

BRAKES

13.1 INTRODUCTION

The primary function of an automobile is to move from one place to another. As the automobile moves from one place to another, it is very important to keep control on its movement. The driver must have complete control on the movement. He/she should be able to reduce its speed whenever needed and also the automobile must stop when desired. Brakes are the control mechanism in an automobile. These are used to control the movement of the automobile. These provide frictional resistance to the motion of the wheel thus causing reduction in the speed or stopping of the automobile. Due to frictional resistance, the kinetic energy of the wheel is converted into heat which is dissipated to the atmosphere suitably.

13.2 REQUIREMENTS OF BRAKES

The brake is an essential control device in an automobile. If anything goes wrong in the brakes it may be disastrous. Following are the requirements of the brakes so that these can function properly.

1. These are required to be strong enough to stop the automobile within minimum distance in case of emergency. Also, the driver must have control on the automobile when the brakes are applied due to some emergency.
2. The brakes are required to be applied repeatedly to reduce the speed of automobile. With repeated application of brakes their efficiency should not decrease. The repeated application of brakes causes a rise in temperature that in turn causes reduction in the coefficient of friction of brake lining material. However, the coefficient of friction is restored when the temperature is lowered. One of the basic requirements of brakes is that its efficiency is not reduced even after repeated use. This is known as 'anti-fade' characteristic. The brakes must have good 'anti-fade' characteristics. As the automobile descends from the hills, there may be constant and prolonged application of brakes. In these conditions also, the effectiveness of the brakes must not be reduced. This can be achieved by providing efficient cooling of the brakes.

13.3 BRAKE EFFICIENCY AND STOPPING DISTANCE

In maximum force of retardation, F applied on wheels depends upon (*i*) the coefficient of friction between tyre and road surfaces, μ and (*ii*) the weight of the automobile on the wheels, W.

The force of retardation, $F = \mu.W$

If $\mu = 1$ then $F = W$

which means retarding force becomes equal to the force experienced by the freely falling body of mass equal to that of automobile. Thus, the deceleration becomes equal to '*g*' and the brake is said to be 100% efficient. Theoretically, this is the maximum limit.

In practice, safety of passengers and goods is most important. If the brakes are 100% efficient, this means that as soon as the brakes are applied the automobile would become stationary. At high speed, if the automobile becomes stationary all of a sudden it would experience a huge jolt and the passengers inside it would get injured. Even if it is a goods carrying vehicle, sudden jolt would cause damage to the goods. Hence the efficiency of the brakes is never kept 100%. The sudden application of brakes always causes a jolt and that must be avoided. It is essential that the automobile stops without causing collision with other vehicles on the road. To ensure this, a safe distance must be kept from the vehicle in the front so that one can stop without colliding with it. The speed of the automobile also plays a very important role. Higher the speed larger is the distance the automobile takes to stop. The distance covered by the automobile before it stops after the application of brakes is known as 'stopping distance'. The stopping distance comprises of the following:

(*i*) The reaction time of each human being is different. It is also known as 'reflexes'. Some human beings may have quick reflexes but others may not have and the time taken by an individual to react to some unexpected situation would be different. It is during this time that an automobile covers some distance that forms the part of stopping distance.

(*ii*) The distance travelled by the automobile during the time when brake pedal is pressed and brakes are actually applied to wheels forms the second part of stopping distance. For any mechanism to work some time period is needed. For an efficient mechanism, the time required would be small and that would reduce the distance travelled.

(*iii*) The third part of stopping distance is due to the speed with which automobile is moving. If the speed high, the distance taken by it to stop would be more.

The safe distance between the two vehicles is larger than the stopping distance. The stopping distance is a very important parameter as far as the safety of passengers is concerned.

13.4 FACTOR AFFECTING THE APPLICATION OF BRAKES

The effectiveness with which the brakes are applied is determined by the following factors:

Pressure

The friction generated between the two surfaces in contact, depends partly upon the force exerted. In automobiles, fluid present in the cylinder exerts force necessary to create friction. This force brings the brake shoes in contact with the drum of the rotating wheel.

Coefficient of friction

The amount of friction between two surfaces in contact depends upon the coefficient of friction. Higher the value of coefficient of friction, higher would be the amount of friction. However, the maximum value of coefficient of friction is limited to one and it depends upon the properties of the material and roughness of the surface. With repeated application of brakes, the surface becomes smooth. The coefficient of friction is reduced and along with it the frictional force is also reduced. If that happens the brake shoes are replaced with new ones which have the rough surface. Coefficient of friction is equal to the force needed to pull an object divided by its weight.

Area of contact

The force of friction depends upon the area of contact. More the area of contact higher is the force. The area can be increased by increasing the size of the shoe brakes and by providing them on all the four wheels.

Heat dissipation

When the brakes are applied, the kinetic energy that is lost is converted into heat. The heat thus produced must be dissipated. If not so, it may damage the brake shoes and other components located nearby. This heat is dissipated with the help of air that is made to pass through the wheel.

13.5 BRAKE LINING MATERIAL

Brake linings are made of friction material and fixed on the brake shoes. These may be riveted or bonded to the shoe. Sometimes these are molded as integral part of the brake shoe. Asbestos has been used as friction material in linings for very long time. Recently, its use has been restricted and even banned in some countries due health hazards.

Asbestos lining material has been replaced by fully metallic, semi-metallic and non-asbestos lining materials. Sintered iron has been used as fully metallic lining material and is more suitable for heavy duty and racing vehicles. These require very high pedal pressure. Also the wearing out of drums is faster.

Semi-metallic lining materials have good anti fade and frictional characteristics. These are made of iron fibre molded with adhesives. Non-asbestos lining materials are synthetic materials. Attempts are being made to improve their life span, friction characteristics, wear characteristics and heat dissipation.

13.6 HYDRAULIC BRAKES

In hydraulic brakes, fluid in liquid form known as brake fluid or brake oil is used. It should have proper viscosity and with rise in temperature the viscosity should not change. One of the advantages of using liquids is that these are non-compressible. The hydraulic brake system is closed system therefore pressure applied on brake pedal is completely transmitted to the brake shoes actuating the brakes.

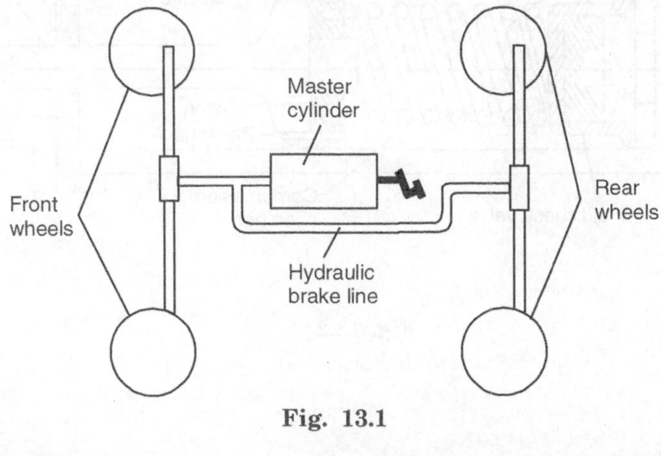

Fig. 13.1

The brakes are actuated in each wheel with the help of brake shoes. The brake shoes are pushed by outward motion of the piston in wheel cylinder and come in contact with disc on which the wheel is mounted or with the inner surface of the rim. The force of friction between two surfaces in contact causes the resistance to motion of the wheel slowing down its movement or making it stationary. The wheel cylinders are connected to the master cylinder through fluid lines. Master cylinder supplies the fluid to the wheels. When the brake pedal is pressed, fluid in the master cylinder goes to the wheel cylinders through these lines. The pressure applied at the brake pedal is transferred to brake shoes consistently as the fluids are non-compressible. The output force can be increased by increasing the cross-section area of the wheel cylinder. Though, this would reduce the piston travel. If the cross-section area of master cylinder and wheel cylinder is same, say 1 cm^2 then same force would be transmitted but if the area of wheel cylinder is increased to 2 cm^2 double output force would be available at brake shoes. Similarly, if the cross-section area of the wheel cylinder is reduced to 0.5 cm^2, the output force would be reduced to half though, in this case the piston travel would increase.

Small pressure of about 0.5 kg/cm^2 is maintained in the fluid lines. This is essential to keep the cups of wheel cylinders expanded. Upon release of brakes this also prevents the air from entering the wheel cylinders.

13.7 MASTER CYLINDER

Master cylinder is the main component of the system. When brake pedal is pressed, the fluid in the master cylinder, under pressure, moves out of it and in the fluid lines which takes it to the wheel cylinders. It is divided into two parts, namely fluid reservoir and compression chamber. As the pedal is pressed, piston at the end of the pedal rod moves towards the left,

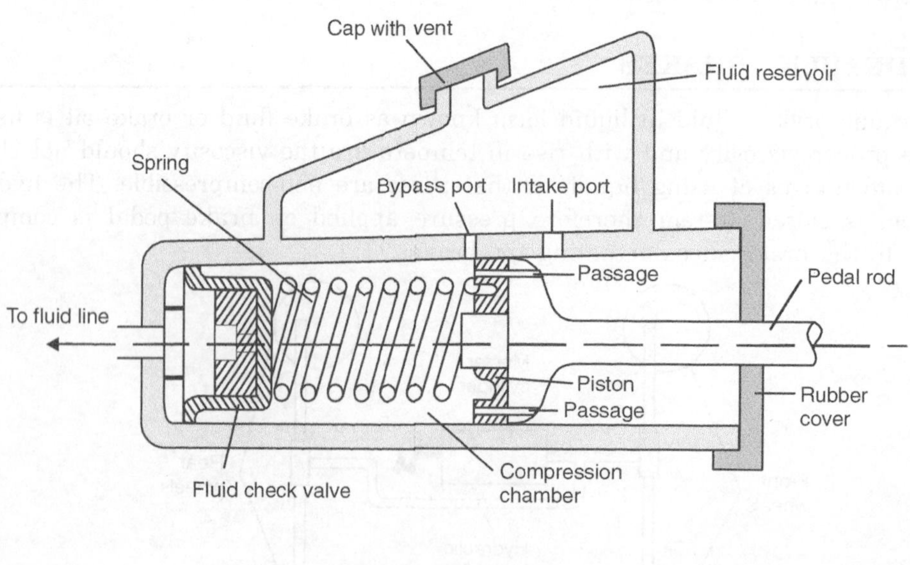

Fig. 13.2

against the force of spring, till it covers the by-pass port. Further movement of piston causes build-up pressure in the compression chamber. With increases in pressure, the inner rubber cup of the fluid check valve is deflected, forcing the fluid to go to lines under pressure. The fluid enters the wheel cylinder and moves the piston inside thereby actuating the brakes. When the pedal is released, spring causes the movement of piston to the extreme right position. The force of spring, also, keeps the fluid check valve pressed on its seat for sometimes. The fluid returns from the lines to the compression chamber and due to inertia some delay occurs. This creates vacuum in the compression chamber which is to be destroyed immediately. When vacuum is created, air tries to occupy the space and affects the working of brakes adversely as air is compressible. As soon as vacuum is created, atmospheric pressure in the fluid reservoir forces the fluid, through intake port and holes in the piston into the compression chamber. This destroys the vacuum. The fluid comes back into the reservoir by lifting the fluid check valve off its seat. The fluid in the compression chamber becomes excessive and this excess fluid goes back to reservoir through by-pass port. The excess fluid, if allowed to remain in the compression chamber, causes the movement of fluid to the lines. The fluid is retained in the lines and even slight application of brakes causes generation of heat. This would cause expansion of fluid in the lines and ultimately jamming of the brakes.

The master cylinder is provided with rubber cover at the pedal rod end. This does not allow the entry of dust in the chamber through pedal rod opening. There is an opening at the top. The opening is used to keep the fluid level as per requirement of the system. The opening is provided with screw cover that makes it air tight.

13.8 DUAL BRAKING SYSTEM

Malfunction in master cylinder would cause failure of brakes in all the four wheels if brakes in all the four wheels are operated through a single master cylinder. If that happens it would not be possible to stop the vehicle. To avoid such situation, it is essential that brake system be designed in such a way that brakes in all the four wheels do not fail simultaneously. Dual braking systems are modified systems where failure of brakes does not occur in all the four wheels. These employ a modified master cylinder, known as Tandem master cylinder. It is a combination of a primary and a secondary cylinder. Each cylinder is connected to two wheels. Dual braking systems can be of two types:

(*i*) Front—Rear split system and

(*ii*) Diagonally split system

13.8.1 Front—Rear Split System

Here, the front wheels are operated by primary cylinder and rear wheels are operated by secondary cylinder. In case of failure either brakes in front wheels will become non-operative or brakes in the rear wheels will become non-operative. Two-third of braking work is taken up by front wheel brakes and remaining one-third is taken up by rear wheel brakes. Two portions in the master cylinder are suitably designed to take up the different brake work. Figure 13.3 shows a Front—Rear split system.

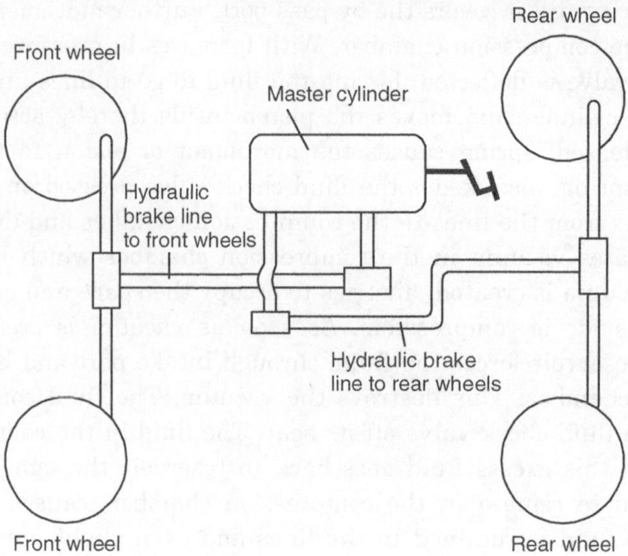

Fig. 13.3

13.8.2 Diagonally Split System

Here one front and one rear wheels are combined and connected to primary/secondary cylinder. As shown, the haydraulic brake lines have been split front to rear *i.e.*, left front to right rear. Each cylinder takes up equal braking work. In case of failure, half of the total braking force can be applied. Figure 13.4 shows a diagonally split system.

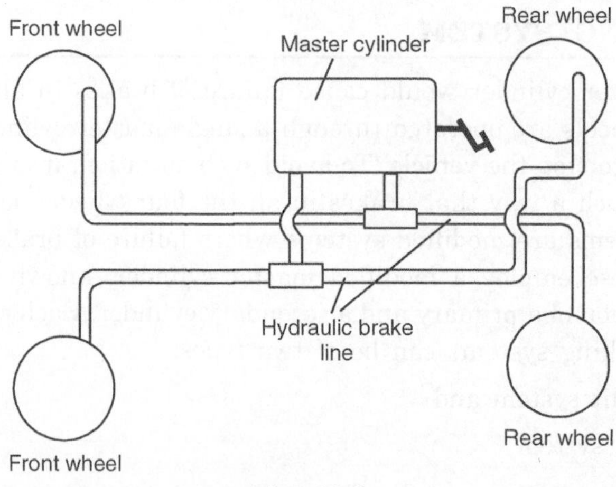

Fig. 13.4

13.9 TANDEM MASTER CYLINDER

A tandem master cylinder is shown in figure 13.5. It has primary and secondary cylinder in one row. These are provided with individual pistons and have their own return springs. The return spring keeps primary piston cup slightly behind the bypass port to keep the cylinder filled with fluid. Return spring also helps to return brake pedal back to its original position when driver removes foot.

There are two separate lines, one connects the front wheels and the other connects to rear wheels. In normal situation, with pushing of pedal rod both pistons move towards the right brake fluid being supplied to both the lines and brakes being actuated in all the four wheels. In case of failure in one line say front line, secondary piston moves till it reaches to the wall of the cylinder. After this, pressure starts building up in space between the two pistons and fluid is supplied to rear wheels. The brakes on rear wheels are actuated. In other situation, where there is flaw in rear fluid supply, the primary piston moves till it comes against secondary piston. Afterwards the two pistons move together supplying fluid to the front wheels thereby actuating brakes in front wheels.

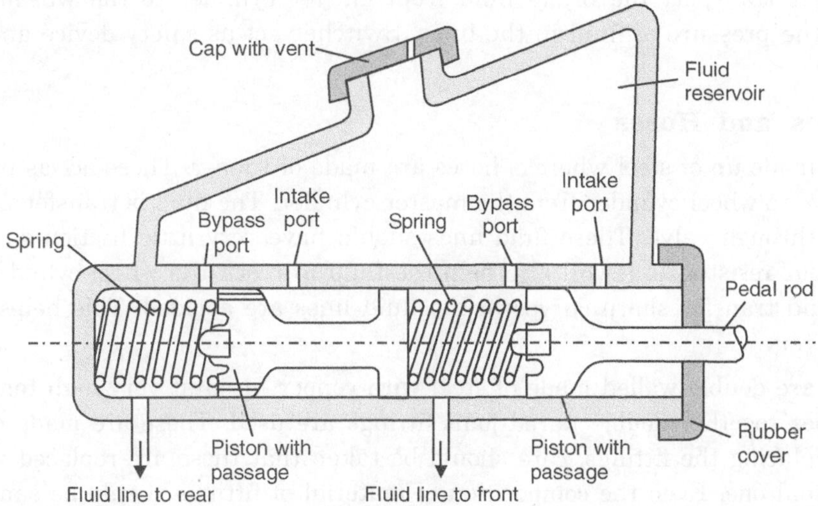

Fig. 13.5

The cylinder is designed in such a manner that force is exerted on front and rear wheels in the required magnitude, as the force required is not the same in front and rear wheels. Although, the system does not fail if malfunctioning occurs in either of the supply line but it is essential that this is indicated to driver so that he may take steps to get the system repaired.

The reservoir of the master cylinder is made of cast iron or plastic. To maintain the fluid in the reservoir at designed level, float valve is provided. The float valve is provided with magnet. The magnet activates the reed switch which causes the glowing of indicator lamp on the dash board. This gives indication to the driver that fluid in the reservoir needs to be checked.

The residual check valve keeps a small residual pressure in the brake lines and in the wheel cylinders. Without this pressure, air may enter the wheel cylinder. This would affect the actuation of brakes adversely. These valves are designed to allow fluid flow from master cylinder to lines when high pressure is built in the cylinder. When brake pedal is released, outlet check valve keeps the hydraulic lines and wheel cylinders filled with fluid.

Lately, cup expanders have been provided in the wheel cylinder. These expanders prevent the air from entering by keeping the cup seal tightly pressed against the walls of the cylinder. Thus residual check valve is no more needed.

Residual pressure is needed to operate the drum brakes but it should have small magnitude. With repeated use of brakes, there is an increase in residual pressure. The temperature of wheel cylinder and brake drum would rise making them hot. The brake shoes would also be heated up and would expand. This pressure causes the contact between brake shoes and drum even after the brake pedal has been released. In this condition these drag against the drum causing further increase in temperature. The building up of excess pressure can be avoided by its release through compensating port.

The displacement of pistons in the wheel cylinder determines the design of master cylinder. Size of the reservoir depends upon the type of brakes. Disk brakes require larger reservoir.

Other components of hydraulic brake system are tubes and hoses, valves and switches. Tubes and hoses transport the brake fluid from master cylinder to the wheels whereas valves control the pressure of fluid in the lines. Switches act as safety device and produce warning.

13.9.1 Tubes and Hoses

Tubes are made up of steel whereas hoses are made of rubber. These act as passage for the fluid to move to wheel cylinder from the master cylinder. The fluid is transferred to tubes after it passes through valve. These fluid lines should have very little friction so that fluid can move without resistance. It will also help the fluid to reach the wheel cylinder quickly. For efficient fluid transfer sharp curves in the fluid lines are avoided. This helps to attain quick fluid delivery.

The tubes are double walled made of steel with copper fused at very high temperature. To join the tubes together double flared joint fittings are used. These are made of steel or brass. While replacing the fittings, care should be taken that these are replaced with same fittings as original one. Even the composition of material of fittings should be same as that of lining as dissimilar metals in lining and fitting may cause corrosion and reduce life.

The hoses are made up of rubber and provide flexible connections. These are used particularly in wheel units. These are made of layers of fabric upon which synthetic rubber is impregnated. These should have high strength and thermal resistance.

13.9.2 Valves

The valves are meant for maintaining required pressure in the linings. Generally rear wheels of the automobile are provided with drum brakes and front wheels with disc brakes. The braking response in the disc brakes is immediate whereas in drum brakes the response is delayed. The valves are meant for metering and keeping a balance between the characteristics of disc and drum brakes.

Metering valve

It is located in the front brake lines and delays the fluid going from master cylinder to front wheel cylinders. This allows the pressure to build up in the rear drums first. After the pressure is built up the valve opens and allows the operation of front brakes. This way the operation of front and rear brakes is balanced. This also prevents the lockup of the front brakes.

Proportioning valve

This valve controls the pressure in the rear brakes. When the pressure in the rear brakes attains the specified value the proportioning valve operates and prevents the flow of fluid in the rear brakes. It regulates pressure in the rear brakes and keeps a difference in pressure between front and rear brakes. This also prevents the locking of rear wheels.

Height sensing valve

This valve controls the pressure on the rear brakes depending upon whether the vehicle is loaded or not. If vehicle is loaded, the actuator lever in the valve moves and permits full hydraulic pressure to the rear brakes. When vehicle is not loaded the pressure is reduced. It consists of plunger, cam, clutch spring and actuator lever. The actuator lever is connected to lower shock absorber bracket. The up and down movement of axle and the frame causes the valve to turn on and off. The clutch spring prevents change in position of the valve if the vehicle moves over a bump suddenly.

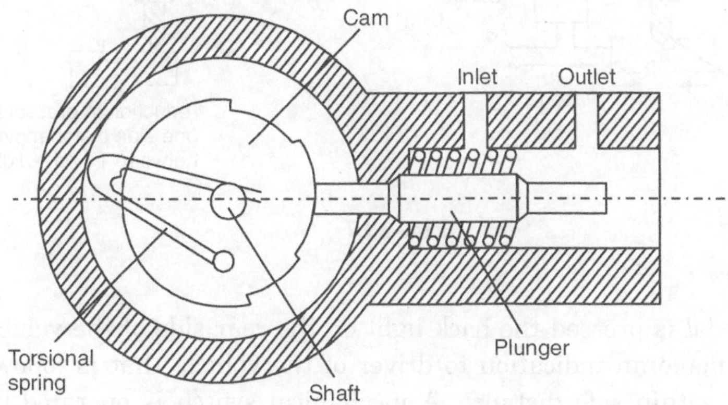

Fig. 13.6

Combination valve

Lately, metering and proportioning valves have been combined together. These are three functions or two functions valves. Three functions valve act as metering valve, proportioning valve and brake warning light switch. The two functions valves have two variants. One acts as metering valve and warning light switch and other has a combination of proportioning valve and warning light switch.

13.9.3 Warning Light Switches

The warning light switches are operated by pressure differential valve. If the pressure is lost in either the front or rear brakes, it is indicated with a warning light on dash board so that driver becomes aware of it. The loss of pressure means that brake pedal travel would increase. Driver will have to apply more force on brake pedal and actuating of brakes would not be that effective.

At designed pressure difference, the valve piston is centrally located and trigger switch is positioned in the groove as shown in the Fig. 13.7. With the reduction in pressure on one side, the piston moves towards that side. The trigger switch attains new position out of the groove. This causes the closing of the switch and warning light glows.

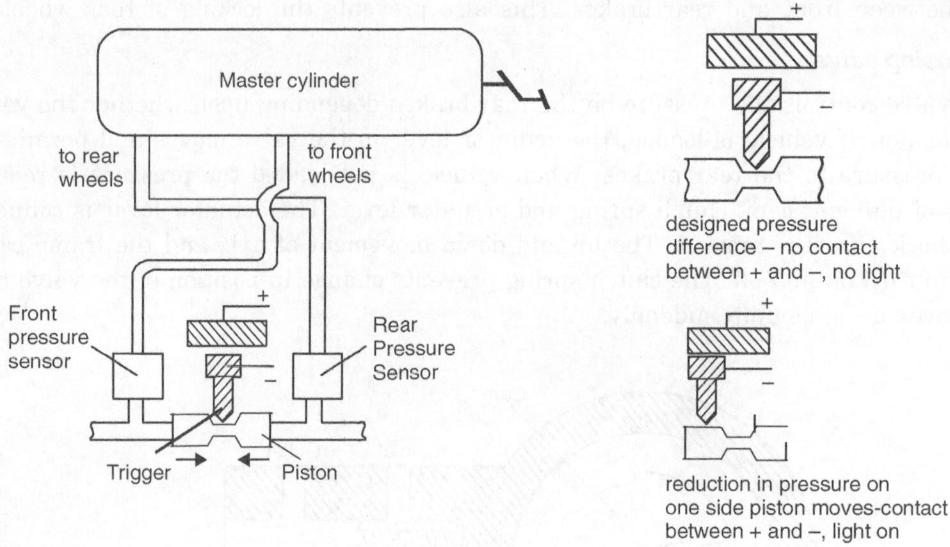

Fig. 13.7

Stop light switch

As the brake pedal is pressed the back light on the rear side of the vehicle should glow. This is essential to make an indication to driver of the vehicle that is following so that he can stop his vehicle within safe distance. A mechanical switch is operated by brake pedal. Alternatively, a hydraulic switch may be operated by pressure developed in master cylinder. In both types, the circuit is open when brake pedal is released. As soon as the brake pedal is pressed the circuit is closed causing glow of back lights.

13.10 POWER BRAKES

Power brakes or power assisted brakes were first utilized in heavy vehicles to help driver to have better control on the motion of the vehicle. The force required on the brake pedal to slow down or stop the vehicle is reduced. Later, these have been used in automobiles also. Brake boosters are incorporated in the system for this purpose.

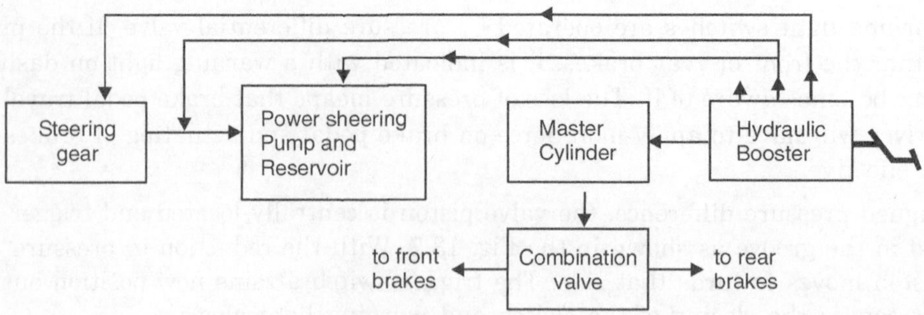

(a)

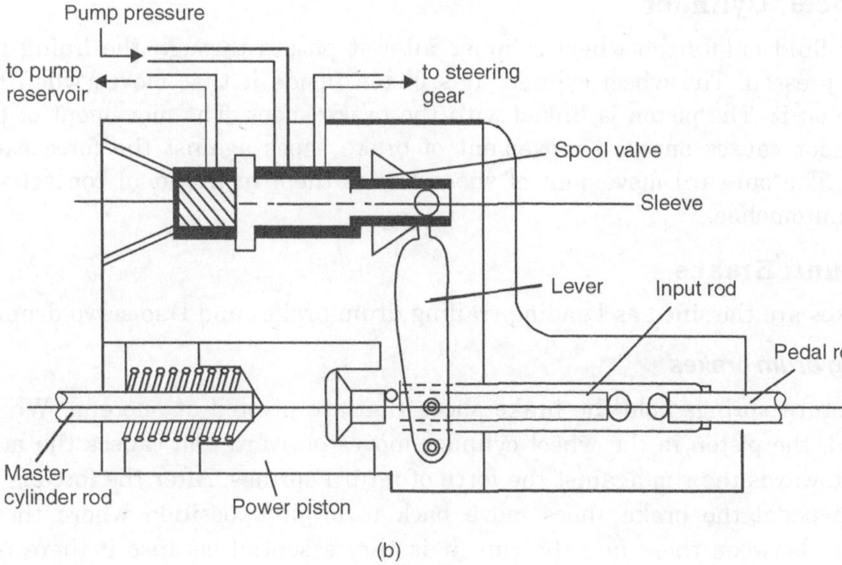

Fig. 13.8

Hydraulic brake booster

Hydraulic brake booster is located adjacent to the master cylinder on the firewall. Power steering pump provides the needed pressure in the booster. In normal condition, when brakes are not applied and engine is running the pump, pressure is transmitted to steering system. When brakes are applied, with pushing of pedal rod, pivot lever connects the booster rod to the spool valve. Forward movement of spool valve provides additional pressure into the space in the main cylinder. The piston in main cylinder is pushed forward and pushes the rod in master cylinder with enhanced force.

Some of the hydraulic boosters are driven by electric pump. The pump operates only when needed. To avoid the heating of the fluid cooler may also be used.

13.11 DRUM BRAKES

Drum brakes consists of brake assembly enclosed in the drum. There are two brake shoes that have curved surface. These shoes move outwards on the application of brakes. The frictional contact between the outer surface of the shoes and inner surface of the rim causes resistance to the motion of the wheel. Earlier, all the wheels used to have drum brakes. Presently, the automobiles are provided with drum brakes in rear wheels and disc brakes in the front wheels.

13.11.1 Brake Shoes

These are made of metal with friction material riveted or pasted on outer side. The friction material is also known as brake lining. If riveted, the riveting is done in such a manner that rivet heads are not projected above the lining material. This way frictional contact between the inner surface of the rim and rivet heads can be avoided. The material used for the lining should be capable of withstanding the heat generated due to friction. Asbestos was used earlier but has now been abandoned as it is health hazard to human beings. Fibreglass or semi-metallic materials have replaced asbestos.

13.11.2 Wheel Cylinder

The brake fluid enters the wheel cylinder after it passes through the lining after the brake pedal is pressed. The wheel cylinder has piston inside it that moves when the fluid exerts pressure on it. The piston is linked with the brake shoes. The movement of piston in the wheel cylinder causes outward movement of brake shoes against the force exerted by return springs. The outward movement of shoes brings them in frictional contact with rim and stops the automobile.

13.11.3 Drum Brakes

Drum brakes are classified as Leading-trailing drum brakes and Duo-servo drum brakes.

Leading-trailing drum brakes

The two return springs hold the brake shoes that are pivoted at one end. When brake pedal is pressed, the piston in the wheel cylinder moves outward and causes the movement of brake shoes towards the rim against the force of return springs. After the force is released from the brake pedal the brake shoes move back to original position where there is no frictional contact between these and the rim. It is very essential because if there is even a slight contact between the two frictional resistance would exist. The engine would have to exert unnecessarily to overcome this frictional resistance. The friction between leading shoe and drum tends to rotate the shoe with drum. This forces the bottom of the shoe against the anchor pin. The leading shoe does most of the braking and therefore wears out fast. For trailing shoe, the drum rotation tends to force shoe away from the drum. The trailing shoe therefore wears less. When the vehicle is moving in the reverse direction and brakes are applied, the leading shoe becomes trailing shoe and vice versa.

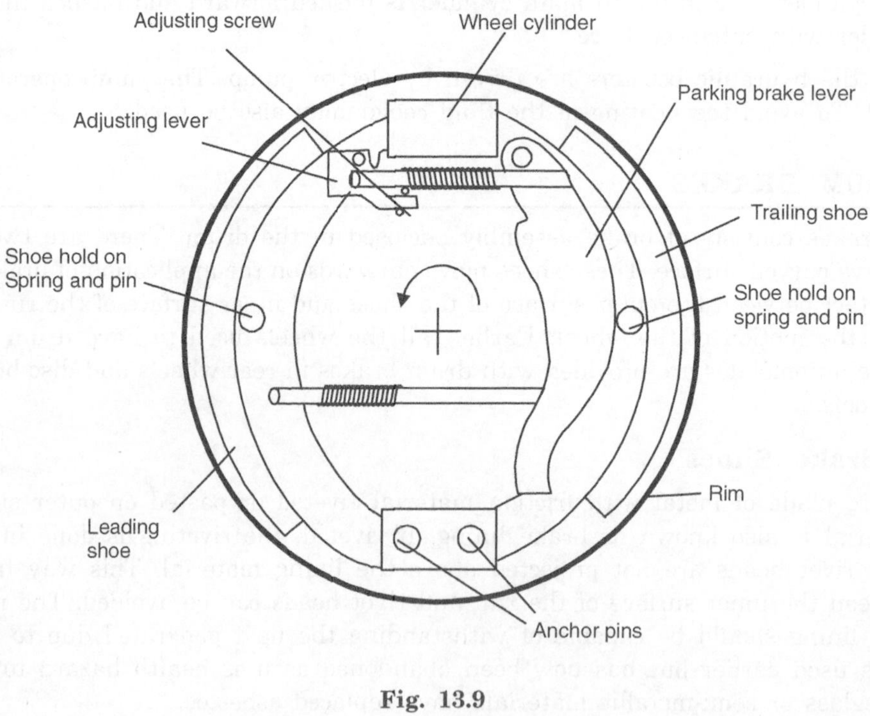

Fig. 13.9

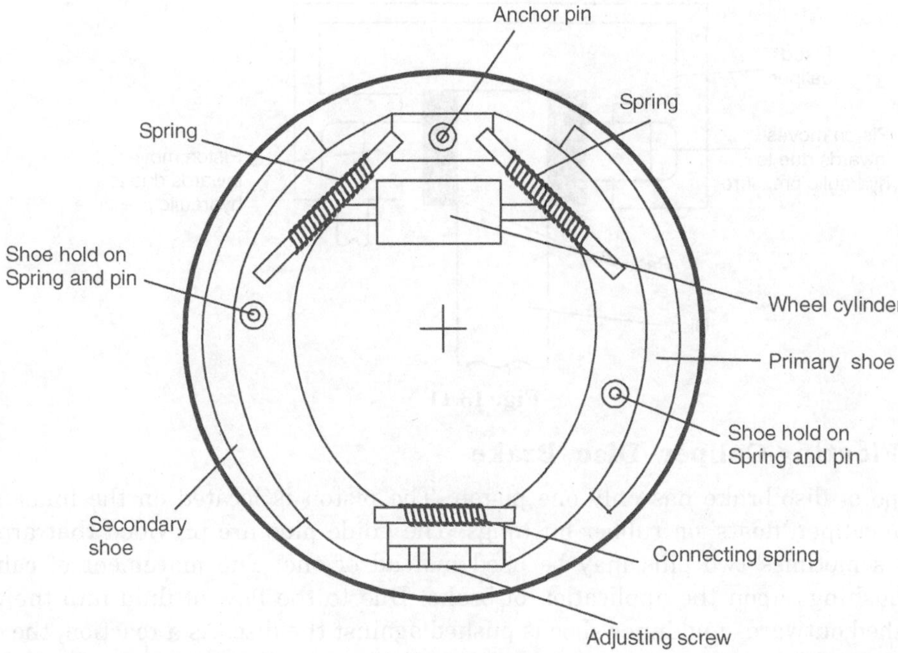

Fig. 13.10

Duo-servo drum brake

The shoes are connected to a single pin. The bottoms of the shoes are linked together with the help of screw. The shoe towards the front of vehicle is known as primary shoe and that towards the rear is termed as secondary shoe. The frictional contact between the shoe and drum tends to rotate the shoe. The upper part of primary shoe tends to move downwards and the lower part of the shoe pushes the adjusting screw backward. The bottom of the secondary shoe is forced against the drum and it is moved upwards against the pin. Shoes are forced against the drum more tightly with further rotation of drum. Therefore, the self-energizing action of secondary shoe is further enhanced. The total braking force is also greater than provided by wheel cylinder. The secondary shoe has longer lining as it provides more braking force than primary shoe.

13.12 DISC BRAKES

The disc brakes have a rotating disc in place of drum. The braking shoes are replaced by pads which are held in calipers. The pads come in frictional contact with the moving disc upon the application of brakes. Due to this frictional contact the resistance to the motion is generated that stops the vehicle. These can be classified into (*i*) fixed-caliper (*ii*) floating caliper and (*iii*) sliding caliper disc brakes.

13.12.1 Fixed-Caliper Disc Brake

It has pistons on both the sides of disc. The number of piston may be one or even two on each side of the disc. The caliper is attached to some stationary component of the vehicle such as steering knuckle.

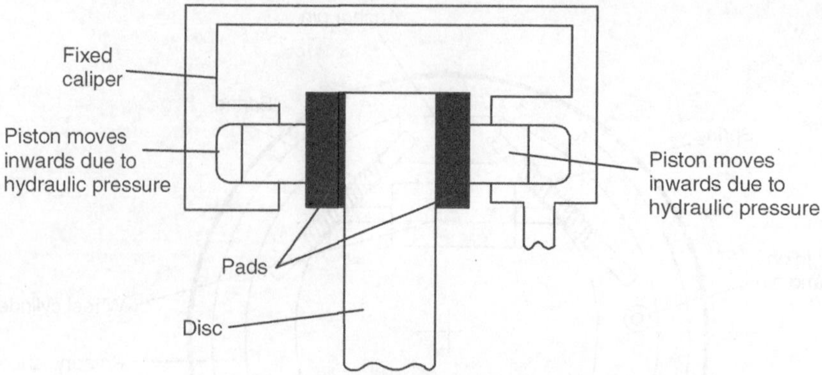

Fig. 13.11

13.12.2 Floating-Caliper Disc Brake

This type of disc brake has only one piston. The piston is located on the inner side of the disc. The caliper 'floats' on rubber bushings. The guide pins are provided that are made of steel and sometimes two pins may be used instead of one. The movement of caliper is allowed by bushings upon the application of brake. Due to the flow of fluid into the caliper piston is pushed outwards and inner shoe is pushed against the disc. As a reaction, the caliper moves slightly and brings outer shoe in contact with the disc. Both the pads clamp the disc and cause resistance to the motion.

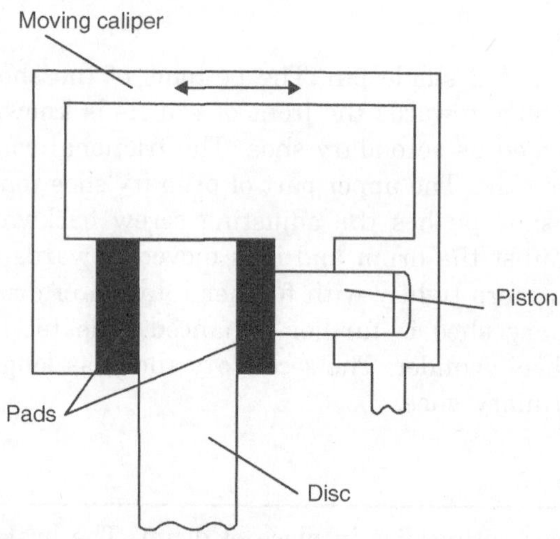

Fig. 13.12

13.12.3 Sliding-Caliper Disc Brake

These are similar to floating-caliper disc brakes. Both the calipers slide on smooth surface on the steering knuckle. The sliding movement is caused when the brakes are applied. There are no guide pins.

13.13 PARKING BRAKES

The parking brakes are applied on the stationary automobile. This is done to prevent its undesirable rolling of automobiles when parked. The possibility of rolling increases when the automobile is parked on inclined surface. The brake is operated by a hand lever. These can be classified as (*a*) *Independent* and (*b*) *Integral parking brakes.*

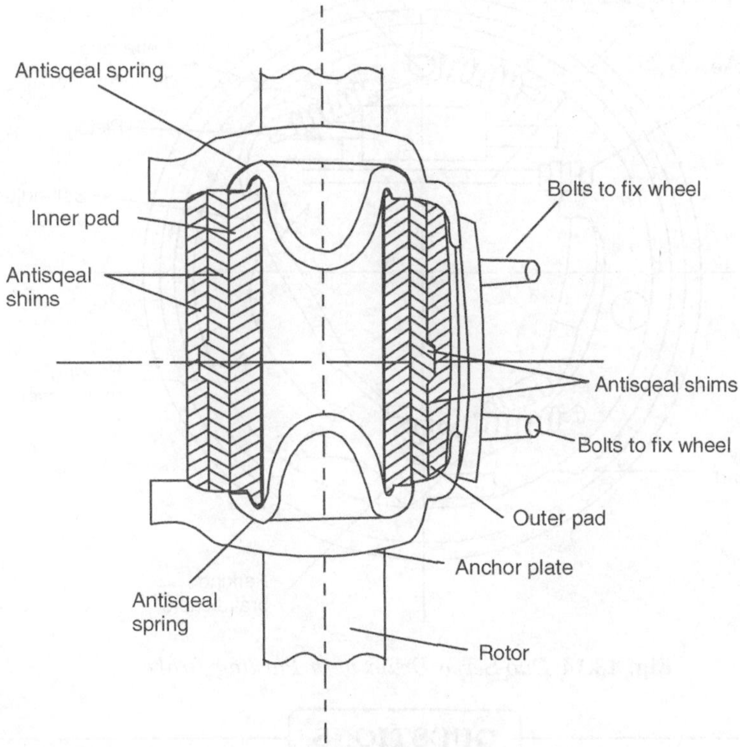

Fig. 13.13

13.13.1 Independent Parking Brakes

These brakes are independent of brakes used during the motion of automobile. These are used in automobiles that have drum brakes on the rear wheels. Small brake shoes are fitted into the hub of the disc. The surface of the hub serves as drum. Another variant of independent parking brakes is transmission mounted. A small drum brake is mounted at the back of transmission and is attached to the output shaft.

13.13.2 Integral Parking Brakes

These are integral with brakes used during the motion of automobile. These have two variants one meant for drum brakes and other for disc brakes on rear side. In rear drum brakes, parking brake lever is attached to rear brake shoe. There is a parking brake link between two brake shoes. When brake lever is pulled, the parking brake lever is pivoted and its upper end forces contact between shoe and drum. The section below pivot forces the link forward and pushes the front shoe into the drum. Thus parking brakes are applied. In rear disc brakes, caliper is provided with a piston that can be operated hydraulically or mechanically. There is a large screw that is screwed into nut and cone assembly which fits inside the piston. When parking brakes are applied the cable pulls the lever. This rotates the screw. The piston is pushed out by nut. This causes contact between shoe and the disc and actuation of parking brake.

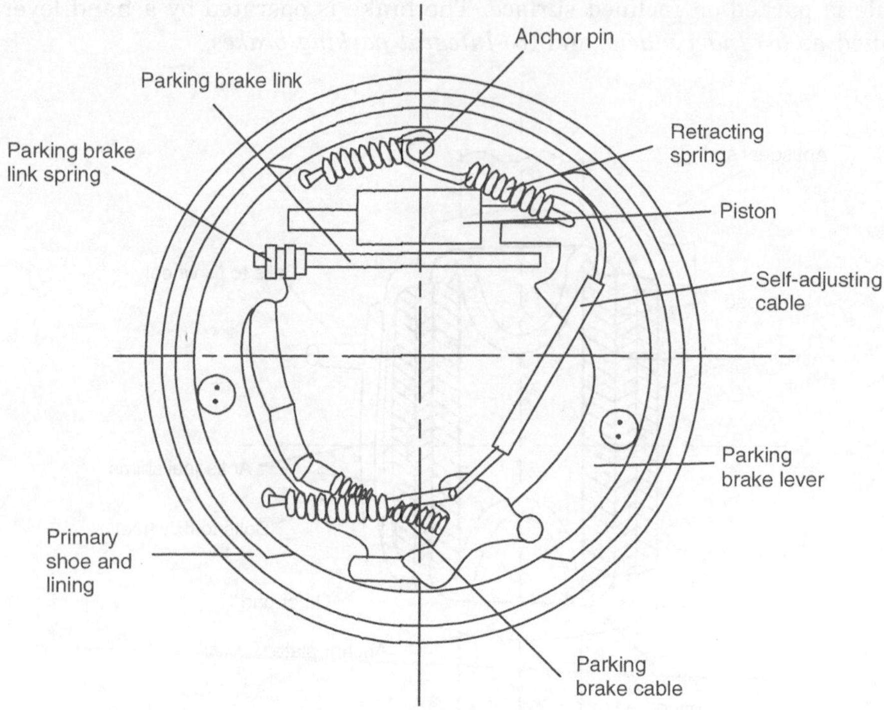

Fig. 13.14 *Duo-Servo Drum with Parking Brake.*

QUESTIONS

1. Explain the different requirements of brakes in automobiles.
2. Describe briefly different components of stopping distance.
3. What are different factors that affect the application of brakes? Describe briefly.
4. Explain the constructional details and working of master cylinder in hydraulic brakes.
5. What is dual braking system? What are its advantages?
6. What is Tandem master cylinder? Explain its constructional details and working.
7. Explain the functions of different valves used in brake system.
8. Explain hydraulic brake booster. What is the advantage of hydraulic brake booster?
9. Explain Leading-trailing drum brakes with the help of a figure.
10. Explain different calipers used in disc brakes.
11. Describe the necessity of parking brakes.

14

ANTILOCK BRAKE SYSTEM

Brakes are an essential controlling device in modern day's automobiles. Automobiles today can run at very high speed and controlling motion at high speed, particularly when brakes are applied, could cause instability. At high speed when brakes are applied, the wheels become stationary and these start sliding instead of rolling on the road surface. As the wheel starts sliding, the steering control is also lost and the vehicle cannot be kept straight. The force of friction is reduced and that also causes instability. It becomes difficult for the driver to keep the vehicle straight and the vehicle skids and goes out of control. This situation is dangerous and an accident can occur. It becomes essential to keep the vehicle under control when brakes are applied. This is achieved by antilock brake system (ABS).

14.1 ANTILOCK BRAKE SYSTEM

When brakes are applied suddenly, the wheels may get locked almost immediately. This causes instability of the vehicle. If the vehicle is moving on loose snow, the locking of wheels causes formation of a small wedge in front of wheel. This helps the vehicle to stop. This is the only situation, where locking of the wheels may be helpful. In all other conditions, it is undesirable.

14.2 PRESSURE MODULATION

To avoid locking of wheels, the brakes are applied and before the wheels are locked, released. The same process is repeated several times. In earlier days, the drivers used to do the same. With introduction of antilock brakes, this is performed by brake system. Application of brakes means building of pressure while releasing the brake pedal means undoing it. This is termed as pressure modulation. Through antilock brake system, modulation of pressure can be achieved even up to 15 times per second. The modulation of pressure maintains the friction between tyre and road surface and provides stability to the vehicle. As the sliding of wheel is avoided the control on steering is also not lost and the vehicle continues to move straight.

Locking of front wheels affects the maneuverability of the vehicle while locking of rear wheels affects its stability. The best results are obtained when wheel are also allowed to slip in between application of brakes. Zero per cent slip means wheel is rolling freely while 100% slip means the wheel is completely locked. It also means that at zero per cent slip the velocity of wheel is same as that of vehicle and at 100% slip the velocity of the wheel is zero and vehicle is moving at its own velocity. Due to slip the velocity of wheel is reduced and vehicle continues to move with the same velocity. It is termed as slip rate measured in per cent.

For example, if the velocity of wheel is 30% less than the velocity of vehicle, it is termed as slip rate of 30%. For antilock brake system the ideal slip rate should vary from 10% to 30%.

Antilock brake systems may differ from each other. These systems could be integral or non-integral. In integral systems, master cylinder, hydraulic booster and their circuitry is combined together into a single unit. In non-integral type of systems, there is vacuum assisted booster and master cylinder. Control unit is also a separate unit. The hydraulic system may be provided fluid from master cylinder in some systems. There is separate hydraulic unit and pump, motor and accumulator. Solenoid valves may be provided to control hydraulic pressure to the wheels.

Antilock brake system may further be classified as single channel or multi channel systems. A single channel system is a two wheel system when pressure is modulated on both rear wheels at the same time and not to the front wheels. The system is provided input from a single speed sensor that is located centrally. The sensor may be positioned on the differential unit. The multi channel system could be two channels, three channels or even four channel system. In a two channel system again pressure modulation is applied to two rear wheels only by each channel. There are two speed sensors for each wheel. Then there are three channel systems. These systems have individual circuits for each front wheels and a single circuit for both the rear wheels. The most effective system is the four channel system that has independent circuits for all the four wheels. There are separate sensors for each wheel. In this system, each wheel receives controlled braking action.

14.3 COMPONENTS OF ANTILOCK BRAKE SYSTEM

The design of an antilock brake system may vary from automobile to automobile. The modern system has hydraulic and electrical/electronic components. Apart from components in normal brakes there are additional components in antilock brake system.

14.3.1 Hydraulic Components

Accumulator

It stores fluid and maintains high pressure in the system. It also provides required pressure for power-assisted brakes. It is charged with nitrogen gas. It has a diaphragm that separates the two compartments. One compartment accommodates brake fluid at high pressure while other has nitrogen at high pressure.

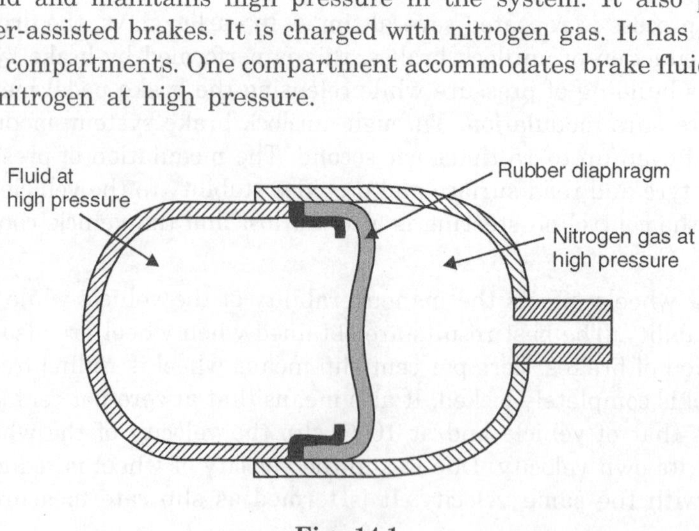

Fig. 14.1

Hydraulic control valve

This valve provides pressure modulation. It builds the pressure in the system when brakes are applied and releases it. This happens several times during the application of brakes. This valve could be combined with master cylinder or mounted separately. In first case it is termed as integral type and in second case non-integral type. For control it is provided with solenoid valves.

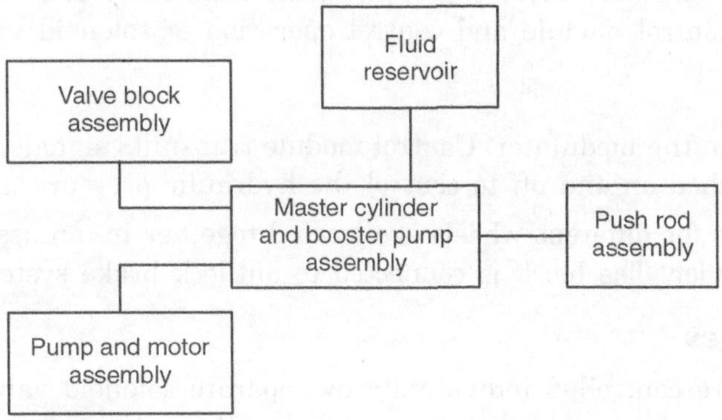

Fig. 14.2 *Block diagram showing components of hydraulic control valve*

Booster pump

It provides fluid at high pressure to antilock brake system. It consists of a pump and electric motor and therefore it is also known as electric pump and motor assembly.

Hydraulic unit

It is an assembly of master cylinder and booster. It modulates the hydraulic pressure in the system with the help of valves and piston. Fluid at high pressure is also provided through hydraulic pump for power brakes application.

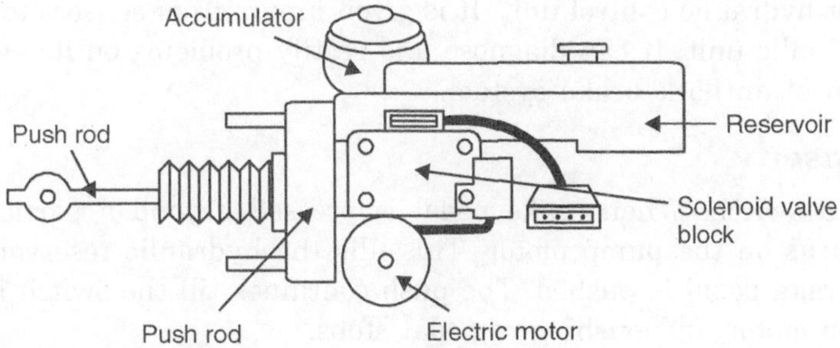

Fig. 14.3

Fluid accumulators

This temporarily stores the fluid from wheel cylinder. This fluid is used to build pressure in brake hydraulic system. There are two accumulators one each for primary and secondary hydraulic circuits.

Valve

Antilock brake system module controls this valve. In open position, brake fluid is sent into the master cylinder from booster circuit at high pressure. This does not allow excessive pedal travel. This is a two position valve.

Modulator unit

This unit controls the flow of fluid to the individual wheel cylinders. It has solenoid valve that controls a number of other valves that control the fluid flow. The unit also has relays that are activated by control module and control operation of solenoid valves.

Solenoid valves

These are located in the modulator. Control module transmits signals that operate these valves. These are switched on and off to control the hydraulic pressure in different wheels.

The solenoid valves for different wheels are located together in .an assembly attached to the side of master cylinder. The block is connected to antilock brake system control module.

Wheel circuit valves

The two circuits are controlled individually by separate solenoid valves. One is meant for controlling the inlet valves and other for controlling the outlet valves. The pressure in the circuit is increased, decreased or maintained at steady level by using the inlet and outlet valves. Control module determines the operation of individual valve. The control module for antilock brake system provides electric supply of 12 volts that operates the solenoid. The circuits are not operated during normal driving.

Apart from these components the system has some electrical and electronic components also. These components make the system more dependable but at the same time complex and costly.

Control module

It is a small electronic unit consisting of micro-processor. The micro-process is programmed and executes instructions according to programme. It is a small unit which can be mounted to master cylinder or hydraulic control unit. It is given inputs from sensors to monitor wheel speed and from hydraulic unit. It can diagnose and rectify problems on its own. It monitors the proper operation of antilock brake system.

Brake pedal sensor

This sensor is a switch. When brake pedal is pressed beyond a particular limit, the switch opens and turns on the pump motor. This fills the hydraulic reservoir with fluid at high pressure and brake pedal is pushed. The push continues till the switch is closed again. This switches off the motor and pushing of pedal stops.

Pressure switch

This switch indicates low pressure through warning light on dash board. The switch activates the pump motor when pressure is reduced below the designed level. When designed pressure is attained the switch again operates and pump motor is stopped.

Pressure differential switch

This switch transmits signal to control module when higher than designed pressure difference occurs in the system. The switch is inside the modulator.

Relays

These are used to switch on the motors and solenoids. These require low current signal to operate and that is provided by control module. These are electro-magnetic devices.

Toothed ring

This wheel is provided with teeth that pass by wheel speed sensor. As this happens the AC signal is generated. The signal fades away as tooth move away. The next tooth that passes by the sensor causes generation of signal. This way pulsing signal is generated. This is sent to the control module. The control module converts it into the wheel speed. The ring is located on axle shaft.

Wheel speed sensor

This sensor is in the form of a coil with permanent magnet in the centre. It sends the pulsating AC signal generated by toothed wheel to the control module.

The system is provided with indicator lights also. These lights act as warning lights in case there is some fault in antilock brake system. Also, there may be another warning light to indicate fault in the basic brake system. Apart from these components, the system has Data Link Connector that acts as a mean to operating conditions, other information about the automobile and diagnostic information. To diagnose the fault in the system numeric identities are provided to different faults. These trouble codes are fed to the control module.

14.4 NON-INTEGRAL ANTILOCK BRAKE SYSTEMS

14.4.1 Two Wheel System

These systems have pressure modulation in rear wheels only. As brakes are applied pressure is transmitted through valves. The control module monitors the speed sensor signal. If it finds that deceleration caused is high and may cause locking of wheels it activates the isolation valve. This stops the building up of pressure on rear wheels that prevents the further deceleration. If still the deceleration is not prevented, control module would activate the dump valve. This continues till the deceleration of wheel and the vehicle becomes the same. When brake pedal is released the control module deactivates isolation valve. This allows the fluid to go back to master cylinder. The control module also controls the solenoid valve. The control module also detects errors in the system. The wheel speed sensors feed the wheel speed to control module continuously. If the automobile is a four wheel drive with an option of two wheel drive the antilock brakes system stops functioning when the operation is shifted to four wheel drive.

14.4.2 Four Wheel System

In this system, there are four channels, one for each wheel. The hydraulic control unit has two solenoid valves for each wheel (Fig. 14.4). Normal braking is assisted by vacuum-power brake system. In another variant, there are three channels, one each for front wheels and one for both rear wheels (Fig. 14.5). The system has two solenoid valves for each channel. Better steering control is available in these systems. The modulation of pressure occurs in all the four wheels which make the operation of brakes very safe and automobile can be stopped in the shortest possible distance.

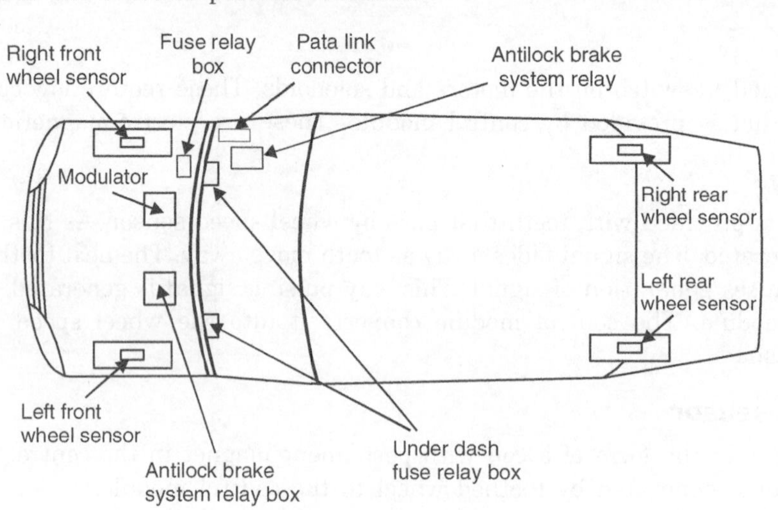

Fig. 14.4

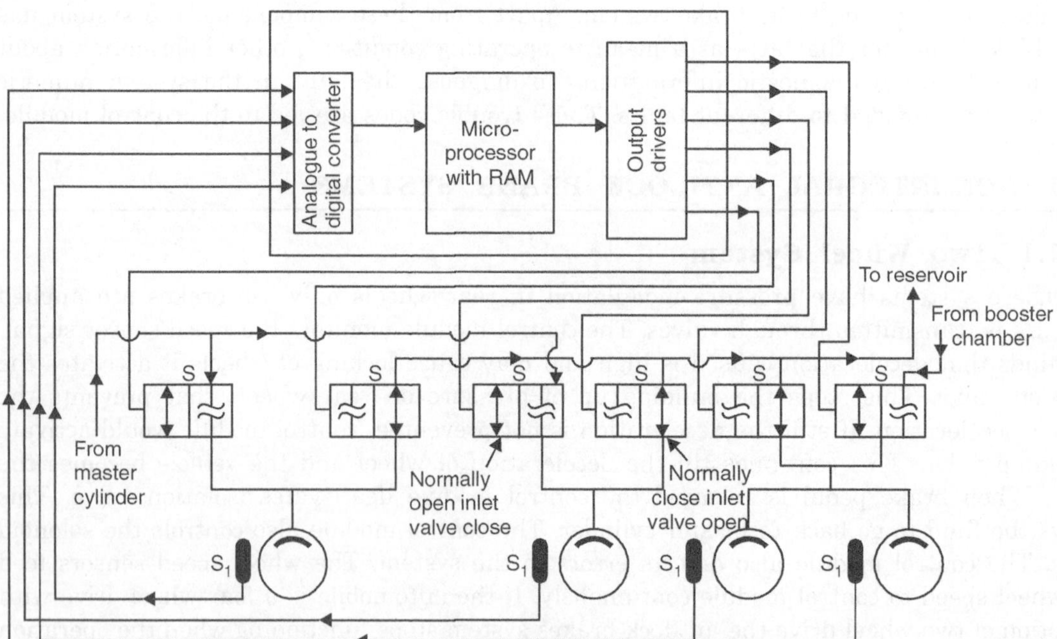

Fig. 14.5

The control module closes the inlet solenoid valve when it senses that due to deceleration the wheel may lock up. This does not allow further entry of fluid in the circuit. If the wheel still decelerates and there is possibility of lock up, the control module opens the outlet solenoid valve. After the brakes are released the control module brings back the inlet and outlet solenoid valves back to original position. The control module also calculates the slip rate of each wheel through wheel speed sensors. If slip rate is high it transmits a control signal to modulator.

Modulator

The modulator consists of inlet and outlet solenoid valves, reservoir, pump and motor (Fig. 14.6). The callipers are relieved of fluid pressure. There are three control modes namely decreasing the fluid pressure, holding the fluid pressure and increasing the fluid pressure. First mode where fluid pressure has been decreased the inlet solenoid valve is closed and outlet solenoid valve is open. The callipers are relieved of fluid pressure as existing fluid in the caliper flows back to master cylinder through outlet solenoid valve.

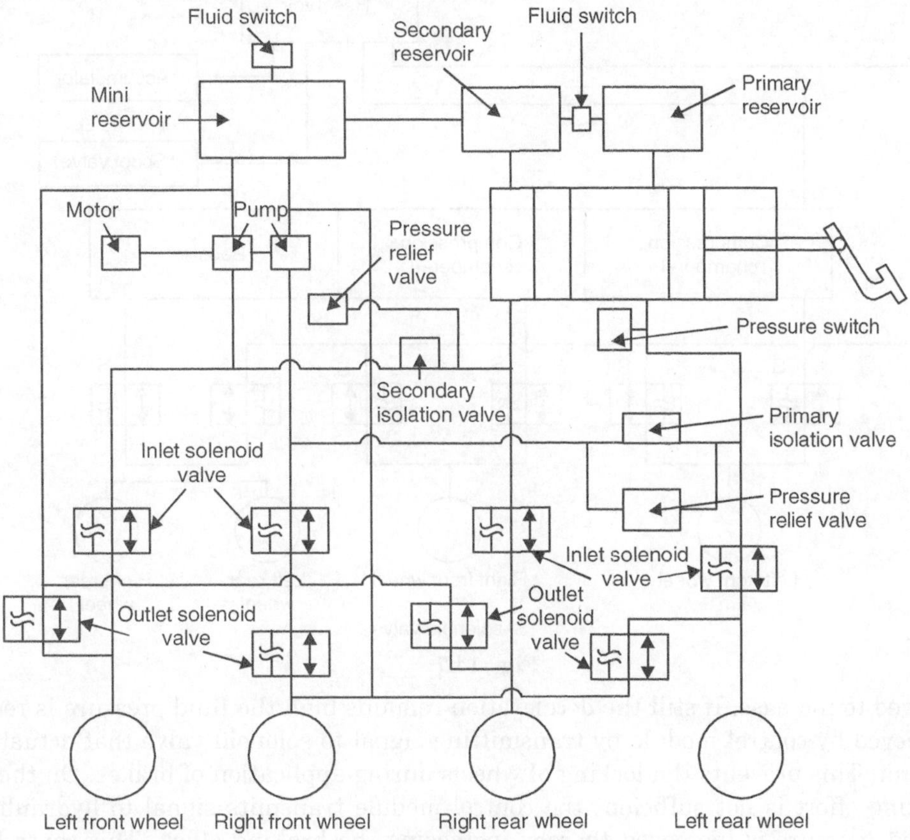

Fig. 14.6

When fluid pressure is increased the fluid at high pressure is pumped to callipers. The inlet solenoid valve is open and outlet valve is closed. In third mode, when fluid pressure is retained, both the inlet and outlet valves are closed.

The pump provides excess fluid needed during antilock operation. The fluid is released to accumulator afterwards that acts as temporary store for it. After the operation the pump drains the accumulator. An electric motor drives the pump. The motor is controlled by the control module. The pump is required to redirect the fluid quickly as the solenoid valves open and close rapidly.

The antilock braking system is self monitoring. If some component does not perform properly it is indicated to the driver through warning light. However, normal application of brakes continues if antilock braking system develops some fault.

14.5 INTEGRAL ANTILOCK SYSTEM

14.5.1 Four Wheel System

Upon applying brakes, the deceleration of each wheel is monitored by the control module. If the deceleration is high and is likely to cause locking of wheel, signal is transmitted to hydraulic unit. The hydraulic unit initially keeps the fluid pressure at wheel constant and it

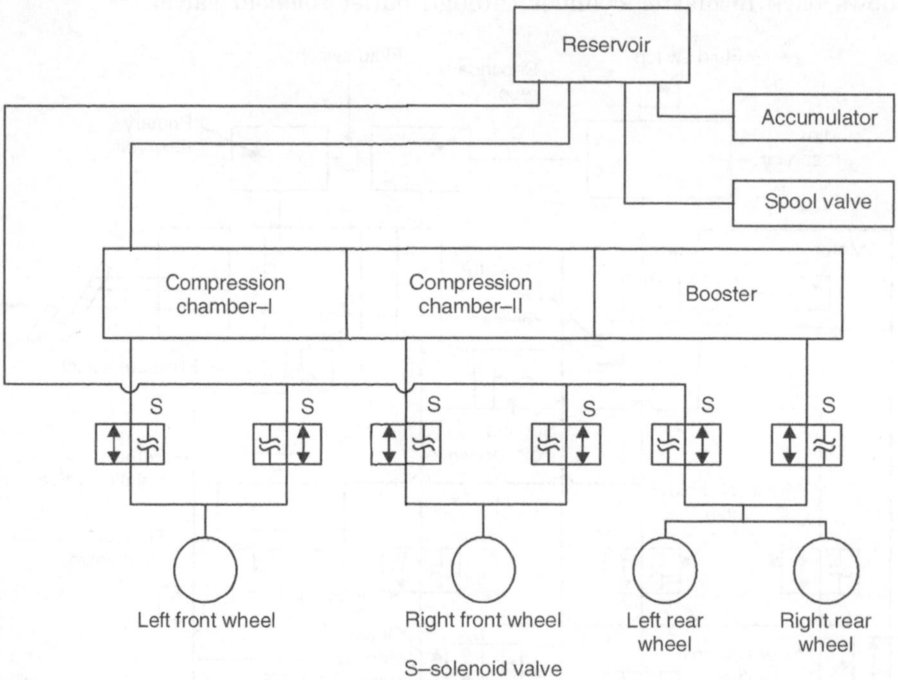

Fig. 14.7

is not allowed to increase. If still the deceleration remains high the fluid pressure is reduced. This is achieved by control module by transmitting signal to solenoid valve that actuates the hydraulic unit. This prevents the locking of wheels during application of brakes. On the other side if braking effort is not sufficient the control module transmits signal to hydraulic unit and the fluid pressure is increased thereby increasing the braking effort. This control cycle is repeated many times depending upon the requirement of brake action.

Once the brake pedal is released, the piston in the master cylinder moves back, the fluid from the booster chamber flows back to reservoir. When the brakes are applied under normal conditions, the pushrod operates a lever. This moves the spool valve and port between booster chamber and reservoir is closed. It also opens partially the port from accumulator. The opening of port is in proportion to the force applied on the brake pedal. The fluid at high pressure moves from accumulator to booster chamber. The fluid pushes the piston forward and adds to the thrust on the push rod. When the control module detects that wheels are locking it opens a valve. The valve supplies the fluid to chambers between pistons in the master cylinders and between the retraction sleeve and piston in first master cylinder. The fluid at high pressure on the retraction sleeve forces the push rod and brake pedal back. In fact, fluid at high pressure is being supplied by accumulator and not through brake pedal action. The control valve also opens and closes solenoid valves to control the brake action on wheels.

When solenoid valves are open, the pistons in master cylinders provide fluid to the front wheels and booster chamber provides fluid to rear wheels. When these are closed, the master cylinder and booster chambers are cut off. The fluid returns from wheels to the reservoir.

While cornering, the brakes are applied and antilock brake system has to work differently in this situation. The control module operates under different programming. A switch gives indication to the control module about the application of brakes while cornering.

14.5.2 Automatic Traction Control

An automobile may loose its track in some situations such as accelerating on wet surface. This creates instability for the vehicle and may be dangerous for the occupants. In

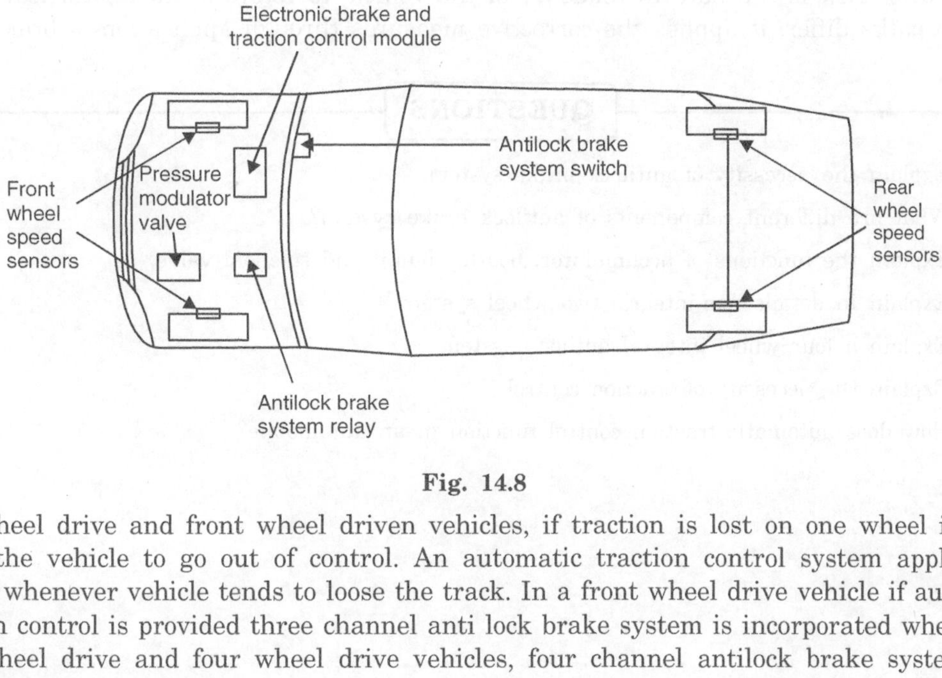

Fig. 14.8

four wheel drive and front wheel driven vehicles, if traction is lost on one wheel it could cause the vehicle to go out of control. An automatic traction control system applies the brakes whenever vehicle tends to loose the track. In a front wheel drive vehicle if automatic traction control is provided three channel anti lock brake system is incorporated whereas in rear wheel drive and four wheel drive vehicles, four channel antilock brake systems are incorporated.

It is essential that the rotary speed of the wheel is in proportion to the vehicle speed. If the rotary speed of the wheel is more than vehicular speed it is termed as 'slippage' and causes the loss of traction. Normally, a slippage of 10% is permissible. Control module in automatic traction control system monitors the wheel speed through sensor. If the loss of traction is sensed due to increase in rotary speed of a particular wheel, control module signals the opening/closing of solenoid valves causing application of brakes to that wheel. The control module and hydraulic valve unit may be different for the automatic traction control system and antilock brake system. In some cases, the two systems may use a single control module and hydraulic unit.

Some automatic traction systems are meant for wet or snow covered roads. These systems operate only at low vehicular speeds. Some automatic traction control systems also control the functions of the engine and work at high vehicular speeds. The functions of engine that are controlled include ignition timing and partially closing the throttle. Even fuel supply to one or more engine cylinders can be cut to reduce the vehicular speed for better traction control. Indicator is provided on the dash board to intimate driver that automatic traction system is operating.

14.5.3 Automatic Stability Control

When the vehicle is moving on the curved path it may 'under steer' or 'over steer'. In case of 'under steer' the vehicle tends to move inwards the curved path and it moves outwards the curved path in case of 'over steer'. Stability control systems can provide stability to the vehicles in these situations and cause desired corrective actions also. In case of over steer, the outside front brakes are applied and in case of under steer inside rear brakes are applied. The system uses the steering angle and speed of the four wheels to calculate the correct path. It also senses the lateral forces and vehicle yaw to calculate actual path followed by the vehicle. Yaw is the natural tendency of the vehicle to rotate about its vertical axis. If the two paths differ, it applies the corrective measures through application of brakes.

QUESTIONS

1. Explain the necessity of antilock brake system.
2. What are different components of antilock brake system?
3. Explain the functions of accumulator, booster pump and solenoid valve.
4. Explain in details non-integral two wheel system.
5. Explain a four wheel integral antilock system.
6. Explain the necessity of traction control.
7. How does automatic traction control function in an automobile?

15

STEERING SYSTEM

Steering system in an automobile controls the direction of motion. It also changes the direction of motion. While moving on the road, there may be a turn or a curve. The automobile must traverse this turn or curve precisely. The change in the direction of motion should occur very accurately. If that does not happen it may cause a collision. Therefore a steering system must provide accurate control on the motion of the automobile when it is taking a turn or when it is moving on a curved path. The turning of the automobile is usually affected by turning the front wheels. The front axle accommodates this turning of wheel through stub axles. The front axle also accommodates the different components of steering mechanism. Because of this, in case of front wheel drive automobiles, the design of the front wheel becomes quite complicated.

15.1 FRONT AXLE

The front axle in a rear wheel drive takes up mainly the load of the vehicle. Hence it is also known as dead axle. Precisely it has to take up the bending load due to the weight of the vehicle and load due to braking torque on the wheels. The axle is needed to be strong enough to take up these loads therefore it is provided with I section in the middle and elliptical section at the ends. The axle is provided with a downward sweep to keep the chassis height low. The main axle beam is connected to stub axles through king pins.

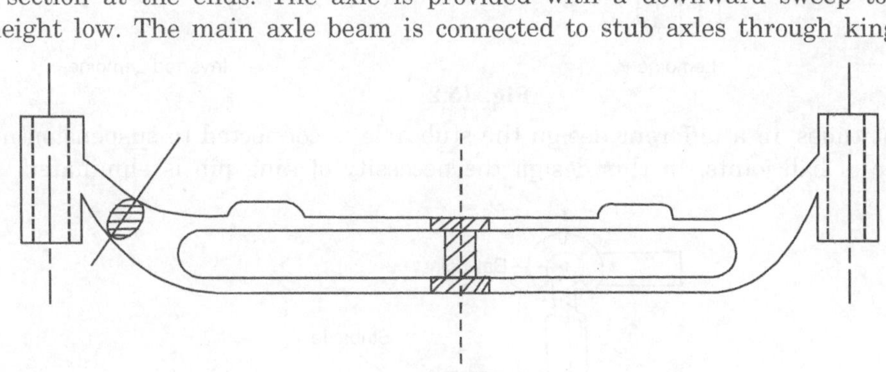

Fig. 15.1

15.2 STUB AXLE

Stub axles are pivoted at the ends of main axle beam. There are different methods to do it. The purpose is to provide proper movement to the wheels while turning and to keep the vehicle stable.

One of the arrangements known as Elliot stub axle there is a swivel pin fixed in forging of the stub axle. The ends of the swivel pin can turn in the forked end provided at the ends of main axle beam. In reverse Elliot stub axle the swivel pin is fixed in the main axle beam. The other two variants are Lamoine and inverted Lamoine stub axles. In Lemoine type, the stub axle is connected to the main axle beam from below it whereas in inverted Lamoine type the stub axle is connected to the main beam axle its the upper side. The wearing of the surfaces occurs on the bearing surfaces and causes 'angular shake' or 'play'. This is not desirable. To keep it minimum the bearing surfaces are kept apart as far as possible.

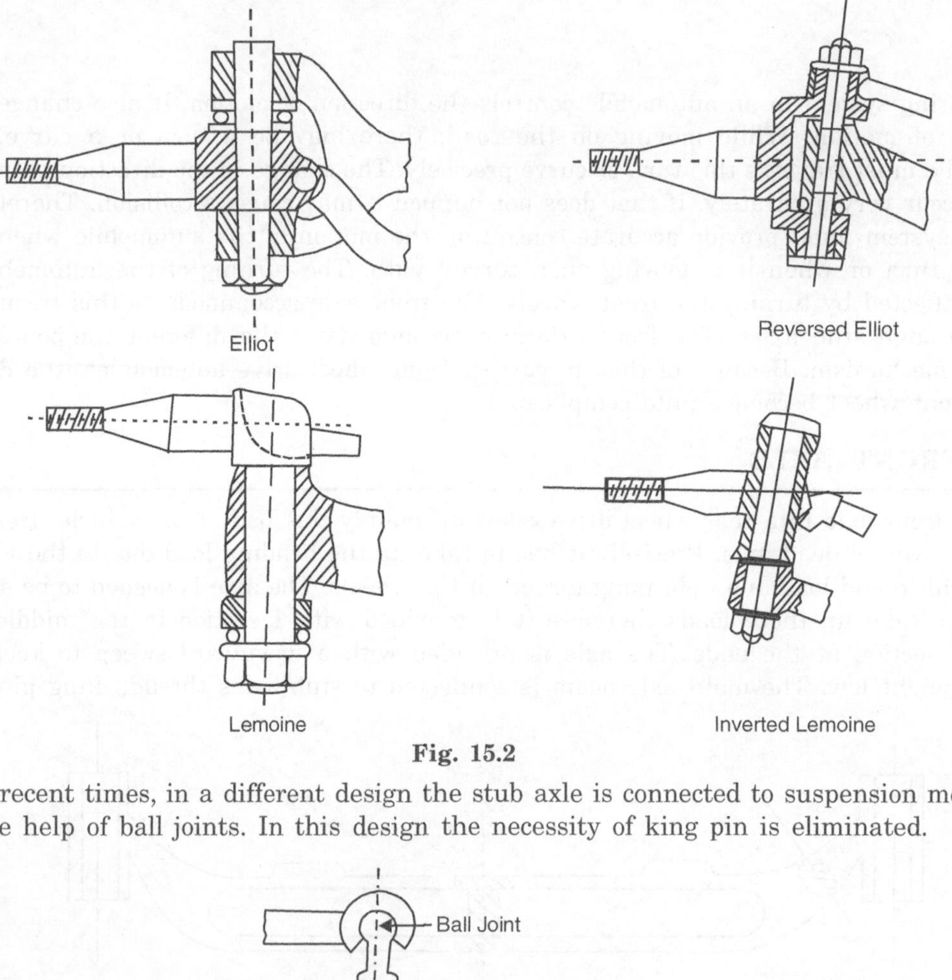

Fig. 15.2

In recent times, in a different design the stub axle is connected to suspension member with the help of ball joints. In this design the necessity of king pin is eliminated.

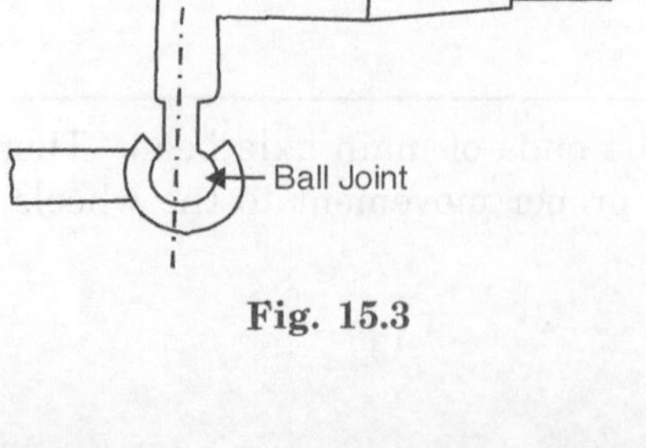

Fig. 15.3

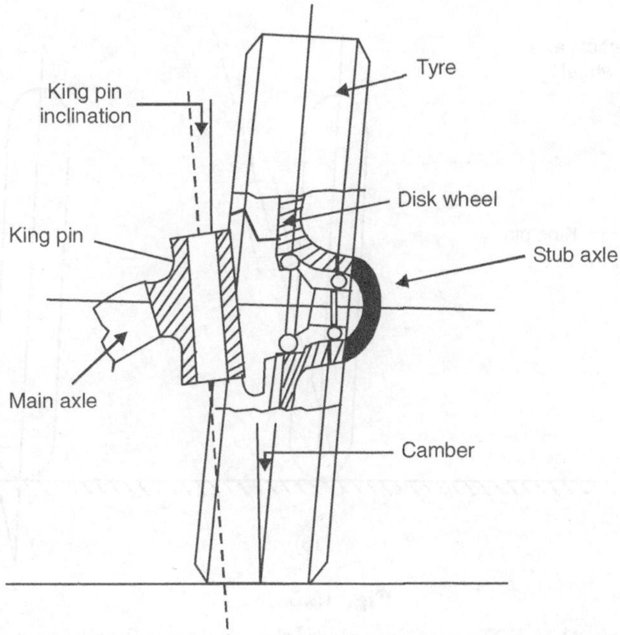

Fig. 15.4

Figure 15.4 represents a front wheel and stub axle assembly. The stub axle is connected to the main axle beam through king pin. The king pin is located in a bush. Lubrication being essential to reduce the friction is also provided. The wheel is attached to the stub axle with the help of two bearings.

As can be seen, the king pin is inclined inwards while the wheel is inclined outwards. The angle made by king pin with vertical is termed as 'king pin inclination'. This angle is kept about 7° to 8°. This is essential to provide directional stability to the vehicle. In some design, where the king pin is replaced by ball joints, the angle is measured between vertical and the steering axis. The steering axis is an imaginary line drawn through the lower and upper steering pivot points.

The angle made by vertical wheel axis with vertical is termed as 'camber'. This angle is kept about 2° to 3°. It is considered to be positive when the wheel is inclined outwards.

Few more parameters related with vehicle stability are included in steering geometry. 'Combined angle' is the sum of king pin inclination and camber. This is the angle between king pin axis (or steering axis) and the vertical axis of the wheel. Its value is about 10°.

The forward tractive force acts at the point on the road surface where king pin axis (or steering axis) intersects it. The road resistance, on the other hand, acts at a point on the road surface where vertical axis of wheel intersects it. The king pin axis (or steering axis) and vertical wheel axis should intersect with road surface at single point. As shown in Fig. 15.5 (*a*). If the intersection occurs above the road surface it causes toe-in in the rear wheel drives [Fig. 15.5 (*b*)]. When the intersection occurs below the road surface, it causes toe-out in rear wheel drives [Fig. 15.5 (*c*)]. In front wheel drives just opposite happens that is intersection above road surface causes toe-out and intersection below road surface causes toe-in.

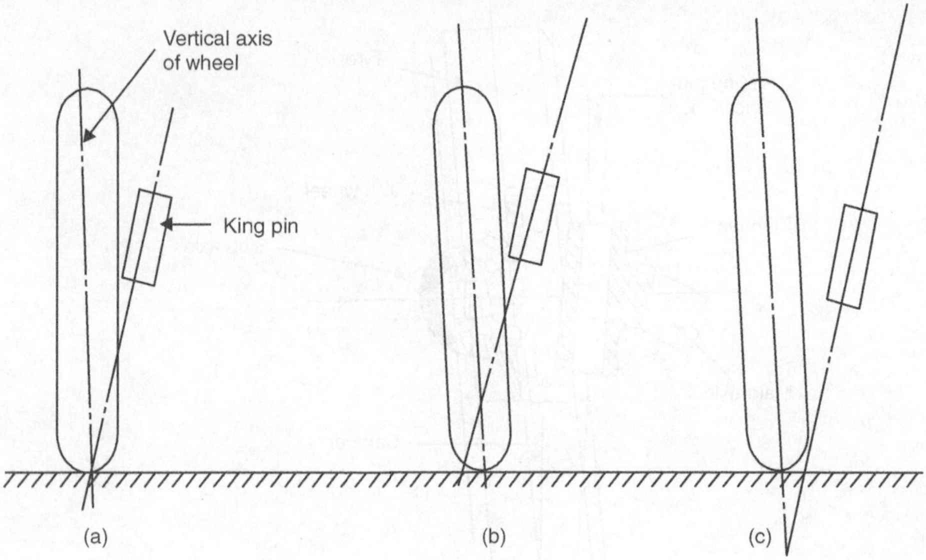

Fig. 15.5

Toe-in and toe-out is the difference between the distance between two front wheels on rear and front side. If the distance is less on front side as shown in Fig. 15.6(*a*) toe-in occurs. If it is vice versa as shown in Fig. 15.6(*b*) toe-out is said to occur. Initially, toe-in is provided as there is an inherent tendency of the vehicle to toe-out. The initial toe-in that is provided has a value less than 3 mm.

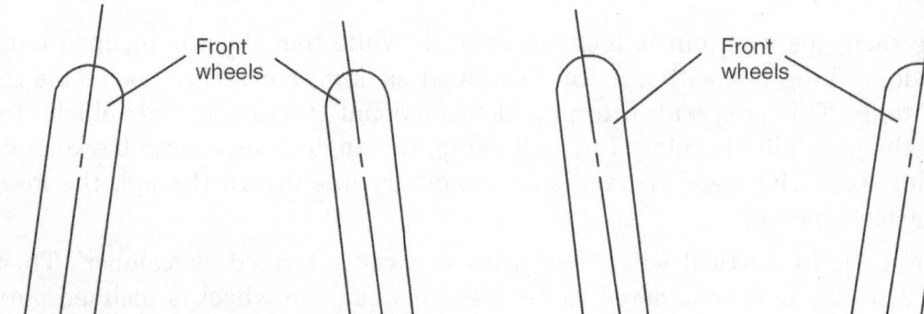

Fig. 15.6

The angle between king pin centre line and the vertical in the plane of wheel is known as 'castor' and is represented in the Fig. 15.7. Castor is considered to be positive if king pin centre line meets ground in front of wheel. The angle is kept about 3° and provides directional stability.

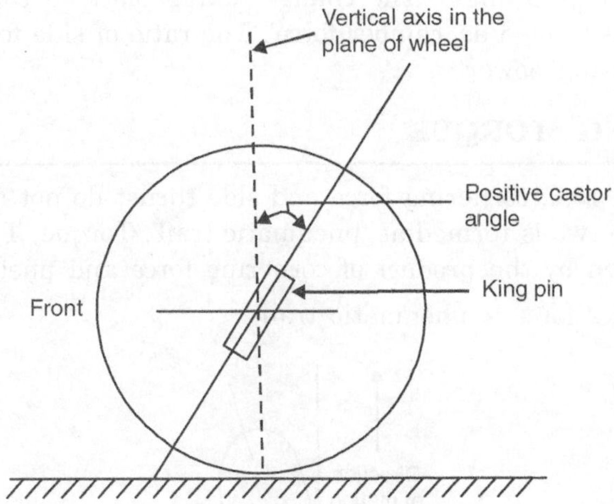

Fig. 15.7

15.3 CORNERING FORCE

While vehicle is taking a turn, the centrifugal force acts on it producing side thrust. To sustain it the plane of wheel must make some angle with the direction of motion of the vehicle. This is attained by the distortion of tyre which is flexible. The angle through which the wheel turns to sustain side thrust is known as 'slip angle'. To counter this side thrust, force acts at right angle to the plane of wheel. This force is known as 'cornering force'.

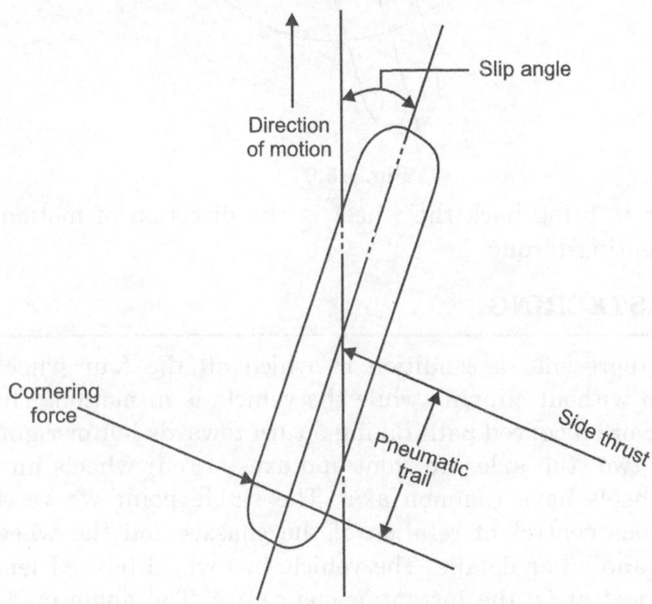

Fig. 15.8

Slip angle depends upon side thrust, the flexibility of tyre, load carried by vehicle, camber and the condition of road surface. The magnitude of slip angle is small at low speed and on less sharp curves. It is large at high speed and on sharp curves. For the same slip angle, a positive camber increases the cornering force. The cornering force is reduced when camber is negative for the slip angle. The change taking place in cornering force due to change in camber angle is termed as 'camber force'. The ratio of side force sustained to slip angle is known as 'cornering power'.

15.4 SELF-RIGHTING TORQUE

As evident from Fig. 15.8, cornering force and side thrust do not act in the same line. The distance between the two is termed as 'pneumatic trail'. Torque, T acting on the wheel and its magnitude is given by the product of cornering force and pneumatic trail.

Torque, T = cornering force × pneumatic trail

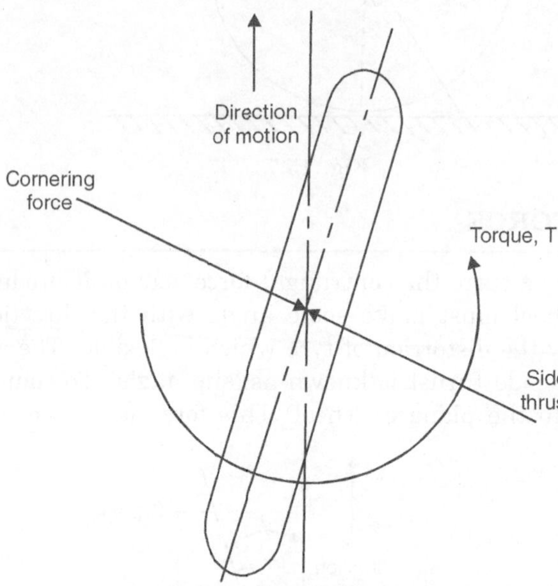

Fig. 15.9

This torque tends to bring back the wheel in the direction of motion. Therefore, it is also known as 'self-righting torque'.

15.5 CORRECT STEERING

Correct steering represents a condition in which all the four wheels of the vehicle undergo rolling motion without slipping while the vehicle is in motion. The vehicle may be moving in straight line or on curved path taking a turn towards left or right. This is possible only when the axes of two stub axles, in front and axes of rear wheels intersect at a single point. In fact, rear wheels have common axis. The single point where these intersect is known as "instantaneous centre" of rotation of the chassis and the wheels. Figure 15.10 represents these axes and other details. The vehicle has wheel base of length, L and track width, W_1. The axes meet at O, the instantaneous centre. The angle made by stub axle of outer wheel is ϕ and that of inner wheel is θ. Inner front wheel turns by larger angle than outer front wheel and therefore $\theta > \phi$. *It* is essential that the two wheels move through different angles for correct steering. Considering the two triangle, Δ DOC and Δ AOB,

$$\tan\theta = \frac{L}{CO} \text{ and } \tan\phi = \frac{L}{BO}$$

Similarly, $$\cot\theta = \frac{CO}{L} \text{ and } \cot\phi = \frac{BO}{L}$$

and $$\cot\phi - \cot\theta = \frac{BO - CO}{L} = \frac{W}{L}$$

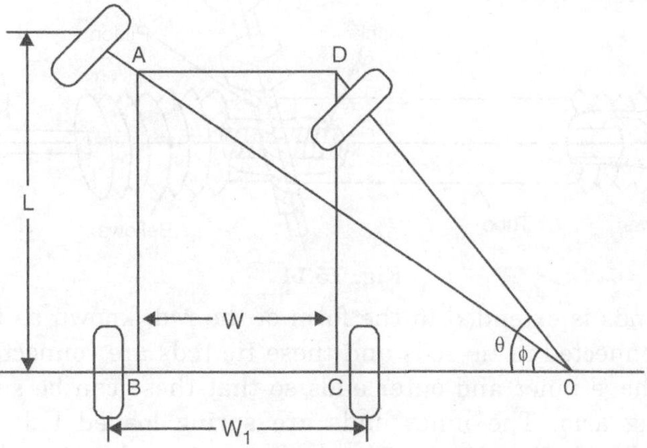

Fig. 15.10

The above equation represents equation for correct steering. The value of angles θ and ϕ are different for different vehicles depending upon its wheel base length. L and track width, W_1. As wheel base length and track width are fixed for a vehicle the equation can be satisfied for only one set of values of angles θ and ϕ. Therefore rolling without slipping is possible only for one particular set of angles. When angles are other than this slipping would occur. Therefore, in practice when a vehicle takes a turn the wheels roll and slip as well. The slipping causes wearing of tyres.

Steering systems may be operated using manual effort only. But to ease the operation and to provide some relief to driver power steering systems are also in use. Power steering systems reduce the driver fatigue and make the steering easy when vehicle is required to turn around for parking.

15.5.1 Manual Steering System

There are three major components in a manual steering system. These are steering linkage, steering gear and column and wheel. When steering wheel is turned, the motion is transferred to the steering linkage by column and wheel.

Steering linkage

There are pivots and other connecting parts between steering gears and steering arms. These are connected to front wheels and change their directions. The steering geometry also depends upon the type of front wheel suspension.

Rack and pinion steering linkage

In case of rack and pinion steering linkage, the movement from steering wheel is transmitted to toothed rack through pinion gear at the end of steering column. It is a worm gear. It meshes with teeth in the rack and rack is moved sideways as the pinion wheel turns. The toothed rack is connected to the tie rods. The rack is a rod with teeth cut along its length. The rack movement pushes and pulls the tie rods. This movement of tie rods makes possible change in the direction of wheels. There is tubular construction provided that encloses the steering rack (Fig. 15.11).

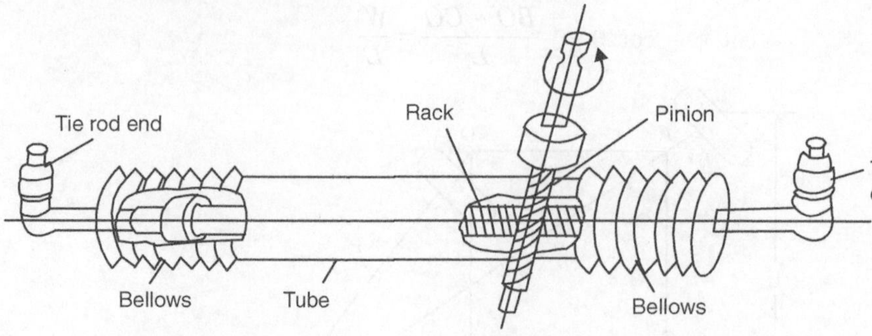

Fig. 15.11

The rack at the ends is extended in the form of the rod, known as track rod. The ends of the track rod are connected to tie rods and these tie rods are connected to steering arms (Fig. 15.12). Tie rods have inner and outer ends so that these can be suitably connected to track rod and steering arm. The inner ends are spring loaded ball sockets. These are screwed onto track rod end (Fig. 15.13). The system is useful for small vehicles.

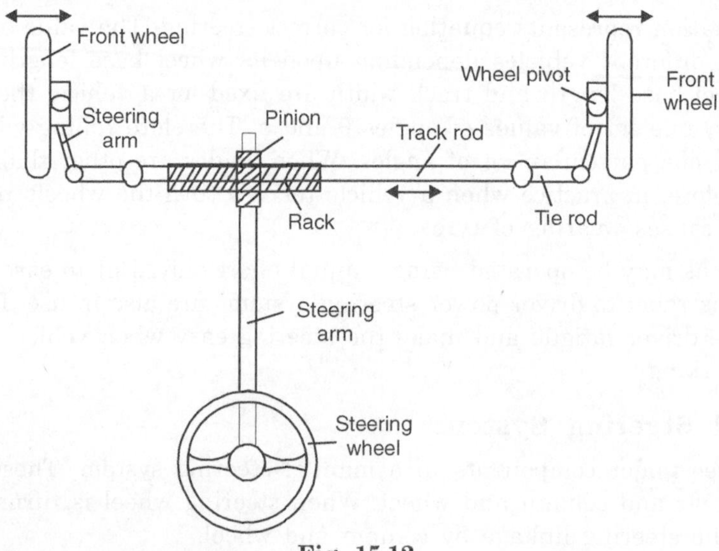

Fig. 15.12

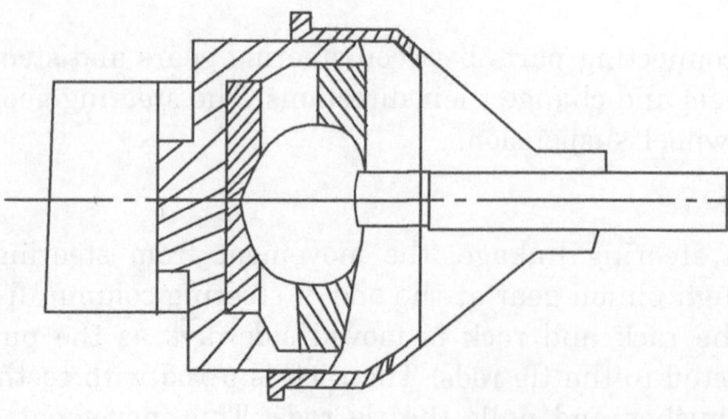

Fig. 15.13

Parallelogram steering linkage

Parallelogram steering linkages are of two types. One type is parallelogram steering linkage located behind the front wheel suspension and the other type is located in front of it (Fig. 15.14 and 15.15). The parallelogram steering linkages are used in conjunction with independent front wheel suspensions.

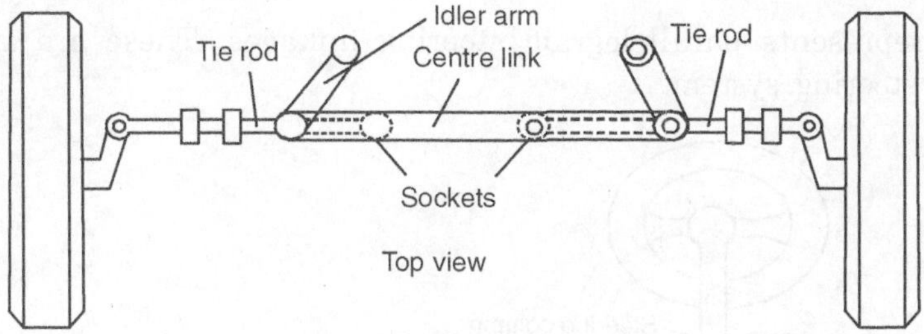

Fig. 15.14

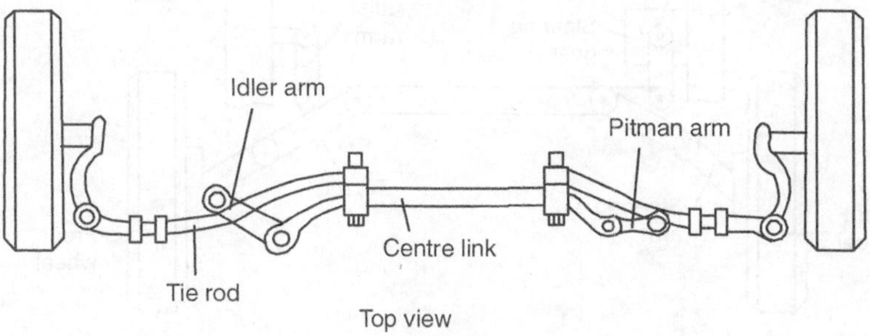

Fig. 15.15

Pitman arm

It connects the linkage to the steering column. This is done through steering pinion located at the end of steering column. The pitman arm transmits the motion to linkage which move left or right and turns the vehicle. The arms have tapered holes or studs at their ends. The arms with tapered holes are known as non-wear arms and that with studs are known as wear arms. Non-wear arms do not need replacement till these are damaged whereas wear arms require periodical inspection as these are deteriorated during normal operation.

Idler arm

These are connected to the car frame at one end. The other end is connected to centre link. The pivoted end of the arm provides sideways movement of the linkage. This also causes more wear. These also keep the centre link at proper height. There may be excessive vertical movement of the idler arm due to worn out bushings or studs in the pivotal assembly.

Links

These are centre links and steering links. These provide sideways movement causing the turning of wheels. These are required to be at the proper height. If not so there may be unstable toe. Unstable toe causes bump steering. The centre links can be wear or non-wear and like pitman arm those with stud or bushing ends are known as wear type and need periodical inspection whereas non-wear type require replacement at the end of their life.

Tie rods

These connect steering linkages with steering knuckles. Adjusting sleeves are provided to accommodate the variation in the length of the tie rods. This is essential for correct toe setting. The tie rods are prone to damage by dust and therefore are provided with dust covers. Moisture too affects the moving parts and protection against moisture is also provided with dust cover.

Figure 15.16 represents parallelogram steering linkages. These are also known as Re-circulating ball steering system.

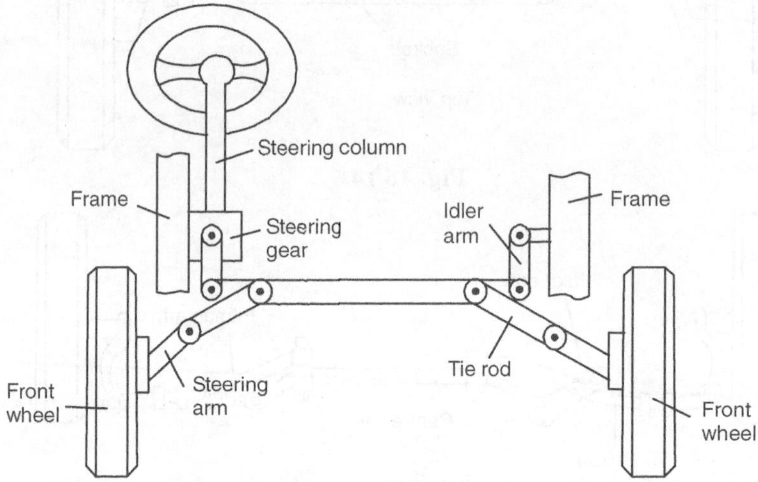

Fig. 15.16

15.5.2 Power Steering System

While steering the vehicle driver may be required to exert a lot of effort particularly at low speed while parking it. In long drives also there is natural fatigue that driver experiences. To make the steering of the vehicle easy and to reduce the fatigue the power steering systems are designed. There are conventional designs of power steering and non-conventional designs which have come into existence after the development in electronics components. In conventional systems hydraulic power supplements the input and therefore the effort required by the driver is reduced. In non-conventional designs, electronic controls and electric motor supplements the input thereby reducing the effort required by the driver.

Following are the main components of the power steering system:

Pump

This is steering pump used to produce hydraulic flow. This flow provides necessary force to operate the steering gear. The engine crankshaft is connected to pump through a belt and therefore it is placed near the engine. There is a reservoir and control valve included in the pump. The pump can be roller or vane type. Any type of pump can be chosen as both the pumps work in a similar fashion.

Valve

The valves act to control the fluid flow and also to control the pressure. The valve controlling the fluid flow is known as Control Valve and that controlling the pressure is known as pressure relief valve. The control valve is connected to the outlet of pump and fluid from the pump passes through an orifice. Due to this there is a pressure drop. There is a spring provided and the spring and low pressure keep the valve closed. In this condition, all the flow goes to steering gear. With increase in engine speed, there is increased flow from the pump that may not be needed in steering gear. The pressure difference on the two sides of the valve becomes high and is capable of overcoming the spring force. This makes the valve open and excess fluid is supplied back to the pump. Thus, the fluid flow to steering gear is controlled.

The pressure relief valve can monitor the pressure at the outlet of the pump and in the hose on the inlet of the pump. There may be a variation of pressure due to variation in engine speed and the valve also controls the pressure at the outlet of the pump.

15.5.3 Steering Gear Box

The steering gear box used in power rack and pinion system is same as gear box used in parallelogram steering linkage system. The only difference is that here it is filled with fluid and a control valve is provided to regulate pressure and flow of fluid. This also supplements the human effort. The pressure exerted by fluid in addition to manual effort causes the movement of rack. This is due to difference in pressure created by valve and the rack moves towards the side of low pressure.

Non-integral piston system

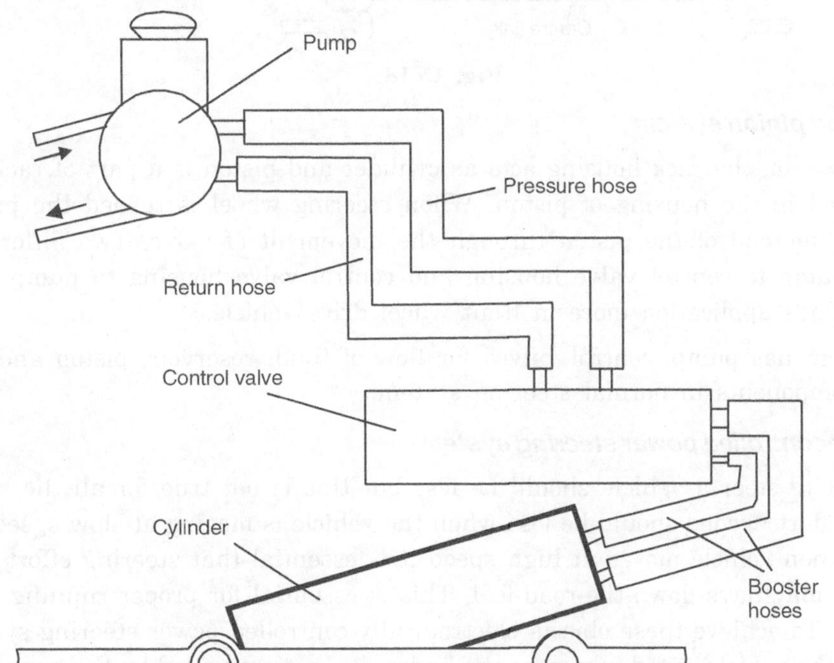

Fig. 15.17

The system consists of a pump, reservoir, control valve, cylinder and hoses. Pump and reservoir form a part of the steering gear assembly. The hose connects the pump to the control valve and fluid at high pressure is directed to the cylinder through booster hoses (Fig. 15.17). The movement of piston in the cylinder provides the force that supplements the efforts made by the driver. This system is also known as external piston linkage system.

Integral piston system

The system also has pump, reservoir, hoses, steering gear, control valve and cylinder as in the previous system. It has steering gears, control valve and cylinder in a single housing. The hydraulic fluid also acts as booster for brakes. Therefore in vehicles with this system power steering pump is used to actuate the brake booster also (Fig. 15.18).

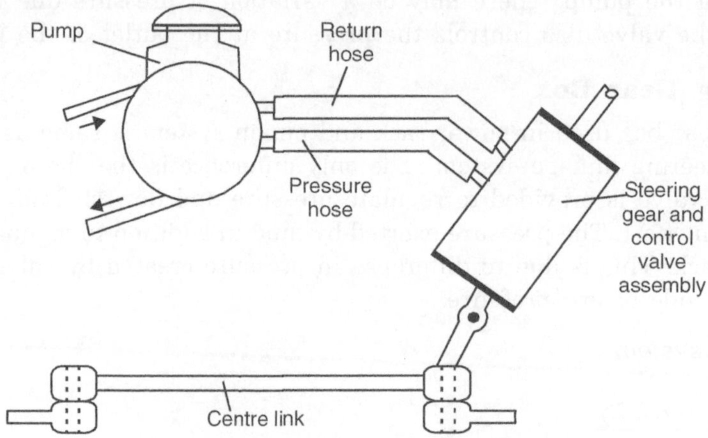

Fig. 15.18

Power rack and pinion system

In this system, the rack housing acts as cylinder and piston is a part of rack. Control valve is located in the housing of piston. When steering wheel is turned the pressure is directed to either end of the piston through the movement of valve. Two different hoses connect the pump to control valve housing and control valve housing to pump reservoir. This system finds application more in front wheel drive vehicle.

The system has pump, control valves for flow of fluid, reservoir, piston and hoses in addition to components in normal steering system.

Electronically controlled power steering system

The effort to steer a vehicle should be less but this is not true for all the situations. The steering effort needed should be less when the vehicle is moving at slow speed or while parking but when vehicle moves at high speed it is essential that steering effort increases so that driver may have down-the-road-feel. This is essential for proper running of vehicle at high speed. To achieve these objects electronically controlled power steering systems are employed. As the vehicle speed increases the hydraulic boost provided by the power steering system is reduced through electronic controls. The effort required to steer in these systems varies from 1 to 3. The effort needed to steer at high speeds becomes three times effort needed to steer at slow speed. Thus, the driver is able to control the steering efficiently at high speed and at the same time steering is convenient at slow speed.

In an electronically controlled power steering system a solenoid valve is used to control the fluid flow to boost steering efforts. This solenoid valve is known as Pressure Control Valve.

15.5.4 Pressure Control Valve

The pressure control valve or solenoid valve has a plunger at the bottom. The plunger as it moves upwards goes inside the electromagnet. The current to this electromagnet is varied and due to this the upward force exerted by plunger also varies. This force is exerted against the force of spring that is provided at the top of valve (Fig. 15.19). The current flow to the electromagnet is dependent upon the speed of the vehicle. Thus, the output force from the valve varies in accordance with the speed of the vehicle. This output force supplements the human force exerted by the driver and thus, the total force for steering the vehicle becomes dependent on the speed of the vehicle.

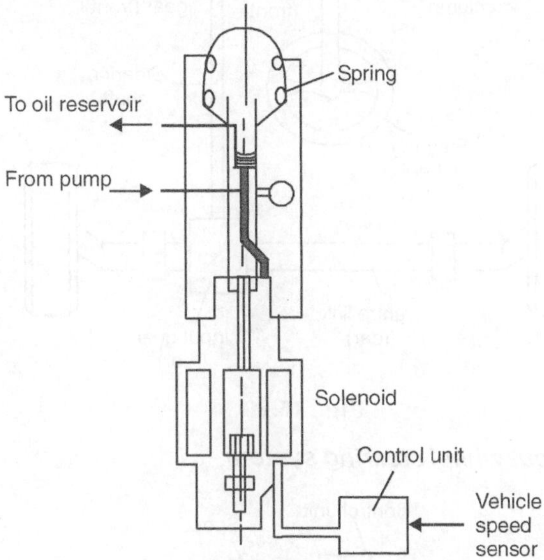

Fig. 15.19

At low speed of the vehicle, the maximum force is provided by the valve and therefore force exerted by the driver is reduced making the steering control convenient. At high speed of the vehicle, the force provided by the valve is reduced and then force exerted by the driver increases which is desirable for proper steering control.

Four wheel steering systems

To improve stability of vehicle as it turns and to have tight turns four wheel steering systems are used. The rear wheels also turn along with the front wheel. These systems are mechanical and in some cases these are electronically controlled.

The turning of the rear wheels occurs according to the speed of the vehicle. At high speed, all the four wheels turn in the same direction whereas at low speeds the rear wheels turn in a direction opposite to that of front wheels.

Mechanical four wheel steering system

The rotary motion of the steering wheel on the front side is transferred to steering wheel on the rear side with the help of a shaft known as steering shaft (Fig. 15.20). The rear wheels also like front wheels can move inwards or outwards on ball joints. At high speeds when the turning is small, the rear wheels turn in the same direction but through a smaller angle. The rear wheels turn in proportion to front wheels. If the front wheels turn through 8° the rear wheels turn through 1.5° in the same direction as the speed may be high but if the front wheels turn through say 30° the rear wheels may turn through 5° in opposite direction because turning the front wheels to this high angle is feasible only at low speeds.

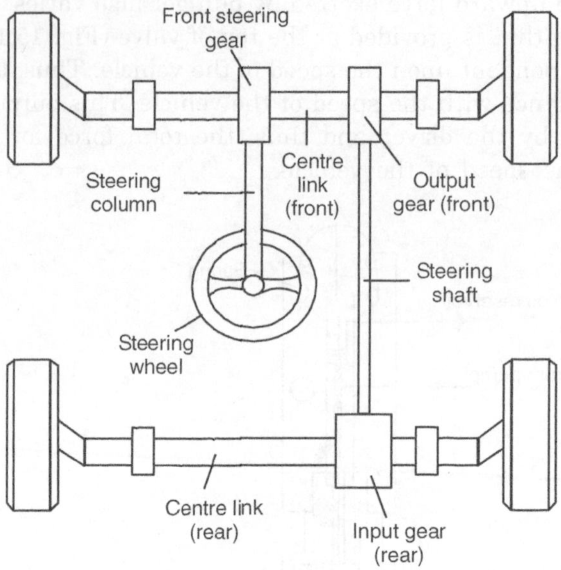

Fig. 15.20

Electronically controlled four wheel steering system

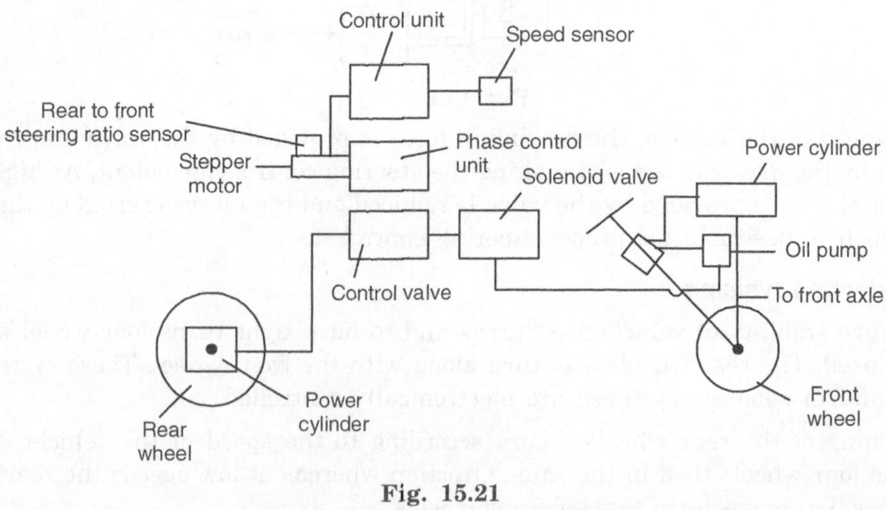

Fig. 15.21

There is flexible toe control link that connect trailing arm to the rear sub frame. This trailing arm carries the rear wheels. When steering wheel is turned, the fluid applies pressure on the front steering gear. This pressure is transmitted on rear control valve. The movement of spool valve directs the pressurized fluid from rear steering pump to one side of the piston in cylinder. The piston turns trailing arms to left or right causing turning of the rear wheels. The fluid pressure depends upon the turning of the steering wheel. The rear wheel turns in the same direction and depends upon speed of the vehicle. Though the maximum limit to which the rear wheel can turn is 1.5°. The system works only at high speeds. Figure 15.21 represents a electronically controlled four wheel steering system.

Electro-mechanical four wheel steering system

In this type of system, there is an electronically controlled module (ECM) that transmits control signals to the rear steering gear. Rear wheel steering varied according to front wheel steering angle and vehicle speed. There is a shaft from front steering gear to rear steering gear through which the motion is transferred on the rear side. The rear and front wheels turn in the same direction at high speed and in the opposite directions at low speeds.

15.6 STEERING RATIO

It is the turning of steering wheel in degree to turn the front wheels through 1 degree. Suppose the steering wheel is turned through 14° and front wheels turn through 1° then steering ratio is 14:1. The ratio is 14:1 for vehicles with power steering and for vehicles with manual steering have steering ratio as 24:1. The higher ratio means it is easier to steer though rotation of steering wheel is more.

In case of parallelogram steering linkage system, the steering gear ratio depends upon relative length of Pitman arm and steering arm. If the two are equal the ratio is 1:1 and if the Pitman arm is twice the length of steering arm the ratio is 1:2. The force applied by the driver at the steering wheel is also multiplied due to steering gears or rack and pinion provided in the steering system. The multiplication is directly dependent upon the gear ratio of the steering gears or rack and pinion. This helps the driver in steering the vehicle.

Variable steering ratio may further ease the steering of vehicle. The higher steering ratio provides better control during high speed driving. The lower ratio provides better control in driving at low speed particularly in city traffic. The ratio varies from 16:1 to 13:1. During first 40° of wheel turning in either direction the ratio remains constant. When wheel turns beyond 40° the ratio is reduced.

Variable steering ratio is achieved by providing bigger tooth in the middle in ball nut of the steering gears. Due to smaller outer tooth, the effective leverage changes. Rack tooth is made smaller in case of rack and pinion steering system to attain variable steering gear ratio (Fig. 15.22).

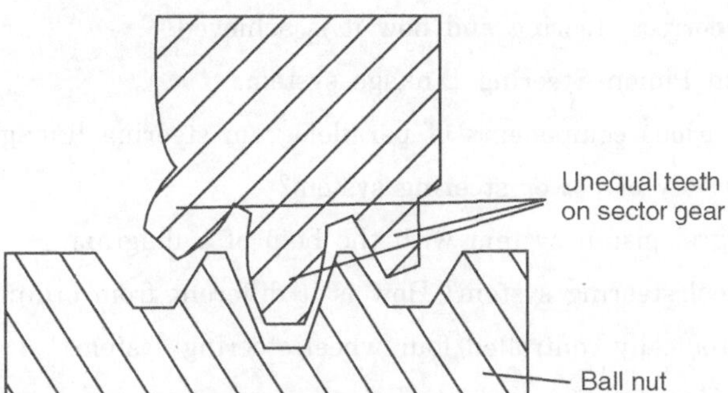

Fig. 15.22

15.6.1 Collapsible Steering Column

In case of collision of the vehicle, there is a possibility that the steering column may pierce through the chest of the driver which proves to be fatal. This is because the driver has to be in front of steering wheel. To avoid such situation the steering column collapses in case of the collision. This is possible by having collapsible steering column (Fig. 15.23). This way the life of the driver may not be put in danger. In some cases, the steering wheel may align itself. There is a hub that absorbs the energy and aligns the steering wheel with chest of the driver during collision. Thus, the largest possible area of the driver's body takes up the impact thereby reducing its intensity.

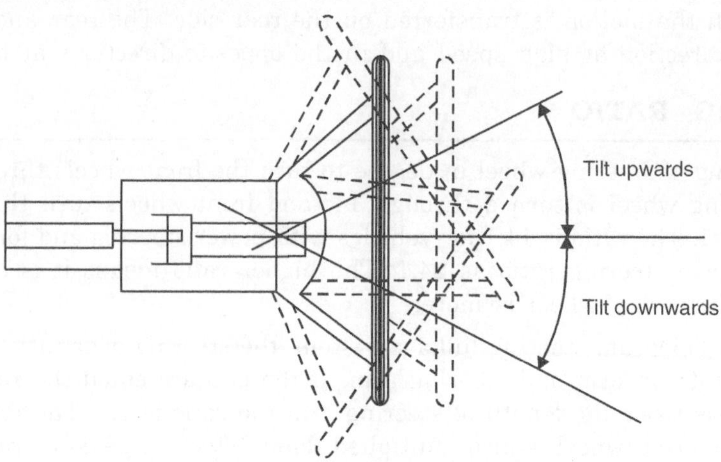

Fig. 15.23

QUESTIONS

1. What is the function of steering system in an automobile?
2. Why middle portion of front axle is provided with I-section?
3. Explain Lamoine type stub axle.
4. Explain the terms camber, king pin inclination, castor, and included angle.
5. What is toe-in and toe-out? What is the initial toe-in that is provided and why?
6. What is cornering force and self-righting torque?
7. Explain what is correct steering and how it is achieved.
8. Explain Rack and Pinion Steering Linkage system.
9. Give details of various components of parallelogram steering linkage system.
10. What is the necessity of power steering system?
11. Explain non-integral piston system with the help of a diagram.
12. What is four wheel steering system? How is it different from ordinary steering system?
13. Explain an electronically controlled four wheel steering system.
14. What is the relevance of steering ratio? Explain.

16

SUSPENSION SYSTEM

16.1 PRINCIPLES OF SUSPENSION SYSTEMS

Irregularities in the road surface cause shocks and vibrations. If these were transmitted directly to the chassis, the structure would be subjected to excessive vibrations. This would cause discomfort to the occupants of the vehicle, if it is a passenger vehicle and damage the goods, if it is a goods carrying vehicle. The suspension system isolates the structure, as far as possible, from shocks and vibrations due to irregularities in the road surface. Also, it must be achieved without affecting the stability, steering control and general handling qualities of the vehicle. The first requirement is met by using the flexible elements and dampers, while the second is achieved by controlling the relative motions between the unsprung mass and the sprung mass. The unsprung mass is wheel-and-axle assemblies which does not get the benefit of spring action. The rest of the vehicle, above the wheel and axle, is the sprung mass as it gets the benefit of spring action.

Figure 16.1 represents the block diagram of an automobile. The tyres are represented in the form of spring S_T. The blocks above these represent the unsprung mass M_{US} which consists of axle and wheel assembly. The block above suspension springs, S_S represent the sprung mass M_{SP} which mainly consists of the body of the automobile.

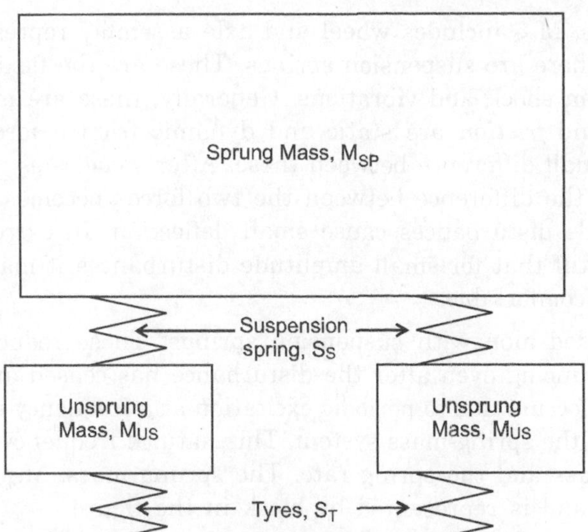

Fig. 16.1

The shock experienced by wheel when it strikes a bump is entirely different from the one which it experiences while passing through the pot hole. In first case the shock is influenced to a large extent by the geometry of the bump and the vehicle speed while in second case it is influenced by the unsprung masses and spring rates apart from the geometry of the pot-hole and also the vehicle speed. The relative motion between the sprung and unsprung masses is controlled by using mechanical linkages. These may be a semi-elliptical spring and shackle or a double transverse link and anti-roll bar or some other such combination of mechanisms.

16.2 HUMAN SENSITIVITY TOWARDS ROAD IRREGULARITIES

The diameter of the tyre, size of contact path between tyre and the road, and weight of wheel and axle assembly affect the magnitude of the shock transmitted to the axle. The tyre while moving acts as spring. The rate of tyre acting as spring affects the magnitude of the shock. Along with it, the rate of suspension springs, damping effect of the shock absorbers, and the weights of unsprung and sprung masses affect the amplitude of up and down motion of the wheel.

Human sensitivity to these disturbances is very complex. It is widely accepted that the vertical frequencies associated with walking speed of about 5 km per hour are 1.5 to 2.3 Hz and cause no discomfort to the human beings. Similarly fore-and-aft or lateral frequencies of the head should be less than 1.5 Hz for human comfort. It may cause dizziness and sickness if the inner ear is subjected to frequencies between 0.5 and 0.75 Hz. It may cause serious discomfort in other important organs if the frequencies range between 5 and 7 Hz.

16.3 SUSPENSION SYSTEM

Figure 16.2, represents a suspension system. It is a modification of figure 16.1. Tyres are the part of an automobile in contact with road surface. These have cushioning effect for smooth movement of the vehicle. This cushioning can be represented by springing effect in the form of spring S_T.

The unsprung mass, M_{US} includes wheel and axle assembly represented by the block above spring S_T. Next, there are suspension springs. These are the flexible elements which isolate the structure from shock and vibrations. Generally, these are multi-leaf semi-elliptical springs. *Stiction* and *friction* are static and dynamic friction forces between leaves. When new, there is a small difference between these. After sometimes with use, the spring becomes old and rusty, the difference between the two forces becomes significant. Due to this, the small amplitude disturbances cause small deflection. In extreme conditions, the spring may become so stiff that for small amplitude disturbances it may not deflect at all. This makes the ride uncomfortable.

Dampers are provided alongwith suspension springs. These reduce the tendency for carriage unit to keep bouncing even after the disturbance has ceased and also prevent the excessive bouncing. This occurs due to periodic excitation at a frequency equal to the natural frequency of vibration of the spring-mass system. This natural frequency is a function of the weight of the sprung mass and the spring rate. The sprung mass, M_{SP} constitutes mainly the body of the vehicle and is represented by block in the figure.

The sprung mass and unsprung mass have different frequencies. The sprung mass will experience a relatively low frequency of 1 Hz to 1.5 Hz. The unsprung mass experiences the frequency of 10-15 Hz which is due to wheel hop. Marginally, the frequency is affected by suspension rate but is independent of motion of carriage unit. To minimise the wheel hop the unsprung mass should be minimum possible.

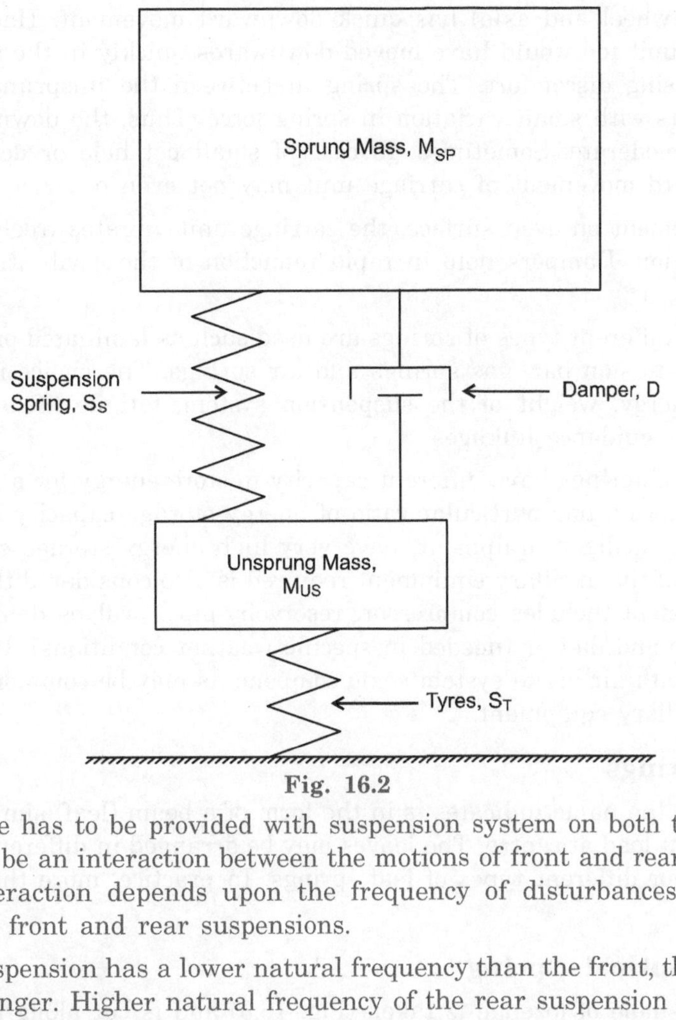

Fig. 16.2

An automobile has to be provided with suspension system on both the front and rear sides. There must be an interaction between the motions of front and rear suspensions. The magnitude of interaction depends upon the frequency of disturbances and the natural frequencies of the front and rear suspensions.

If the rear suspension has a lower natural frequency than the front, the pitching motion tends to persist longer. Higher natural frequency of the rear suspension causes less severe the initial pitching motion. Thus, the natural frequency of the rear suspension is normally higher than that of the front.

Before going into the details of rear and front suspension systems it would be proper to know about the details of suspension springs and dampers which isolate the body of the automobile from shock and vibrations.

16.4 SPRINGS

The necessity and importance of the springs in suspension system could well be imagined. Considering the movement of the wheel over a bump, it would have risen very rapidly. Now, without a spring, shock experienced by wheel would have been transmitted completely to the carriage unit. This would have been undesirable as it could cause harm to the user. With spring in between wheel and carriage unit, and wheel moving over a bump, the spring would get compressed. The force which is required to compress the spring and would be experienced by carriage unit, has a very small magnitude.

Similarly, considering the situation when wheel passes through the pot hole. This would be, as if the wheel is passing through a depression. The relatively light unsprung mass (consisting of wheel and axle) has quick downward movement. Had there been no spring the carriage unit too would have moved downwards quickly in the same manner as unsprung mass causing discomfort. The spring in between the unsprung mass and the carriage unit, deflects with small variation in spring force. Thus, the downward movement of carriage unit is moderate. Sometimes, in case of small pot hole or depression, due to inertia, the downward movement of carriage unit may not even occur.

In case of movement on even surface, the carriage unit vibrates freely on the springs with small acceleration. Dampers help in rapid reduction of these vibrations and provide comfortable ride.

In automobiles, different types of springs are used such as laminated or leaf, taper-leaf, rubber springs, coil, torsion bar, gas springs and air springs. The choice is based on cost, capacity to store energy, weight of the suspension system, fatigue life and requirement regarding location or guidance linkages.

Different types of springs have different capacity to store energy for a particular stress level. Each type of spring has particular ratio of energy storage capacity to its mass. The gas springs, without ancillary equipment, have very high energy storage capacity per unit mass. But if weight of the ancillary equipment required is also considered this becomes low. The ancillary equipment includes compressor, reservoir, pipes, valves dampers and filters apart from gas drier and de-icer (needed in specific weather conditions). In case the automobile is equipped with air brake system some components may be common between brake system and the ancillary equipment.

16.4.1 Leaf Springs

Leaf springs, as the name indicates is in the form of a beam (leaf) simply supported at each end having point load at centre. The leaves may be arranged in different ways. Loading pattern also varies for different types of leaf springs. In practice, more than one leaf constitute a spring.

16.4.2 Semi-elliptical Spring

A sheet in the shape of lozenge is taken (Fig. 16.3) and is cut along the dotted lines. The innermost strip is numbered as 1. The two leaves adjacent to it above and below are numbered 2. These have width equal to half of width of 1. When combined the width of 2 becomes equal to width of 1. The length of 2 is less than that of 1. Next to 2, on both sides, is 3. Again when put together its width becomes equal to that of 1. The length of 3 is less than 2. Similarly there are 4, 5, 6, 7 and 8. When combined, 8 forms the leaf of the shortest length.

All the leaves are put one above another, generally the longest leaf, 1 at the top and shortest leaf 8, at the bottom. These are clamped so that they act as a single unit.

The leaf spring made, as explained above, will have uniform bending stress along its length. This also provides the optimum use of material. Under deflection, when the stresses in the outer layer of all the leaves rise to elastic limit, the energy stored in the spring is much more as compared to other springs of same shape. As the stiffness of beam is proportional to the cube of its depth multiplied by its width, the leaf springs are more flexible. The depth, in the above case, is that of one leaf only and not of the total depth of assembled spring and width is the mean of that of original lozenge shape. Friction between the leaves opposes the relative movement between leaves. This motion is essential for its working. The friction is reduced by putting suitable material with low coefficient of friction and by wrapping and sealing the springs to prevent entry of dust and other materials.

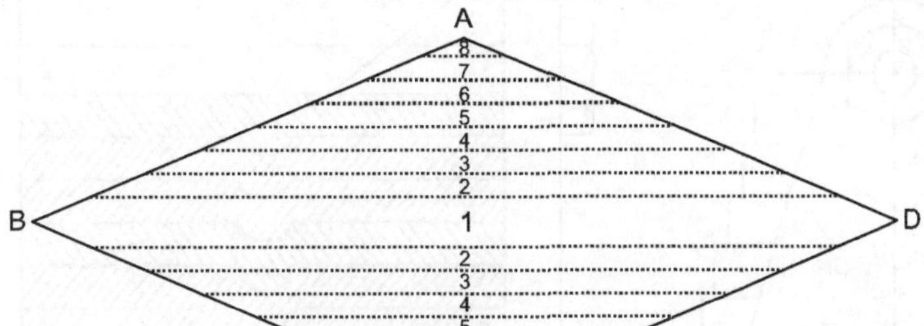

Fig. 16.3

The master leaf, *i.e.,* the largest one has bends provided at the ends, to form eyes to accommodate the pivot or shackle pin to hold the spring. Rubber bushes are used in the eyes of springs in light vehicles and for heavy vehicles metal bearing bushes are used. Provision for lubrication is also made which is necessary for very heavily loaded springs. To increase the effective bearing areas and improve capability to retain lubricant, bushes and pins of threaded form are used. Rubber bushes do not require lubrication and effectively prevent transmission of vibrations but are prone to squeak and deterioration once these become loose. Rubber if replaced by nylon or other plastics and composite self-lubricating bearings provide improved life.

The second leaf is sufficiently long to cover the eyes of the springs partly. This provides safety in case the master leaf is broken. A single bolt passes through all the leaves and holds them together. To keep the leaves together laterally, these may be provided with pips or projection on the upper surfaces. The lower surfaces of leave too have corresponding recesses to accommodate the pips and projections (Fig. 16.4). Ribs or grooves provided along their whole length is alternative to pips and projections.

In case of reversal of the load on the spring, the separation of leaves may occur. This causes the rebound load on the master leaf alone which may damage it. To take up the rebound load, uniformly on all the leaves, the leaves are clipped together. Clips cause the distribution of load over the whole length of spring in the rebound condition. Alternatively, short rebound leaves may be fitted on top of the master leaf so that these can take up the rebound load. In another form, the leaves are symmetrically arranged above and below the master leaf and clamped together.

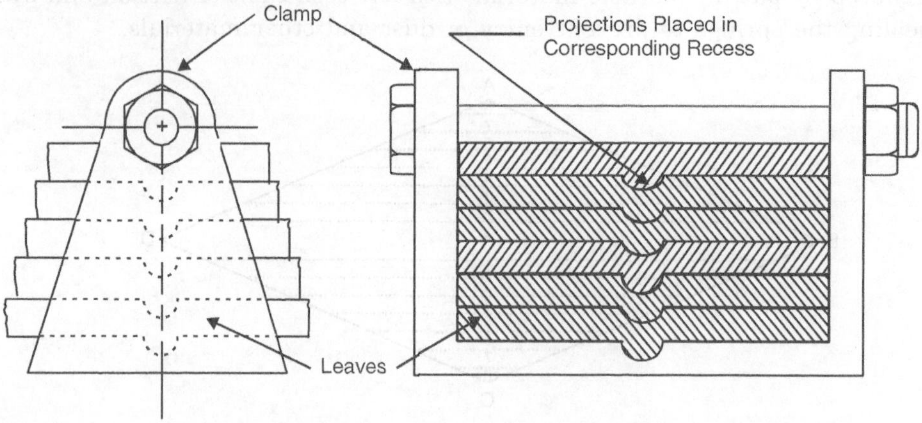

Fig. 16.4

The variation in length of spring, when deflected upwards is accommodated by shackle. The shackle can be replaced by a slider block or by shackles made of rubber. The slider block (Fig. 16.5) has the ends of two top leaves of spring projected into a slot in the cylinder, C. Where they are free to slide. Inside the cylinder C, a block in the form of circular shaft is provided. It has diameter equal to inner diameter of cylinder and length equal to that of cylinder. This block has a slot to accommodate the leaves. Also, it is free to rotate inside the cylinder. The leaves can have sliding movement, as shown in the top view of Fig. 16.5 and angular movement along with block as shown in the front view of the Fig. 16.5. This angular movement is limited to few degrees which is sufficient. Proper lubrication is provided for free movement of block and the spring leaves.

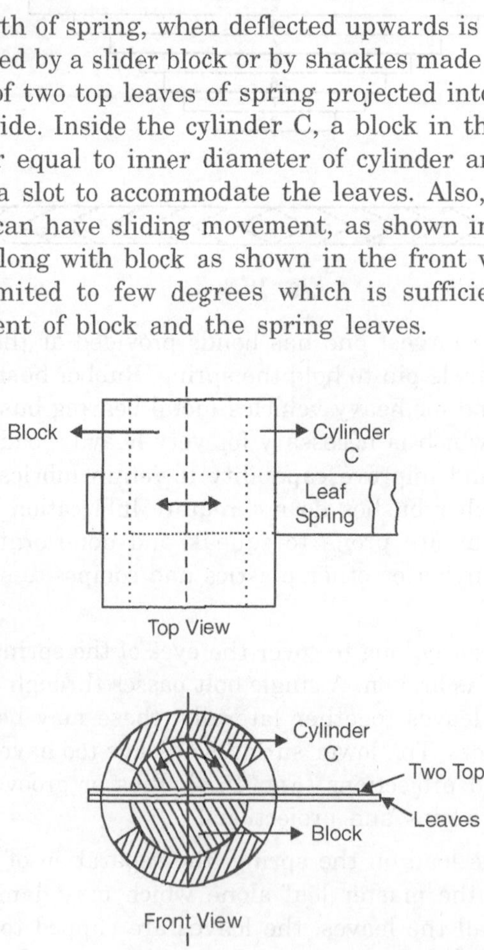

Fig. 16.5

A semi-elliptical spring arrangement is as shown in Fig. 16.6. As the axle is fixed with the help of U-bolts beneath the centre of this spring, the shorter leaves are placed on the lower side. In case of suspension arrangements, where sprung mass is secured to the centre of springs and the axles to its outer ends, the shorter leaves are on upper side.

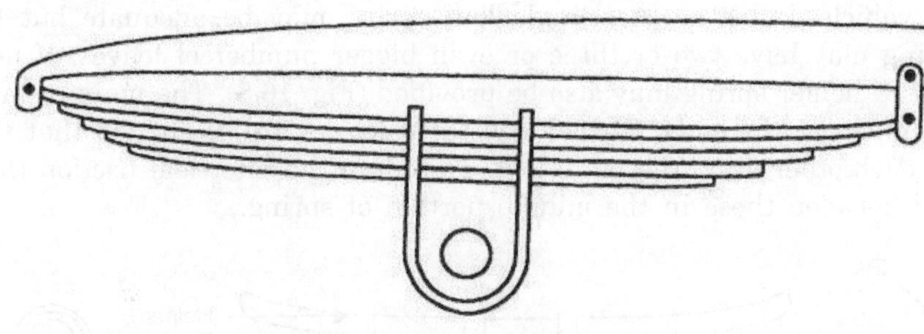

Fig. 16.6 *Semi-elliptical springs*

The quarter-elliptical spring is shown in (Fig. 16.7). It is cantilever multi-leaf spring which has its thick end bolted rigidly to the frame. The other end being either pivoted or shackled to the axle.

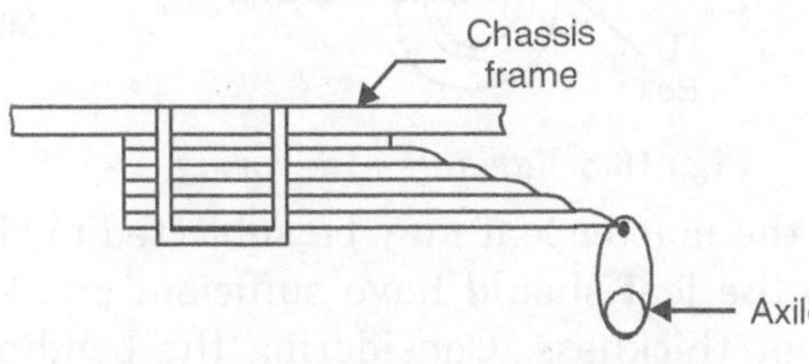

Fig. 16.7 *Quarter-elliptical springs*

Another variant of semi-elliptical spring is *full-cantilever spring.* It is pivoted at its centre. One end of the spring is again pivoted on frame and the other end is either pivoted or shackled to the axle, depending upon whether or not the spring is to take thrust loads and guide the axle.

16.4.3 Taper-Leaf Spring

Considering the stress distribution in a semi-elliptical spring, it has been found that maximum stress occurs in the middle portion, near the U-bolt. The U-bolt is provided near middle portion of the spring and axle is clamped to them. The material of the spring adjacent to the middle portion experiences the maximum stress. Material in the remaining portion of the spring is relatively lightly stressed. This brings forth the fact the excessive material is being used in the semi-elliptical spring where all the leaves are of the same width.

To avoid this, in early sixties, a modified design of semi-elliptical springs was introduced. This spring had tapered leaves and therefore it is named as Taper-leaf spring. Also, it is called parabolic leaf or minimum leaf spring.

Considering the design part of leaf spring, for a maximum permissible stress level and width, the stiffness of a semi-elliptical spring is proportional to nt^3, where n is the number of leaves and t is the thickness of each leaf in mm. If the number of leaves are reduced by half, the corresponding increase in the thickness would be $3\sqrt{2}$ *i.e.,* about 1.26. Assume a spring with four leaves of 8 mm thickness each. If the number of leaves is to be reduced to two, for the same stiffness, from the above expression, the thickness of each leaf comes out to be 10 mm, when rounded off. This means an increase of 2 mm thickness in each leaf. But the total thickness of spring is reduced from 32 mm to 20 mm. For efficient utilisation of material, width could be kept maximum at the centre and tapering could be provided to each end. In this manner, the springs weight could be reduced by 30%.

For lighter vehicles, such as car, single leaf spring may be adequate but for heavy vehicles the spring may have two or three or even bigger number of leaves. If needed, an additional spring or helper spring may also be provided (Fig. 16.8). The main or master leaf has eyes or slider pad at the ends. Second and third leaves are so shaped that they bear, at ends only, on each other or on the main leaf. To reduce the inter-leaf friction the spacers may be provided between these in the middle portion of spring.

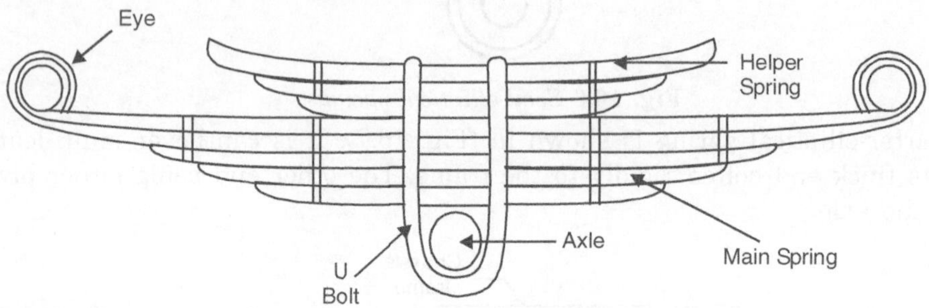

Fig. 16.8 *Variable rate spring*

In case, the vehicle rolls, the master leaf may be subjected to shear as well as torsional loads. To take up these loads the leaf should have sufficient cross-sectional area. In other words, it should have sufficient thickness. Considering the bending loads, the stress is a function of square of the thickness of leaf. Therefore, inverse square law is followed while providing the taper. This type of spring is known as *parabolic leaf springs.* The leaves of the spring are provided with tapered width to have sufficient space for tyre (Fig. 16.9). But if the width is insufficient, the leaves may twist under combined vertical and lateral or torsion loads. It may also cause instability. All these considerations are essential to decide about the width of the spring, while designing the leaf.

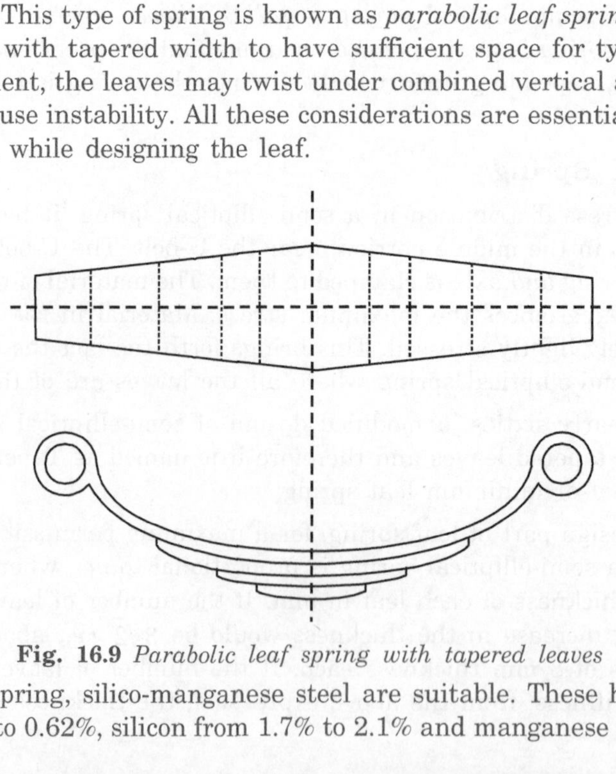

Fig. 16.9 *Parabolic leaf spring with tapered leaves*

For multi-leaf spring, silico-manganese steel are suitable. These have carbon contents varying from 0.53% to 0.62%, silicon from 1.7% to 2.1% and manganese content varying from

0.7% to 1%. Because, the thickness is very less, the needed 80% martensite structure at the centre can be obtained. For higher leaf thickness, up to 28.5 mm, chromium steel, with carbon contents varying from 0.6% to 0.9%, is used. For thickness beyond this, chromium-molybdenum steel is used. High quality steels are essential for taper-leaf springs. Due to their higher yield point (620 N/mm^2), weight of the spring can be reduced further. Surface strength of spring remains unaffected as no carburisation occurs in good quality steel. Corrosion or inter-leaf fretting causes the loss of fatigue strength. To avoid this, scragging is done where spring is loaded beyond yield point. This leaves residual compressive stresses in the surfaces. These surfaces are in tension during deflection. To improve the fatigue resistance, shot-peening is done while the surface is under load. This is also known as strain-peening.

16.4.4 Variable-Rate Springs

The rate of spring is the increment of static load it will carry per unit deflection measured in kg/cm. This remains constant in normal range of deflection. In variable rate, the rate increases with the deflection of spring. The advantages being (*a*) when the vehicle is lightly loaded, statically or dynamically, a high degree of isolation of the carriage unit from the wheels is obtained (*b*) the spring is not deflected excessively under heavy load and (*c*) resonance will not cause an excessive increase in amplitude of vibrations. On the other hand, this increase in amplitude of vibration would imply a change in natural frequency and system will not attain the state of resonance.

A *helper spring* is provided which causes the rising rate and acts only towards the end of the upward deflection of the main spring (Fig. 16.8). In coil spring suspension, a rubber spring or another coil spring which is shorter than the main spring acts as helper spring. It is mounted beneath the chassis frame so that the axle comes up against it only in the latter stage of upward deflection. In case of leaf-spring suspension system, a second semi-elliptical spring with smaller radius of curvature is centrally clamped on the top of it. The arrangement is such that its end come up against stops on the chassis frame only on the latter stage of deflection of main spring. In this manner, the rate of helper spring increases with deflection too. As it flattens, the slender sections of its ends slide outside the span between to stops or supports as the case may be. The rubber springs provide still better progressive rate. Because the gases can be compressed adiabatically, the gas springs have progressive rate as their inherent characteristic.

16.4.5 Composite Leaf Spring

Composite materials to manufacture the leaf springs were used in early seventies. The spring was designed as shown in the Fig. 16.10. Kevlar fibres were laminated longitudinally to reinforce the polyurethane resin of the leaf. But being costly these were replaced by glass fibres.

These springs are lighter by 50% when compared with multi-leaf springs and by 30% when compared with taper leaf steel spring. In case of failure, in a steel spring, the fracture is transverse which causes loss in the axle location. But in case of a composite material spring the failure causes a split longitudinally along its lamination. This is because fibres are continuous from one end to other. Due to this, the location of axle is not lost. Also, composite leaf springs do not have high stress concentration at any point. At the centre of the leaf, polyurethane pads form a resilient platform for the axle and spread the loading uniformly over the contact area of spring. For locating the axle a boss can be moulded into the polyurethane pad or a steel pin inserted into a hole in it. Apart from this composite leaf springs are resistant to fatigue, wear and corrosion, have god noise and vibration damping properties and as they are light they contribute towards a comfortable ride.

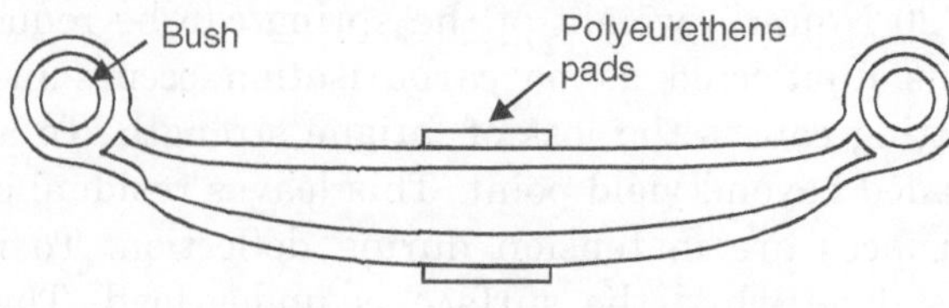

Fig. 16.10 *Composite leaf spring*

16.4.6 Rubber Spring

Rubber springs are used in different forms in commercial vehicles. As precise control of ride cannot be achieved, which is essential for comfort, these are not suitable for use in passenger carrying vehicles.

The rubber springs can easily be given a rising rate. For this, their fixing brackets need be so designed that initially these are loaded in shear and then progressively changed to compression. The reason being that rubber is much stiffer in compression than in shear.

Another advantage is that rubber has best fatigue strength in compression and the worst in tension. In tension the cracks formed may open up and impurities may go inside the cracks. There has to be a check strap or a tie to prevent the tensile loads from acting. The rubber springs do not require any maintenance which is an advantage.

16.4.7 Coil Springs

Both these types are mainly used in independent suspension systems. Coil springs are widely used in live and dead axles both. These occupy lesser space. The coil springs (as shown in Fig. 16.11) experience, both, the bending stress and torsion.

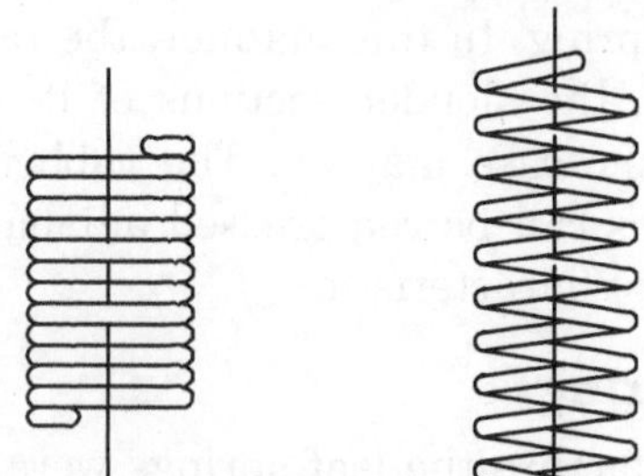

Fig. 16.11 *Coil spring*

16.4.8 Air Springs

Air contained in a cylinder fitted with piston or in a flexible bellows can act as a spring, as shown in Fig. 16.12. Under the static load, the air is compressed to a predetermined pressure, and subsequent motion of the piston either increases or decreases the force acting on the piston. Plotting this force against the travel of piston gives curve similar to that of compression curve of an engine indicator diagram. So, with increase in air pressure, the rate at which the force varies with piston travel becomes greater. This also indicates that the rate of air spring is not constant. This varying rate is advantageous in the sense that a low rate can be obtained for small deflections from the mean riding position while keeping the total rise and fall of the axles within reasonable limits.

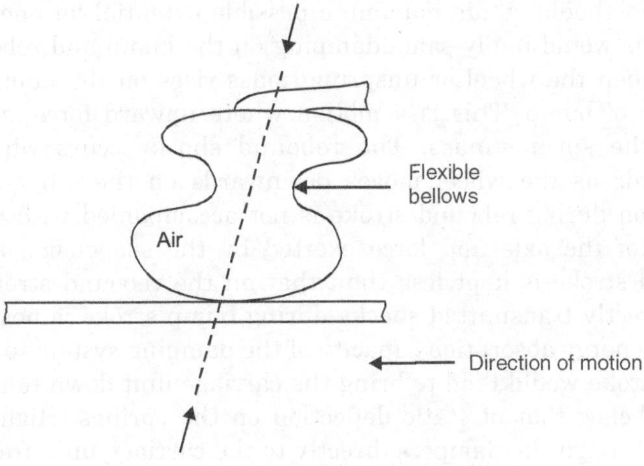

Fig. 16.12 Air springs

Air springs are quite favourite with automobile manufacturers in Continental Europe and USA. Their high cost, complex ancillary system for compressed air and risk of break down are some disadvantages. For cars they are not suitable as they are bulky and too complex. They are suitable only for tractors, heavy vehicles and passenger buses.

16.5 DAMPING

The vibrations, in an automobile, are main source of discomfort. These may be equally harmful to passenger and goods being carried in the vehicles. Vibrations occurring in an automobile are required to 'disappear' gradually. This is better known as 'die-away' of vibrations. The variations may occur randomly or periodically or at the natural frequency of the suspension system. When they occur at natural frequency they introduce state of resonance. The state of resonance is totally undesirable as it may lead to the damage to vehicle. Dampers, or shock absorbers cause a die-away of any vibrations. To achieve this, the dampers apply a force in a direction opposite to that of the instantaneous motion of the suspension. Earlier, there used to be friction dampers or semi-rotary vane type hydraulic dampers. These were abandoned and replaced by modern dampers that are either telescopic hydraulic struts or lever type hydraulic units.

The telescopic hydraulic strut type dampers are interposed between carriage unit and axle or, in other words, between the sprung and unsprung mass. In lever type hydraulic units, the body, which is a hydraulic cylinder, is mounted on the carriage with actuating lever connected to the axle.

Damping is effected by the damper piston. There can be a single piston or more than one piston to force the hydraulic fluid at high velocities through small holes. The fluid absorbs the energy which is converted into heat. The heat is partly dissipated through conduction to the surrounding structure of the vehicle and remaining passes into the surrounding air. The amount of energy absorbed and dissipated, for any given rate of energy input, is a function of the volume and viscosity of the fluid and the numbers, sizes and geometry of the holes through which it is forced. The resistance to deflection of the damper is a function of the square of its velocity of motion. Thus, the slow movement of wheels can occur with relative freedom but as the velocity of motion increases the resistance to deflection increases rapidly.

The damper design should obtain maximum possible potential for energy absorption for any given size, and this would imply same damping on the bump and rebound strokes. The bump stroke occurs when the wheel or unsprung mass rises on the bump in the first half till reaches at the top of bump. This is a motion where upward force is exerted violently causing a shock to the sprung mass. The rebound shock occurs when the unsprung mass moves downwards as the wheel moves downwards on the other side of the bump (Fig. 16.13). The motion during rebound stroke is not accompanied with violent force. This is due to the weight of the axle and force exerted by the suspension spring. Therefore, damping on the bump stroke is kept less than that on the rebound stroke. To relieve the carriage unit of all directly transmitted shocks during bump stroke is not practical. If done so it would reduce the energy absorption capacity of the damping system to half and damping only on the rebound stroke would tend to bring the carriage unit down to a mean level. This mean level would be below that of static deflection on the springs. High frequency, small amplitude vibrations through the dampers directly to the carriage unit are prevented by the interposition of the rubber bushes or blocks between their end fittings and their anchorages on the axle and carriage unit.

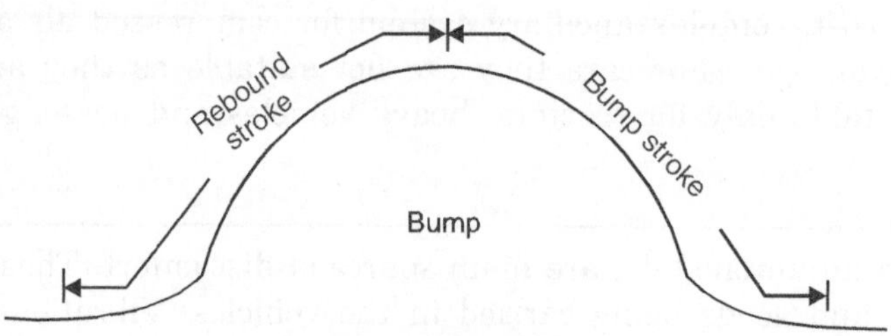

Fig. 16.13

16.6 DAMPERS

In all telescopic dampers, as the piston moves inside the cylinder its whole area is effective in transmitting the load but as it moves outwards the effective area is reduced. This is because some space is occupied by the piston rod and the effective area is reduced by the area of piston rod. For damping to be equal on both the sides, pressures at which the valve opens the small holes in the piston are adjusted to different values. These pressures force the fluid through these holes which provides the damping. Also, the total cross-sectional areas of the holes for the flows in the two directions must differ. This can be obtained by use of simple plate valve to close some of the holes during motion in one direction only.

Also due to piston rod, the space available to accommodate fluid differ on the two sides of the piston. This can be compensated by incorporating a flexible element in the cylinder, so that total volume within it can be adjusted automatically, as required. This flexible element can be an elastic sphere containing an inert gas, or a free piston with an inert gas between it and closed end of the cylinder. An alternative to this being the use of double-tube design.

16.6.1 Double Tube Damper

Referring to Fig. 16.14, a double tube damper consists of a cylinder C to which the head H is welded. The head is screwed into the outer tube T. To the outer tube T a pressed steel cap and eye E are welded. Through E the cylinder C is secured to the axle or wheel assembly. Piston A, in the cylinder C, is secured to the piston rod R. This rod, at its upper end has got an eye welded to it. This eye is attached to the frame of the vehicle. The part of the piston rod emerging out of the cylinder C is protected by a cover which is welded to the fixing eye. There is a gland, G provided to prevent leakage where the piston rod passes through the head, H. The fluid scraped off by the gland packing passes down the drain hole to the space between cylinder C and the outer tube T which acts as a reservoir.

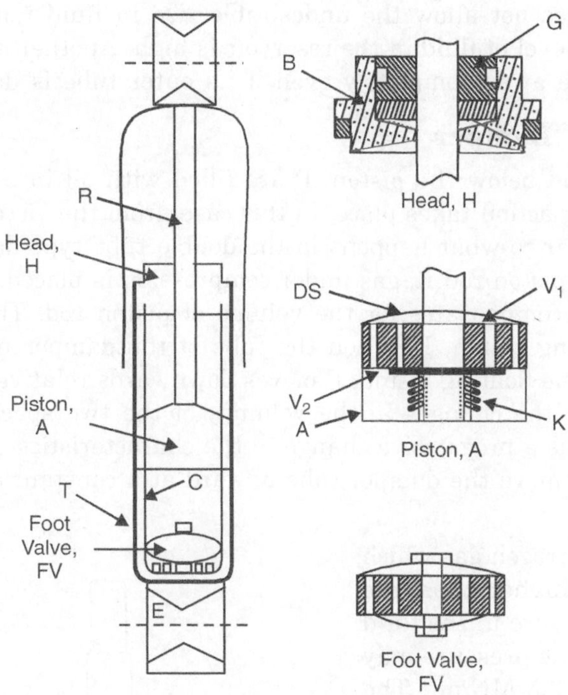

Fig. 16.14 *Double tube damper*

Piston A is provided with two concentric rings of holes. The outer ring is covered by a disc valve V_1 which is held in position by star shaped disc spring DS. The inner ring is covered by the disc valve V_2 held up by the coil spring K. The foot-valve assembly FV, which exists at the bottom of cylinder C, is similar to that in the piston except the lower disc valve is held up by a disc spring instead of a coil spring. This reduces the dead length of the shock absorber. The dead length of the shock absorber is the length which is not available for the working stroke.

The cylinder C is completely filled with fluid whereas the space between cylinder C and T is only partly filled. With the upward movement of eye E, fluid is displaced from beneath to above the piston A. This fluid passes through the outer ring of holes, by lifting the valve V_1 against the spring DS. Due to presence of piston rod the increase in volume on upper side is less as compared to decrease in volume on the lower side. The difference in the two being equal to the volume of the portion of piston rod that enters the cylinder. Due to this, the fluid will also be displaced through the inner ring of holes in the foot valve. This causes a rise in the level of fluid in the reservoir.

The size of the passage opened by the valves in the piston and foot valve determines the pressure set up. Also, it depends upon the square of speed at which the cylinder is moved upwards. When the cylinder moves downwards, the valve V_2 is forced off its seat and fluid is displaced from the upper end to the lower end. This flow occurs through the inner ring of holes in the piston. But as the piston rod moves, its volume is removed from the cylinder. This causes the fluid drawn from the reservoir space. This occurs through the outer ring of holes in the foot-valve.

The fluid displaced from cylinder to the reservoir carries heat with it. This heat gets conducted away. This does not allow the undesirable rise in fluid temperature. The heat transfer is greater if the level of fluid in the reservoir is high. Another advantage being that the damper does not give away completely even if its outer tube is damaged.

16.6.2 Single-Tube Damper

The spaces above and below the piston, P are filled with oil in a single tube damper (Fig. 16.15). The damping action takes place, in this case, from the viscous losses that occur in the orifices. It is similar to what happens in the double tube type dampers. However, to compensate, the effect of piston rod R, gas under compression is placed at the bottom of the tube. The volume of gas compensates for the volume of piston rod. The fluid and the gas are separated by a floating piston, F. When the tube of the damper moves up, the gas is further compressed and the floating piston, F moves downwards relative to it by the amount required to accommodate the changes in the volumes of the two spaces. Due to the compression of the gas there is a progressive change in the characteristics of the damping, such that the forces needed to move the damper tube upward at a constant speed will rise at an increasing rate.

These damper use nitrogen gas which must be at a pressure higher than the maximum operating pressure in the fluid below the main piston. The pressure may be of the order of about 2.5 MN/m^2. The total spring-rate of the suspension is increased by an amount equal to the gas pressure multiplied by the effective cross-sectional area of the piston rod.

As a variant, in this type of dampers, no floating piston is provided. Instead, these use inert gas which is free in the cylinder. The gas tends to emulsify the fluid. Although the oil and the gas separate when the vehicle is stationary, re-emulsification occurs at fast pace due to the large flow rates inherent in the design. In this type of dampers if there is a leakage of fluid past the piston, jacking and subsequent bending of piston rod does not occur. As there is no free piston the dead length of the damper is small. The performance of emulsion type dampers is less affected by

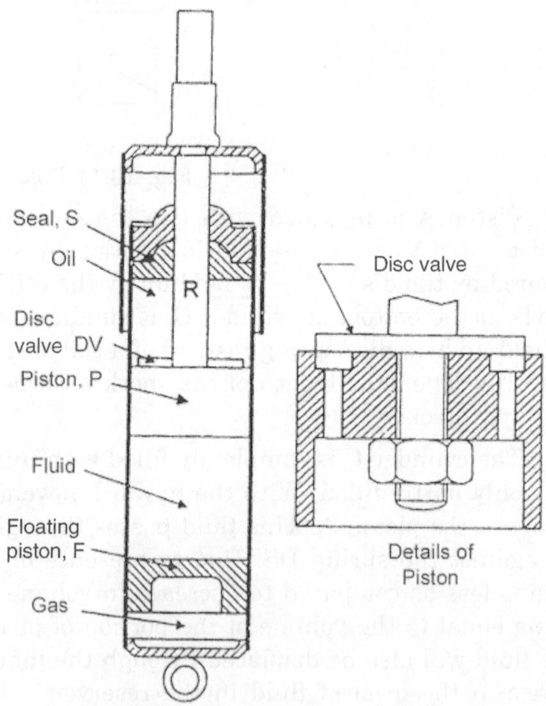

Fig. 16.15 *Single tube damper*

change in temperature whereas the affect is significant in case of fluid type dampers. In addition to this, for a given overall diameter, the piston area is greater in this type of dampers. Due to high built-in pressure the dampers are heavy which is disadvantageous.

16.6.3 Lever-Arm Type Damper

In this type of damper, the two pistons P_1 and P_2 move in bores B_1 and B_2. These bores are connected to space through orifices provided as shown. The pistons are actuated by the ends of a double lever LL carried on a shaft S (Fig. 16.16). The shaft being coupled externally to the axle. Damping arises, in this case also, from the viscous losses that occur in the orifices. The lever arm acts as one of the links of an independent suspension system.

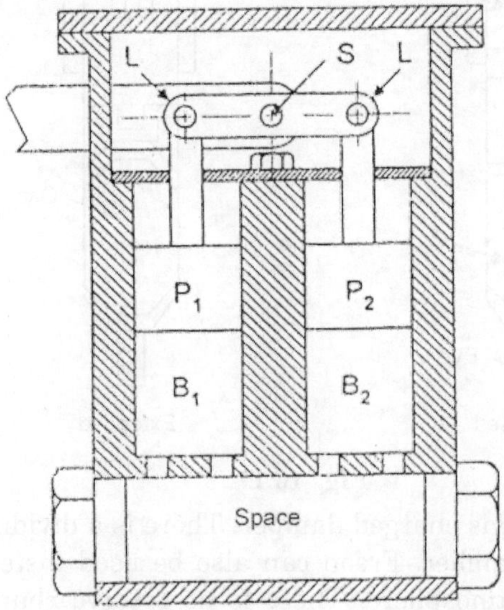

Fig. 16.16 *Lever arm type damper*

The working of dampers has been improved by incorporating certain changes in design. There are few more variants of dampers as described below.

16.6.4 Shock Assist Damper

These dampers can be used for the front or rear suspension. These can also enhance the load carrying capacity of the springs. There is a coil spring fitted on a conventional tube damper. The upper end of the spring being fixed on the upper tube and lower end of the spring fixed to the lower tube. The spring is under some tension in normal position and is compressed when damper is compressed due to up and down movement of axle.

16.6.5 Gas Charged Damper

These dampers work on the same hydraulic principle applied in conventional dampers. The double tube is replaced by a dividing piston that creates two chambers. One chamber is used for fluid and other chamber is used for gas. The gas is utilized to separate air from mixing with damper fluid. On rough roads the possibility of air mixing with damper fluid is higher and causes skip in the damper's action. A bag filled with Nitrogen gas seals the air passage and damper fluid comes in contact with gas only.

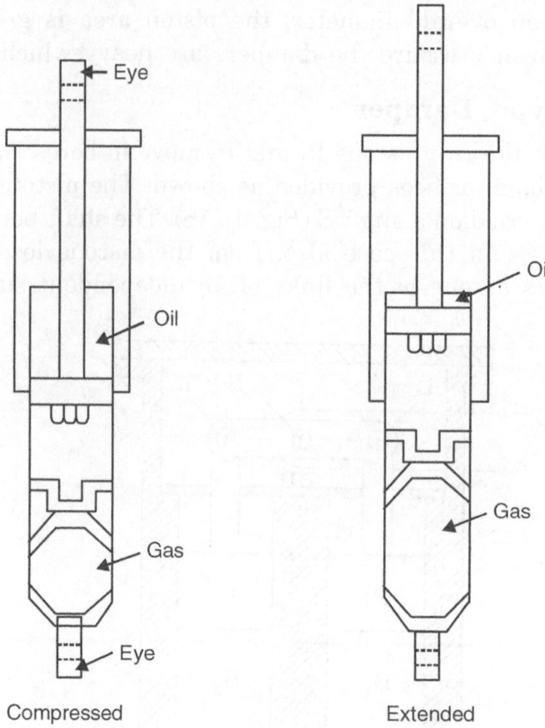

Fig. 16.17

Figure 16.17 represents a gas charged damper. There is a dividing piston that separates the gas chamber from fluid chamber. Freon can also be used instead of nitrogen. The gas is at a pressure of about 25 atmosphere. There is no reserve chamber and double tube.

When piston moves down, the fluid is displaced as in double tube damper. This causes the piston to compress the gas. When piston rod returns, the gas at high pressure causes the downward movement of piston and it comes back to its original position. The pressure in the fluid chamber is so maintained that pressure behind the piston is low and does not allow the gas to escape from gas chamber.

There is provision to adjust the dampers. The adjustments could provide firm or soft ride. Sometimes there can be an adjustment in between also termed as medium. These three options can be achieved manually. The variable damping is achieved by varying the size of metering orifices in the damper. The adjustment can also be done electronically by rotating a control rod inside the damper. The rod is actuated by a small motor. The driver can operate the system whether it is manual or electronic. Sometimes electronic systems are part of computer controlled suspension systems.

16.7 TORSION BAR

Torsion bar is located between vehicle frame and lower control arm. It is made of alloy spring steel which is heat treated. As the wheel moves up the lower end of the control arm moves up and causes twisting of torsion bar and absorption of shock. The natural resistance to twisting restores the bar to original state returning the wheel to road. The torsion bars are pre-stressed to give them desired fatigue strength. The pre-stressing is directional and therefore these are identified as 'left' or 'right' so that these can be used on left or right side of suspension system. A damper is placed between the lower control arm and frame to dampen the twisting motion of the torsion bar.

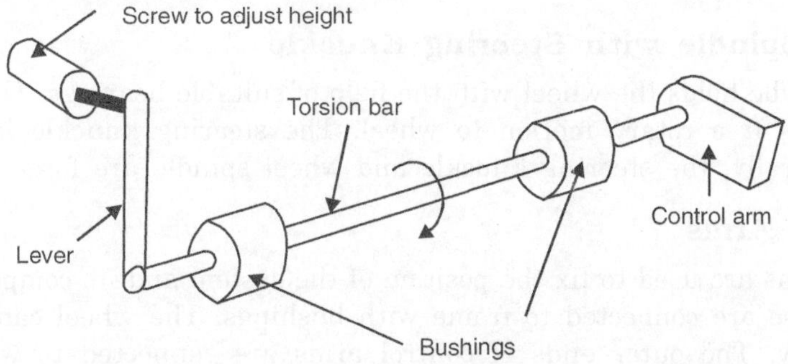

Fig. 16.18

The automobiles are provided with suspension systems on both front and rear side. According to requirements, the design of front suspension system differs from design of rear suspension system.

16.8 INDEPENDENT FRONT SYSTEM

The design of the front suspension systems is quite complicated. They keep the wheels firmly positioned while allowing them to steer left or right. The front suspension system also absorbs almost the whole of the braking torque. Therefore it is required to be stable.

The components of independent front suspension systems include coil springs, leaf springs, air springs, torsion bars and dampers that have already been described. These are assembled in a particular manner for independent systems.

One of the types of independent system had parallelogram design. The upper and lower control arms were hinged at the frame and at the top and bottom of spindle support. The hinging was done through pins or bushings. These had two control arms of the same length. This kept the wheel camber constant when vehicle bounced on rough road. In this design, the limitation was that contact of the tyre surface with road surface was not uniform and this caused instability and uneven wearing of the tyre.

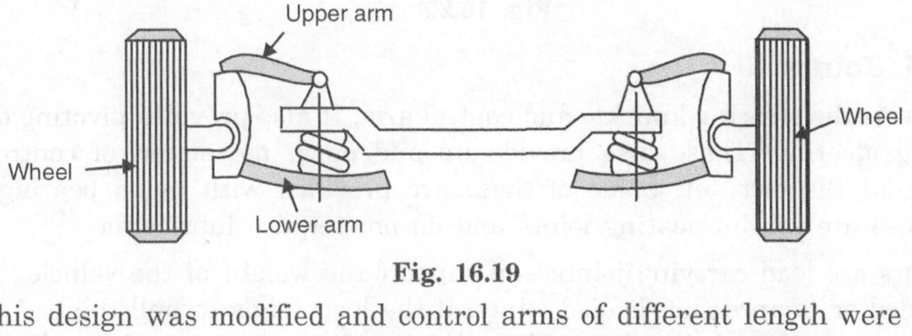

Fig. 16.19

Later this design was modified and control arms of different length were provided on upper and lower side. The upper arm was made short and lower arm was long (Fig. 16.19). Due to shorter arm done by upper arm the top of wheel moved in and out slightly but provided a uniform tyre surface contact with road surface. The steering control was also improved in this design.

16.9 DOUBLE WISHBONE SUSPENSION

Double wishbone suspension system consists of wheel spindle with steering knuckle, control arms, ball joints, damper and springs.

16.9.1 Wheel Spindle with Steering Knuckle

The wheel spindle holds the wheel with the help of suitable bearings. The bearings are essential to provide free rotary motion to wheel. The steering knuckle is connected to control arms. Generally, the steering knuckle and wheel spindle are forged together.

16.9.2 Control Arms

The control arms are used to fix the position of the system and its components relative to the vehicle. These are connected to frame with bushings. The wheel can move up and down independently. The outer ends of control arms are connected to wheel assembly through ball joints inserted into steering knuckle (Fig. 16.20). The arms have wishbone shape. Alternatively, these could be double pivoted or single bushing control arms. The wishbone type control arm provides better lateral stability but requires more space.

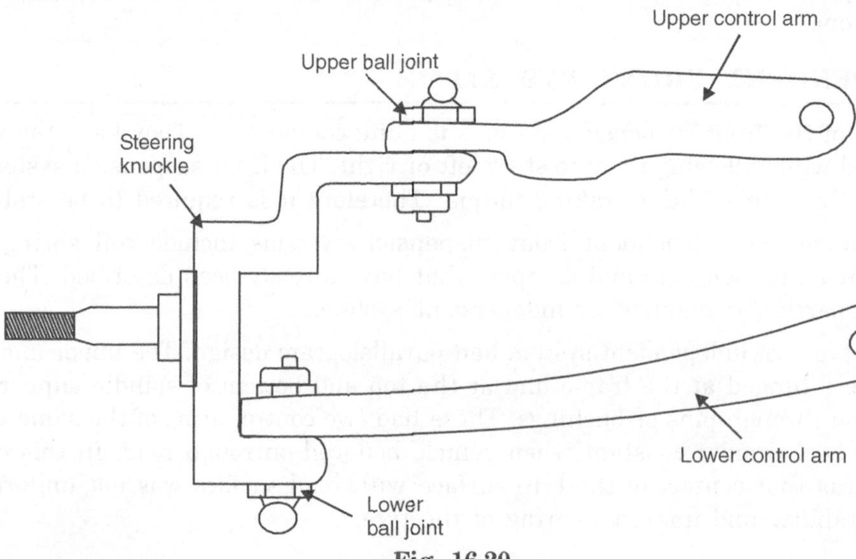

Fig. 16.20

16.9.3 Ball Joints

This connects the steering knuckle and control arm. It also provides pivoting of steering knuckle during steering. These also provide up and down movement of control arm. A rubber seal keeps the dirt out. Some of these are provided with nylon bearing that are lubricated. These are self-lubricating joints and do not require lubrication.

These joints are load carrying joints and support the weight of the vehicle. These can be tension loaded or compression loaded joints. If the load tends to pull the ball out of the socket these are tension loaded and if load tends to push the ball into the socket these are compression loaded joints (Figs. 16.21 and 16.22).

The ball joints may not carry load *i.e.,* the weight of the vehicle. Then these are called follower ball joints. These mount on control arm. The control arm does not provide seat for spring.

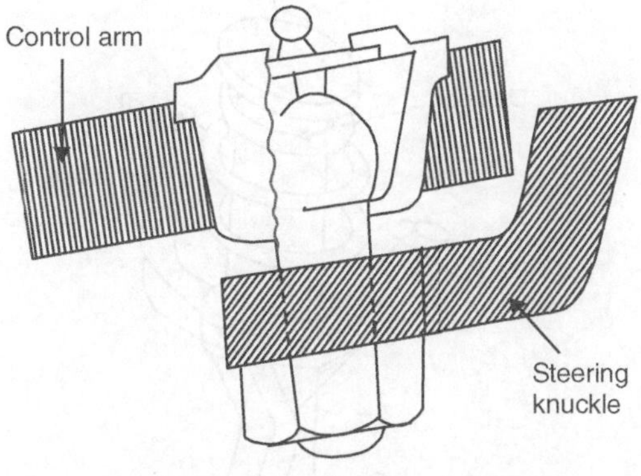

Fig. 16.21

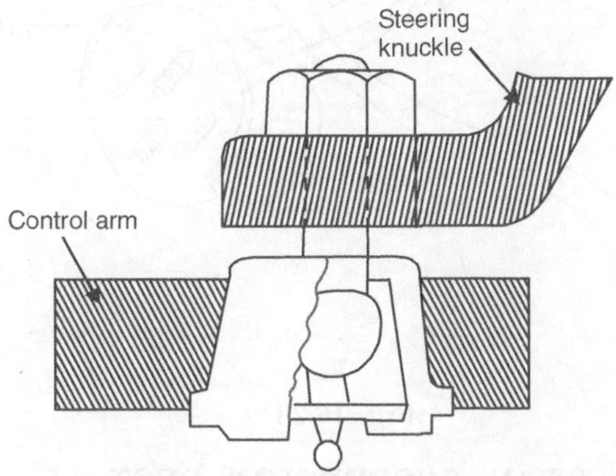

Fig. 16.22

Some ball joints are provided with wear indicator. With wear, the grease fittings recede and as the shoulder of the fitting flushes with housing the joint is replaced with new one.

16.10 BUSHINGS

Bushings made of rubber or polyeurethane are used in different components of suspension systems such as control arms and strut rods. These absorb shocks and reduce noise entering the vehicle. The bushes made of rubber deteriorate fast but Polyeurethane bushings are stronger and take up load betterly. These provide better steering ability and have longer life.

16.10.1 Stabilizers

These are used for additional stability. They are metal bars placed between the opposite ends of lower control arms. As one wheel moves up or down due to uneven road surface this movement is transferred to other wheel through these bars. The bars can be one piece or have a U-shape. These are fastened to end rods with rubber bushings.

The design with unequal control arms was further modified by using double wish bone. This can also be used in front wheel drive vehicles. The components are not affected by torques due to braking and cornering. This type of suspension can be readily adapted in front wheel drive vehicles and provide comfortable ride (Fig.16.23).

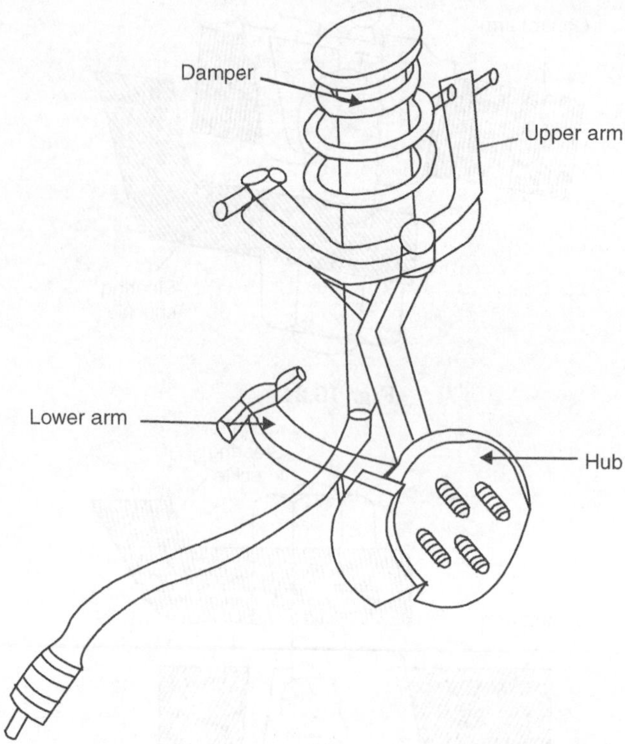

Fig. 16.23

16.11 INDEPENDENT REAR SUSPENSION SYSTEM

The independent suspension systems provide better traction and comfortable ride. This is because the wheels move separately on the road. In these systems, A-shaped control arms are used. These control arms can be placed differently. One of the arrangement has the apex of A-shaped control arm pointing towards the rear of the car (Fig. 16.24). These are known as trailing arms. In another arrangement, the A-shaped control arm is mounted at an angle and are known as semi-trailing control arms (Fig. 16.25). Coil springs are used above the control arms and support the vehicle body. The damper is connected to spindle or sometimes control arms. The semi-trailing arms with lateral links also control the wheel angles.

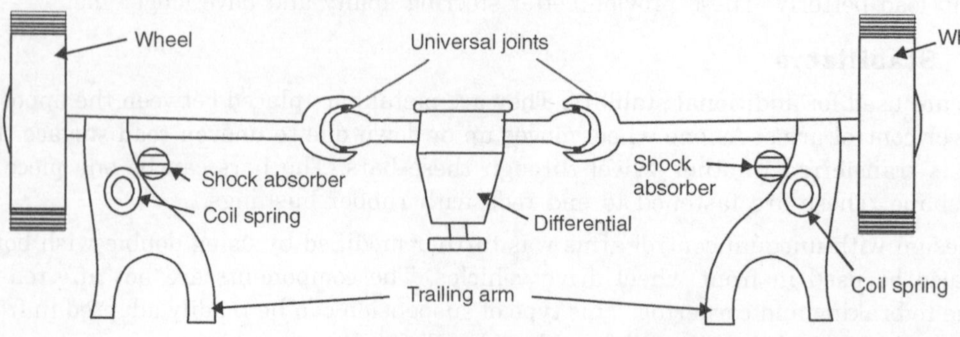

Fig. 16.24

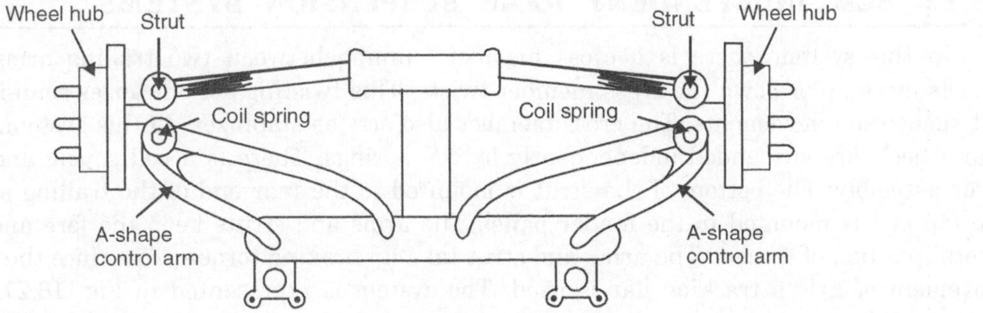

Fig. 16.25

Some modified designs of independent rear suspensions, use lower control arm. It has open driving axle. There is cross member provided that supports the control arms. The upper ends of dampers are mounted to the body of the vehicle. There are coil springs provided between the top and the bottom cross members.

In some other designs, upper end control arms are replaced by a wishbone shaped sub-frame. There are no control arms at the top. There are two torque arms that transfer the rear end torque to wishbone shaped sub-frame. In some cases, the dampers have been replaced by struts (Fig. 16.26). Here, the rear wheel, rear ends of the tie rods, outer ends of two control arms and strut are secured through a spindle. The control arms have small bushings at the end which is connected to the body centre line and the end with larger bushings is connected to the spindle. This is known as MacPherson strut system.

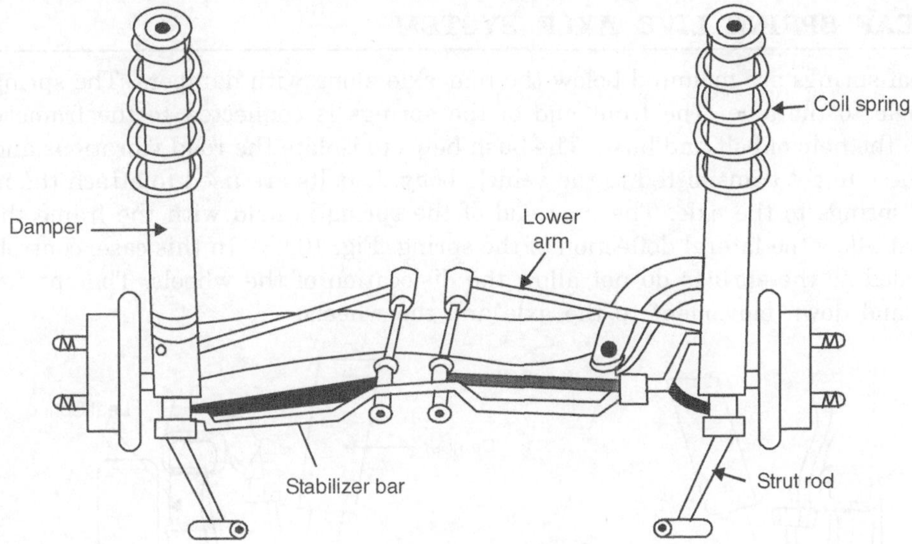

Fig. 16.26

In a further modification, another system was designed that was readily adapted by the vehicles with front wheel drive. Here, the coil spring is mounted between the lower control arm and side rail. The spindle supports the rear wheels and also acts as a location where outer ends of the control arm and rear ends of the tie-rod are attached. Chapman strut finds place in some designs instead MacPherson strut. This strut is not directly involved in vehicle's steering system. It can be used with conventional leaf springs. Chapman strut does not carry any load and acts as damper also.

16.12 SEMI-INDEPENDENT REAR SUSPENSION SYSTEMS

In this system, there is a cross member running between two trailing arms. As the wheels move up or down the cross member twists. This twisting action causes semi-independent suspension movements. The cross member also acts as stabilizer. In this system, both the rear wheels are suspended independently by coil springs. There is a coil spring and damper strut assembly. The bottom of the strut is mounted to the rear end of the trailing arm while the top end is mounted to the fender panel. The arms and struts keep the fore and aft and lateral position of wheels. The arms and strut take up braking torque. To reduce the sideways movement of axle a tracking bar is used. The system is represented in Fig. 16.27.

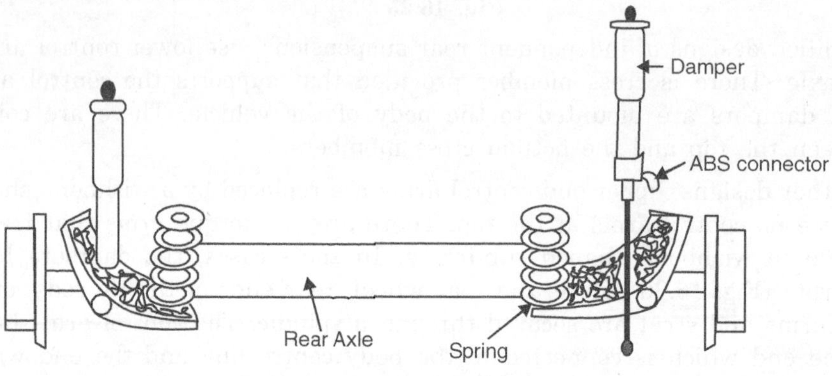

Fig. 16.27

16.13 LEAF SPRING LIVE AXLE SYSTEM

Two leaf springs are mounted below the rear axle along with dampers. The springs are at right angle to the axle. The front end of the springs is connected to the frame of the vehicle with the help of bolt and bush. The bush helps to isolate the road vibrations and does not allow these to get transmitted to the vehicle body. U-bolts are used to attach the middle part of the springs to the axle. The rear end of the spring is held with the frame through shackles that allow the lateral deflection of the spring (Fig. 16.28). In this case, control arms are not needed as the springs do not allow the dislocation of the wheels. The springs also provide up and down movement to the axle and the wheels.

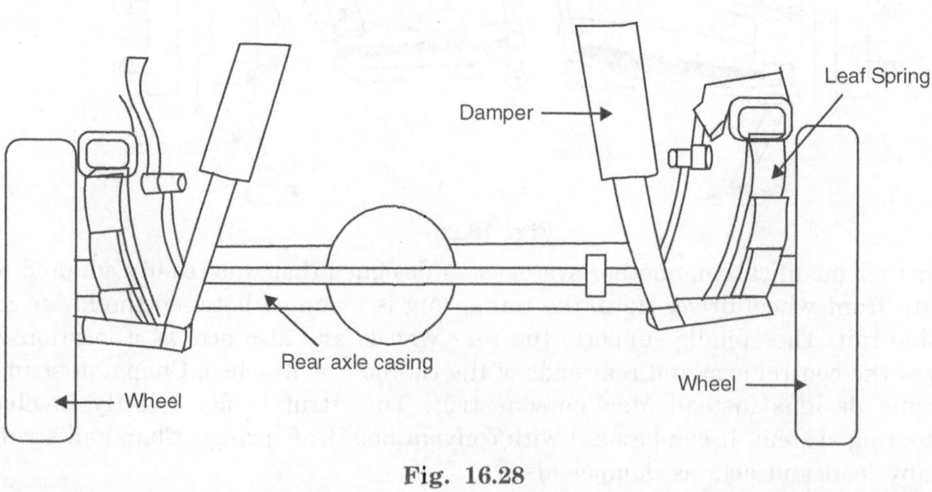

Fig. 16.28

The system has certain disadvantages. It is instability due to solid axle. Also, there is poor traction. When vehicle is accelerated, rear axle may jump up and down rapidly as the torque is absorbed by the springs. This may damage the dampers and spring mounts. To avoid this, dampers are mounted on front and rear side of the axle.

16.14 COIL SPRING LIVE AXLE SYSTEM

In this design, there are two coil springs at the rear live axle (Fig. 16.29). Forward and lateral control links are provided to keep location of the axle. The coil springs only take up the load of the vehicle. The control arms are channel beams mounted on rubber bushings. These arms take up all the torques including accelerating, driving and braking torques. Rubber bushings are provided for mounting the arms to reduce the noise and transmission of shocks.

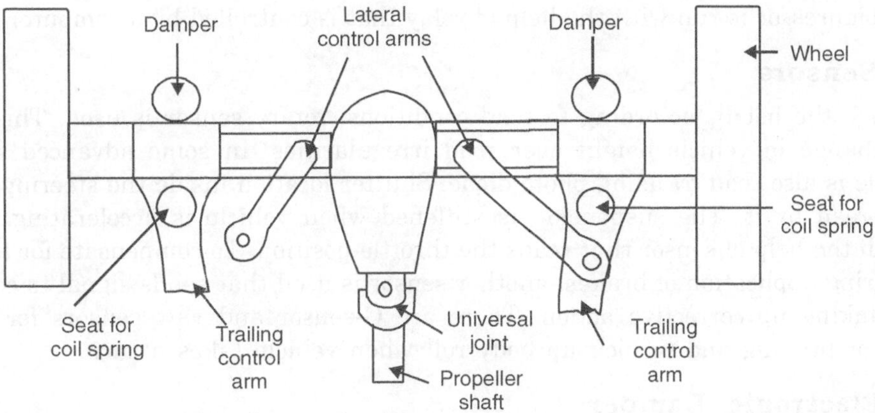

Fig. 16.29

16.15 ELECTRONIC SUSPENSION SYSTEM

In the recent past, due to the development of sensors and computer controlled technology, it has become possible to have improved suspension systems. The application of computer controlled technology has made it possible to programme the system in such a way that these can respond to various operating conditions perfectly. The systems are capable of adjusting the height of the vehicle as per requirement. It is also possible that, on a curve, the height of outside actuators is increased and height of the inside actuators is decreased so that the vehicle leans like a motorcycle.

These systems are provided with air adjustable dampers. A compressor provides the compressed air. Some more advanced systems are capable of changing the ride height and damping height continuously. The systems have sensors that feed data to computer directly and commands are transmitted by computer. These systems have high pressure hydraulic actuators that work more efficiently that air actuators. These hydraulic actuators take the weight of the vehicle and conventional springs are not needed.

16.16 ADAPTIVE SUSPENSIONS

Electronic sensors monitor the vehicle height, its speed, steering angle, braking force, throttle position and ignition switching along with dampers status and even door position.

This data is fed to a computer, which after analysing it, switches the suspension system to desired operating mode. Adaptive system has electronic dampers with variable valve timings. These have variable rate air springs.

16.17 COMPONENTS OF ELECTRONIC SUSPENSION SYSTEM

To control the flow of air in the system, air compressor, sensors, computer controlled module and solenoid valves are used. These are connected with each other through nylon tubes. Practically, there can be several designs that have been developed by different manufacturers.

16.17.1 Compressor

It provides compressed air that acts as working fluid for the system. It may be run by battery that is provided in the vehicle. The moisture contained in the air is removed by air dryer. The compressor is run with the help of relay that is controlled by a computer module.

16.17.2 Sensors

To change the height according to road conditions, rotary sensor is used. This sensor allows the change in vehicle height over road irregularities. In some advanced systems, steering angle is also read by using photo diode. Shutter location inside the steering column can also be read by it. The suspension is stiffened when vehicle is accelerating. This is achieved with the help of sensor that reads the throttle position. To compensate for the front nose dive during application of brakes, another sensor is used that sends signal to computer module for taking up corrective action. There are G-sensor and yaw sensors for sudden acceleration or braking and to pick up body roll when vehicle takes a turn.

16.17.3 Electronic Damper

Variable shock damping is achieved through electronically controlled dampers. The input to the computer module determines the degree of damping. This input is based on vehicle speed, steering angle, and brake application. Solenoid actuated damping is used for real time damping. Solenoids allow instantaneous changes in the operation of valve and the time taken is few milliseconds. This allows instantaneous reaction to a bump under the wheel.

16.17.4 Electronic Struts

Air springs or dampers are replaced by electronic struts that are similar to electronic dampers (Fig. 16.30). The fluid pressure is controlled by variable orifice. The input is provided by different sensors monitoring different parameters. The variable orifice along with a deflected disc valve can control the fluid flow effectively in all the conditions. Moderate damping force is provided in normal mode when unit is set to balance the flow between the small orifice and deflected disc valve. Minimum damping force is provided through large orifice in comfort mode. The fluid flows through deflected disc valve in firm mode.

The height of the vehicle can also be controlled through electronic struts. This improves the aerodynamic characteristics of the vehicle. The increase in speed causes the reduction in the height of the vehicle in the front side to reduce the wind resistance. When the speed is reduced the front height of the vehicle comes back to normal.

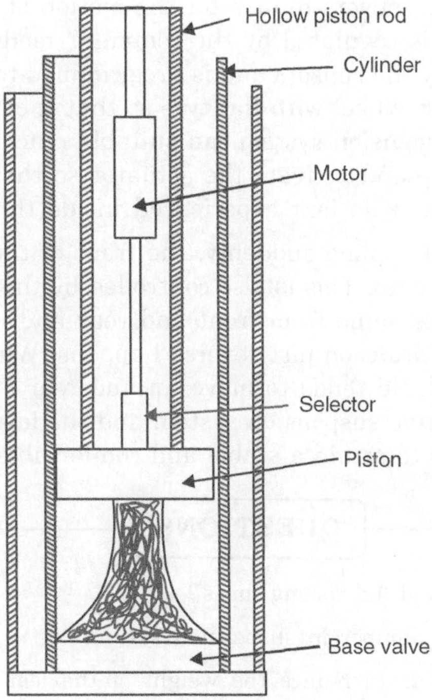

Fig. 16.30

16.17.5 Computer Control Module

The motor for air compressor, compressor vent solenoid, and air spring solenoids are controlled by the computer module. The valves in the struts and operation of electronic damper are also taken up by this module. The sensors provide input to the module. The module is programmed to carry on diagnostic tests. It also controls the warning light, an indication to driver if anything in the system goes wrong.

16.17.6 Electronic Level Control

The height of the vehicle can be adjusted by electronic level control. Height sensors are provided to provide input to the system. These sensors can sense the load on the vehicle as the passengers ride or get down.

16.18 ACTIVE SUSPENSIONS

These systems are fully automatic and are provided with hydraulic actuators. These are capable of changing ride height and damping height continuously. They can reduce body roll and provide very comfortable ride.

Each wheel is provided with actuator which is a double-acting hydraulic cylinder. Each actuator acts as its own damper. It also acts as spring. Each responds to road conditions that it experiences. This is done by controlling the pressure in the actuator.

There is computer module and information to the module is fed by different sensors. Up and down movement of each wheel is sensed by linear displacement sensors and acceleration sensors in the actuators.

The steering sensor provides the information about turning of the wheel. There are roll sensors and lateral acceleration sensors to monitor the motion of the vehicle. The hydraulic pressure within each actuator is regulated by the computer module. The computer module works on the inputs provided by the sensors and is programmed to act. These active systems can also be programmed to raise wheel with flat tyre so that it can be replaced. The module can be so programmed that suspension system can suitably react to infinite number of road conditions. It can sense a bump and activate the actuator so that pressure is released and wheel rises upwards and there is no jerk experienced inside the body of the vehicle.

Similarly when brakes are applied suddenly, the front of the vehicle tends to dip and rear of the vehicle tends to move up. This is also controlled by the active suspension system. The level of the vehicle remains same from front and rear and it is easy for the driver to control the vehicle. In another situation just reverse happens. When the vehicle is suddenly accelerated the front of the vehicle tends to move up and rear of the vehicle tends to dip. This is also taken care of by active suspension system and the level of vehicle remains same from front and rear. This helps to attain a stable and comfortable ride in all the situations.

QUESTIONS

1. What is 'sprung mass' and 'un-sprung mass'?
2. What is the function of damper in suspension system?
3. Describe different methods to reduce the weight of the leaf spring without affecting its strength.
4. Explain with the help of a diagram a variable rate spring.
5. Explain the constructional details and working of double tube damper.
6. Describe a gas charged damper.
7. Describe briefly different components of a double wishbone suspension system.
8. Explain a semi-independent rear suspension system.
9. What is electronic suspension system? How is it better than conventional suspension system?
10. Explain briefly different components of electronic suspension system.

17

IGNITION SYSTEM

The ignition system is responsible for the flow of current across the two electrodes of the spark plug. This is essential to create spark. The spark, thus produced, causes the ignition of fuel. This ignition is converted into flame and causes the combustion of fuel in the cylinder. The process becomes quite complicated as, in automobiles, multi-cylinder engines are used. Considering a four strokes, four cylinder engine (that is very common in an automobile) and say it is running at 4000 revolutions per minute. This means 2000 cycles per minute. Four cylinders mean total 8000 cycles per minute. Therefore, the ignition system is required to produce 8000 sparks per minute. The spark in each plug is required to be produced at right time. It must generate correct amount of heat. Also, the variations in engine operating conditions make the function of ignition system more complex. The improper functioning of ignition system would mean poor fuel economy and engine performance.

17.1 FUNCTIONS OF IGNITION SYSTEM

The ignition system should generate enough heat, through electric spark, to ignite the fuel air mixture. Every fuel has got a particular ignition temperature and until that temperature is attained ignition is not possible. Hence spark is required to generate enough heat so that ignition temperature is attained.

Sometimes it is required for the combustion to occur and therefore the duration of the spark should be sufficient to allow for combustion to occur. If the spark is not produced for sufficient time the combustion may not occur.

The hot gases produced due to combustion force the piston to move downwards during expansion stroke and combustion process requires sometime to complete, the beginning of combustion begins slightly before the piston has reached top dead centre that is towards the end of compression stroke. This means the spark should be produced towards the end of compression stroke.

The time when the spark occurs, before the compression stroke, varies with the engine speed. It has been found experimentally that if for a particular engine running at 1200 RPM the spark should occur 18° before the piston reaches top dead centre position then at a speed of 3600 RPM the spark should occur 40° before the piston reaches top dead centre position.

However, as the speed of the engine increases, the turbulence in air-fuel mixture also increases. This requires the ignition must occur slightly later. Hence two opposite requirements are to be fulfilled. This means that the best possible timing must be decided by taking into account both these factors.

The load upon engine varies. It depends upon so many things including weight being carried and resistance experienced by the vehicle. When load is high, there is resistance to the motion of crankshaft and that causes resistance to the movement of piston also. The throttle is fully open and quantity of fuel-air mixture entering the cylinder is more. This causes low vacuum in manifold and rapid combustion. This demands retardation in ignition timing.

When vehicle is running under light load, the throttle is partially open. It causes high vacuum in the intake manifold. The quantity of fuel-air mixture entering the cylinder is small. This makes combustion slow and therefore ignition timing is advanced.

In multi-cylinder engine, it is not enough to consider ignition process in single cylinder. It is essential that ignition process occurs in different cylinders in a particular sequence. This sequence is determined by firing order of the engine. In order to get uniform output from the engine it is essential that output from individual cylinder comes in a particular manner during the cycle. This has been explained in detail in chapter 4. Ignition system should be capable of monitoring the relative position of piston in each cylinder and should act at the proper time towards the end of compression stroke.

The electronic ignition system has been developed for automobiles. Before that, the ignition systems had distributors so that spark could be created in different cylinders. The electronic ignition systems do not require a distributor and can work without it.

17.2 IGNITION SYSTEM WITH DISTRIBUTOR

These systems are still being used in some automobiles. The system has battery, ignition switch and coil, distributor with capacitor, secondary wiring and spark plug.

17.2.1 Battery

The battery is essential to provide electric current. It can provide voltage of 12 volts and desirable amount of current. A rechargeable battery is used. It can be recharged and has long life.

17.2.2 Ignition Switch

The ignition switch turns the system 'on' and 'off'. This is done with the help of ignition key. When key is turned it moves the actuator rod and actuator rod operates the switch. The starting motor is operated with this switch. It operates the steering wheel lock to make the vehicle immovable. Audio system in car and heater motor gets power through this switch. In some, vehicles, electric fuel pump is also connected to battery through this switch. The five positions in the ignition switch indicate audio system, lock, off run and start (Fig. 17.1).

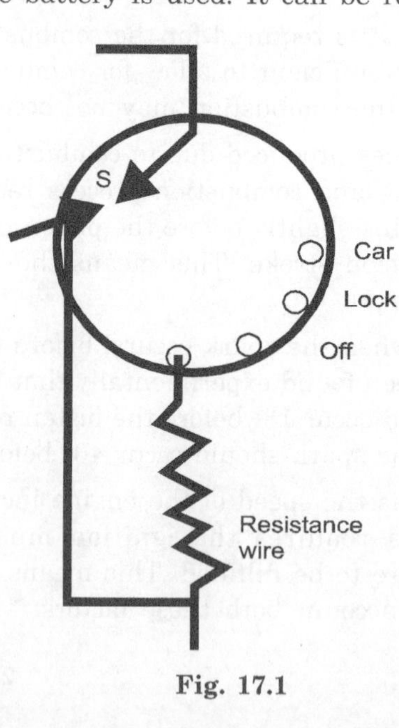

Fig. 17.1

Figure 17.2 shows the inner details of ignition switch. The steering wheel is also locked through the same switch. The plunger moves inside the notch of the disc. The disc is mounted on steering shaft and this way the rotary motion of steering shaft is restricted and steering wheel is locked. The locking occurs when ignition key is brought to 'lock' position. Moving the ignition key away from 'lock' position moves the plunger downwards outside the notched disc thereby permitting the rotary movement of steering shaft.

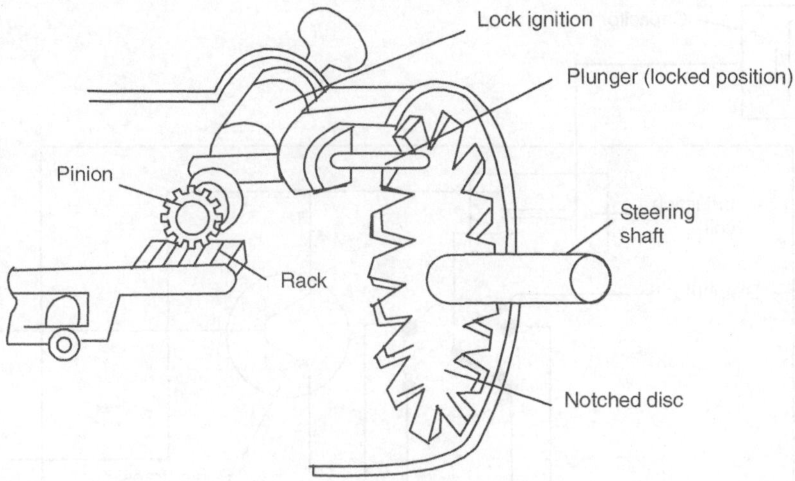

Fig. 17.2

17.2.3 Ignition Coil

The battery can provide a voltage of 12 volts. To produce spark between the electrodes of the spark plug, a high voltage of about 25000 volts is required. The voltage from the battery is 'stepped up' to the desired level through this coil which is a step up transformer. Figure 17.3 represents an ignition coil.

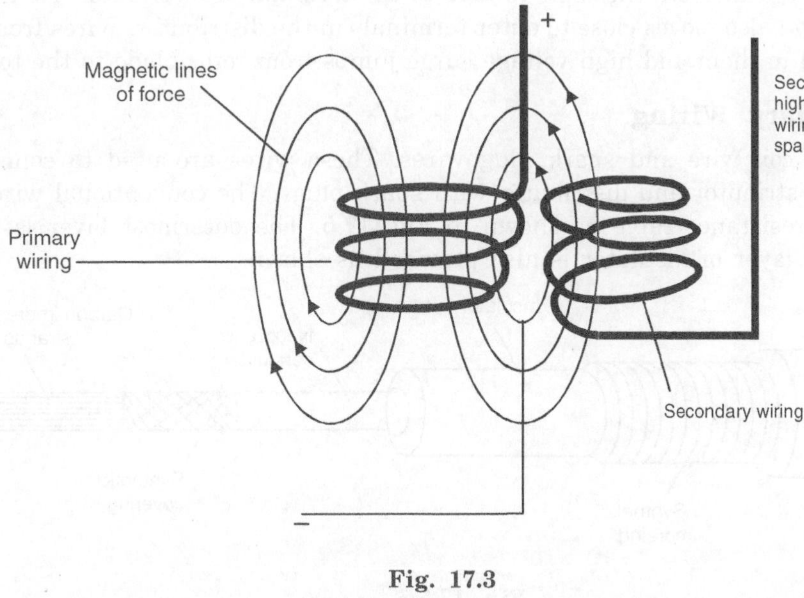

Fig. 17.3

17.3 DISTRIBUTOR WITH CAPACITOR

The distributor is provided with a set of contact points (Fig. 17.4). This is a switch that acts very fast. When the points are close, the current passes through the coil and when the coil does not produce high voltage surge. For this purpose, the distributor is accompanied by capacitor. It also helps in the collapse of magnetic field and prevents the burning away of points.

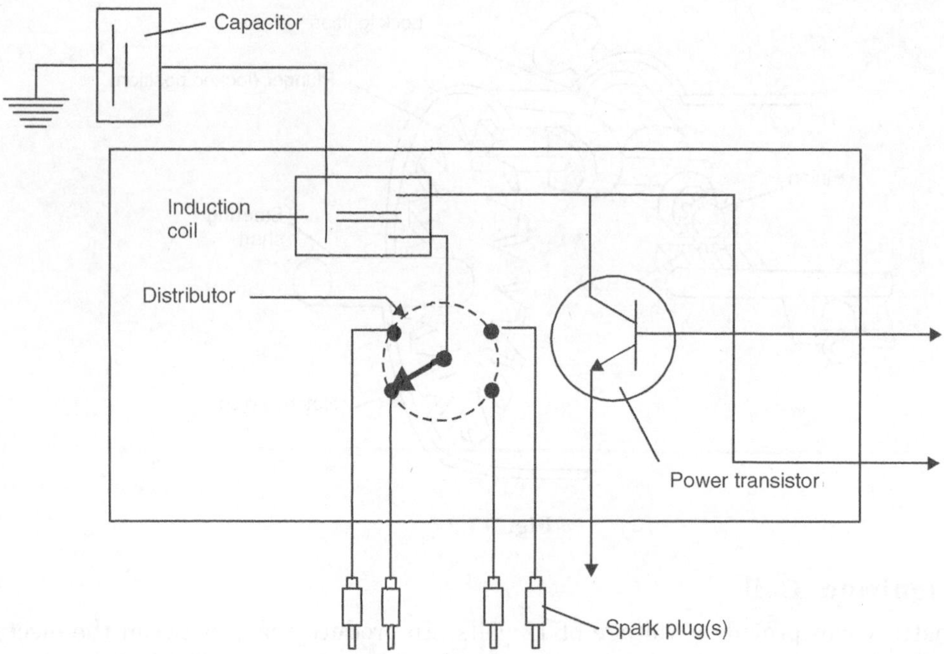

Fig. 17.4

The distributor also provides high voltage surge to spark plugs in the order of firing. The wire from ignition coil delivers the high voltage to the terminal of distributor. As the rotor moves, the other end also moves close to outer terminals in the distributor. Wires from spark plug are connected to them and high voltage surge jumps from rotor blade to the terminal.

17.3.1 Secondary Wiring

This includes coil wire and spark plug wires. These wires are used to connect the ignition coil and distributor and distributor with spark plugs. The conventional wires have been replaced by resistance cable as shown in Fig. 17.5. The outermost layer is silicone insulator. Another layer of insulator is also provided as shown.

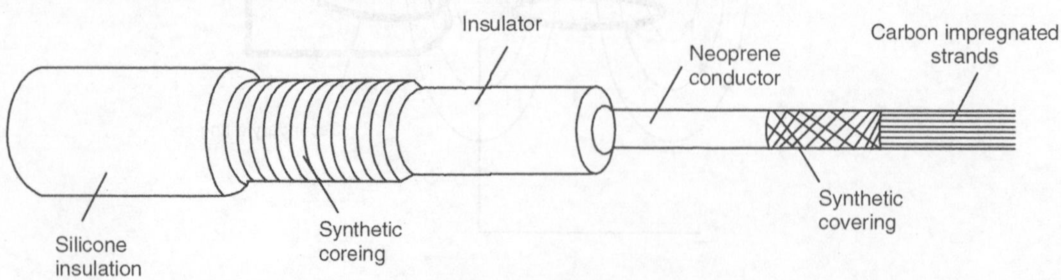

Fig. 17.5

17.3.2 Spark Plug

It has two electrodes which are made of metal conductor. One electrode is centre electrode and other is grounded electrode. There is a suitable gap between the electrodes. The spark jumps the gap and ignites the fuel-air mixture. There is outer metallic shell with ceramic insulator. Inside the ceramic insulator, electrode is located. The ignition coil supplies high voltage current to this electrode. The ground electrode is given a specific shape to provide suitable gap as shown in Fig. 17.6. The gap varies from 0.9 mm to 2 mm. The bigger the gap higher is the voltage required. The spark plug is provided with a resistance built into center electrode that does not allow television or radio interference. This interference is due to ignition. The mounting of the spark plug is done with help of screwed end. A gasket between the plug and seat makes it leak proof. Tapered screwed ends provide leak proof seat without gasket.

The material used to make electrodes is nickel and chrome alloy. This alloy is corrosion resistant. Sometimes copper core or platinum tip on electrodes is provided for longer life.

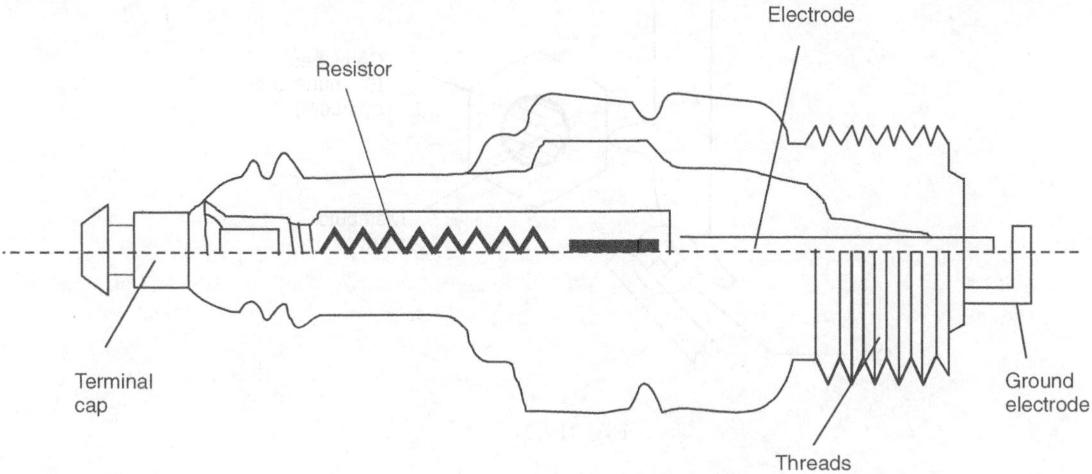

Fig. 17.6

There is a primary circuit and a secondary circuit in the ignition system. The primary circuit handles low voltage while the secondary circuit handles high voltage. The circuits have windings. Primary winding consists of less number of turns, say in hundreds, while secondary winding has large number of turns, in thousands. When switch is in 'on' position, current flows through primary winding and produces magnetic field. When switch is in 'off' position, current flow stops and magnetic field crumples and cuts the secondary winding. This produces high voltage that is supplied to spark plug through the secondary circuit and spark is produced across the electrodes of spark plug.

The contact points are located on the breaker plate of the distributor. There is a cam provided. The number of lobes in the cam is equal to number of cylinders. With movement of cam, the points close and open and act as mechanical switch to make and break the primary circuit.

One point of the switch is stationary and is mounted on grounded breaker plate. The other point is located at the end of insulated movable arm (Fig. 17.7). The arm is pivoted at one end and swings about this pivot. The cam lobe causes the swinging action of the arm. When contact is created the current flows, in the primary circuit and magnetic field is built up in the coil. When contact is lost the current flow stops and magnetic field crumples. The process is repeated each time lobe pushes the arm. The number of lobes being equal to number of cylinders ensures that each cylinder gets high voltage to produce spark in the spark plug. The duration for which current flows is known as 'dwell'. The distance when points are separated is 'gap'. The cam is mounted on shaft driven by camshaft which in turn is driven by crankshaft. Therefore, the point, where contact and high voltage occurs (thus producing spark) is determined by the position of crankshaft. The crankshaft position is determined by the position of piston in the cylinder. Hence spark is produced according to position of piston in the cylinder towards the end of compression stroke.

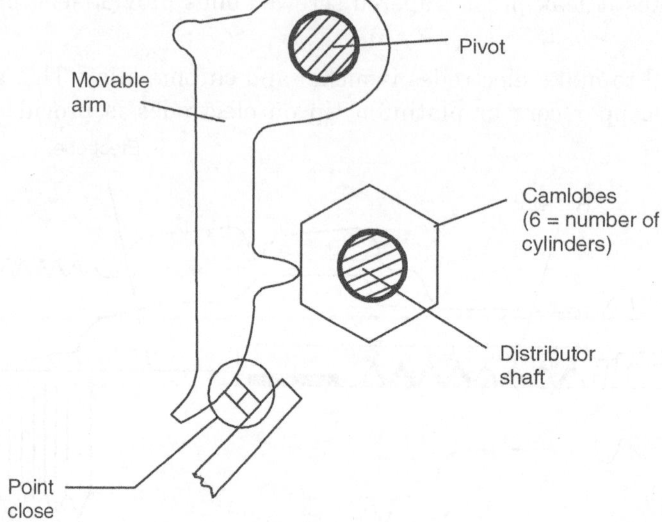

Fig. 17.7

17.4 RESISTANCE IN PRIMARY CIRCUIT

To prevent the burning of contact points, it is essential that flow of excessive current is not permitted. This is achieved with the help of resistor that is placed in series in the circuit. To supply the enhanced amount of current needed during starting, this resistance is by passed. After the start of engine, the current passes through resistance. This reduces the coil voltage to about 8 volts.

Voltage in the secondary circuit is very high. The pattern of the voltage has been viewed on an oscilloscope. The pattern is reproduced in Figure 17.8. Initially switch is off and crumpling of magnetic field causes high voltage in the secondary winding, A to B, in the figure. This voltage causes the spark. Next, the voltage drops from B to C because to sustain the spark less voltage is required. The spark continues till D. The duration of the spark measured in terms of crank shaft rotation is 20° or 1 millisecond approximately. Next up to E, there is voltage ripple when the amplitude of voltage is low. At E, the circuit closes and current flows in primary circuit. The magnetic lines of force expand and pass through the primary winding. This is represented by ripples and finally smooth curve till F. The dwell period is shown as EF. At F the points open and cycle begins once again as shown at A.

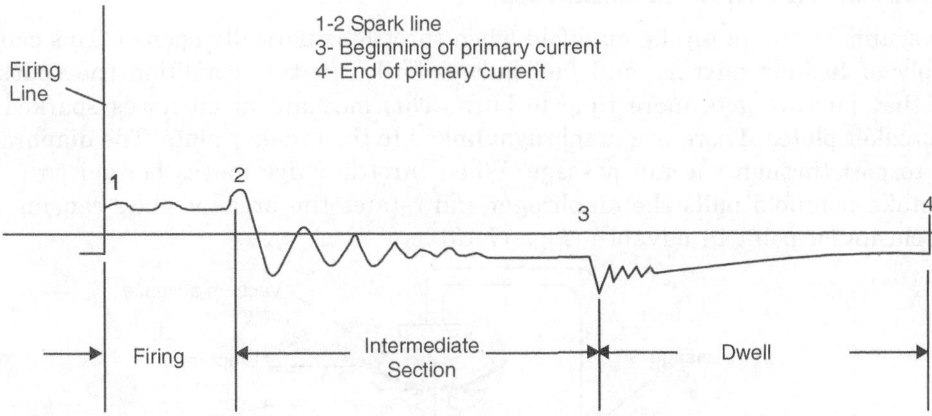

Fig. 17.8

The different running conditions determine the positions when spark should occur. This means the timings when spark occurs are different. At higher speeds the spark should occur earlier. While at low speeds or when engine is running idle, the spark may occur later that is when the piston is almost at the top dead centre towards the end of compression stroke. This is also known as 'spark advance'. This is achieved with the help of either centrifugal advance mechanism or through vacuum advance mechanism.

17.4.1 Centrifugal Advance Mechanism

There is a breaker cam that is pushed ahead with increase in engine speed. There are two weights, two springs and cam assembly. The assembly has breaker cam and an advance cam that has oval shape. At low speed, the weights are held in by springs. With increase in engine speed, due to centrifugal force, the weights are thrown out after overcoming the force exerted by springs (Fig. 17.9). This pushes the cam assembly ahead and spark is advanced.

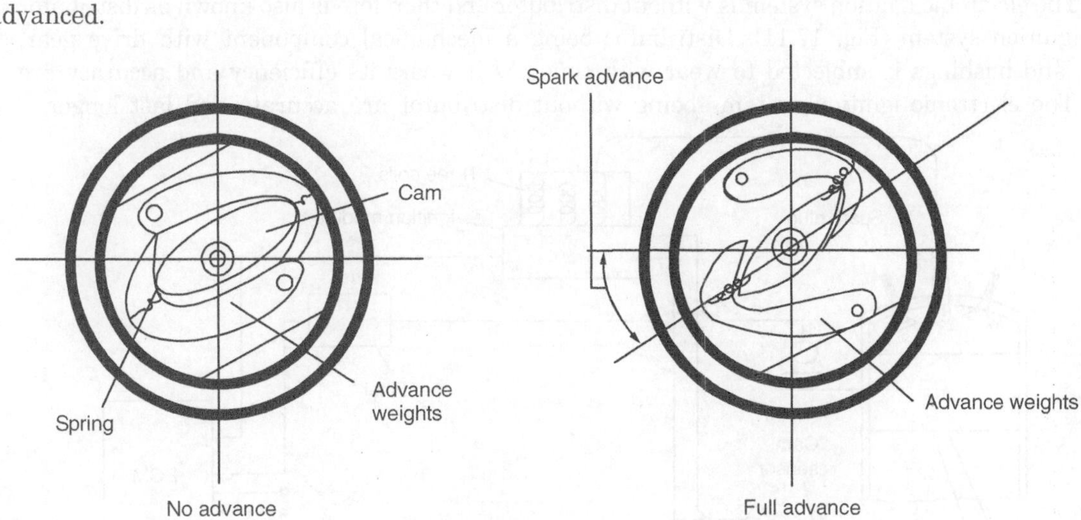

Fig. 17.9

17.4.2 Vacuum Advance Mechanism

Partial vacuum occurs in intake manifold when throttle is partially opened. This causes reduced supply of fuel-air mixture and fuel burns slowly. In this condition the spark is advanced so that mixture gets more time to burn. This mechanism advances sparked by shifting the breaker plates. There is a diaphragm linked to the breaker plate. The diaphragm is connected to port through vaccum passage. When throttle valve moves beyond port, the vacuum in intake manifold pulls the diaphragm and rotates the breaker plate causing the opening and closing of point in advance (Fig. 17.10).

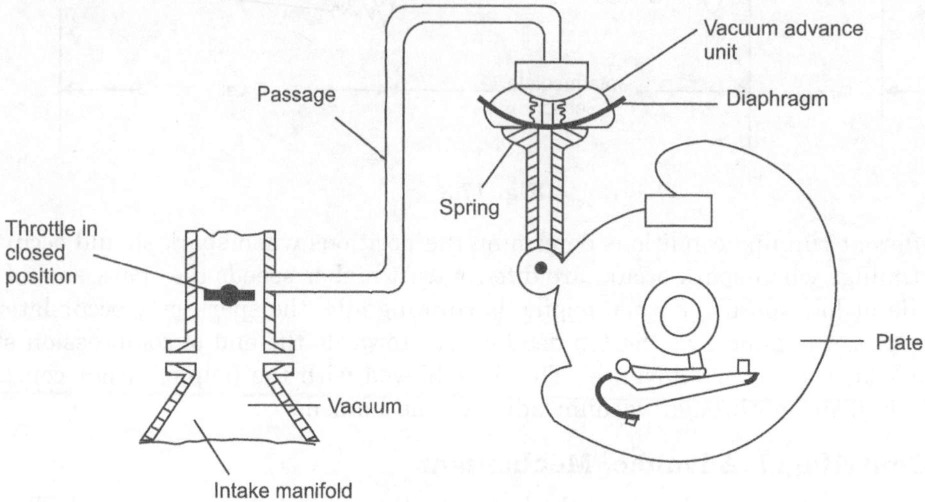

Fig. 17.10

17.5 ELECTRONIC IGNITION SYSTEM

The electronic ignition system is without distributor and therefore is also known as distributor less ignition system (Fig. 17.11). Distributor being a mechanical component with drive gear, shaft and bushings is subjected to wear with time. As it wears its efficiency and accuracy are lost. The electronic ignition systems being without distributor are accurate and last longer.

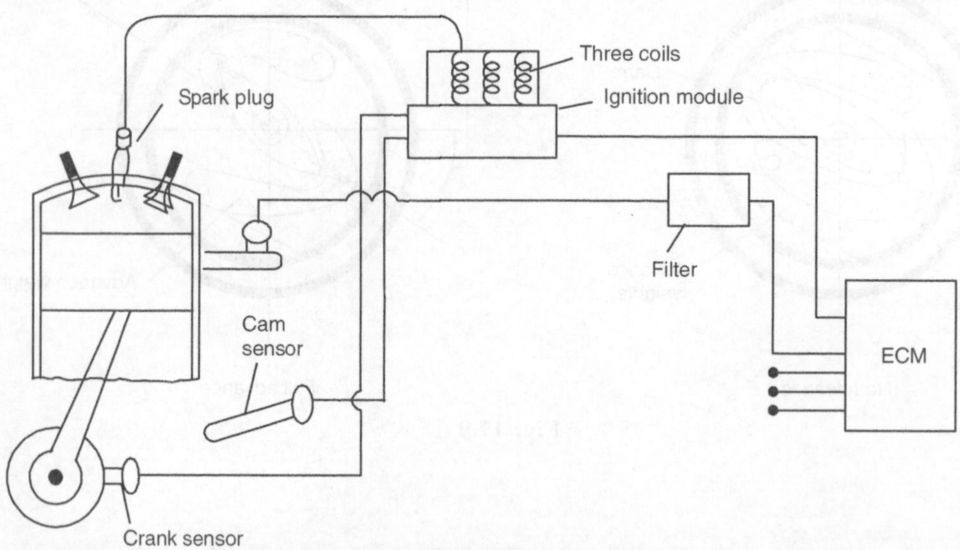

Fig. 17.11

17.5.1 The Ignition Control Module

The system has ignition control module connected to main computer control system. The module controls the firing order, spark timing and advance. The ignition system is activated by crank sensor that monitors the position of crankshaft. These systems have fewer moving components which mean less wear. This also reduces the maintenance. There is no problem of radio frequency interference. The timing is adjusted by electronic module. Transistors act to control the primary current. Each ignition coil is provided with transistor that acts as switch. Switching on means flow of current in primary circuit. When the current is interrupted secondary voltage is induced and spark occurs. There is a triggering device that controls the sequence and timing of ignition. The module also limits the dwell time and spark plug timings during starting of engine.

17.5.2 Coil Pack

There is more than one coil in electronic ignition system. Generally one coil is provided for each spark plug. The coil assembly is termed as coil pack (Fig. 17.12). The coil is saturated with high voltage as there is time gap between firings. This is helpful in providing high voltage to system when needed. But in case it is not needed this would produce heat that would be harmful. The ignition control module can be programmed to allow saturation only when needed by the system.

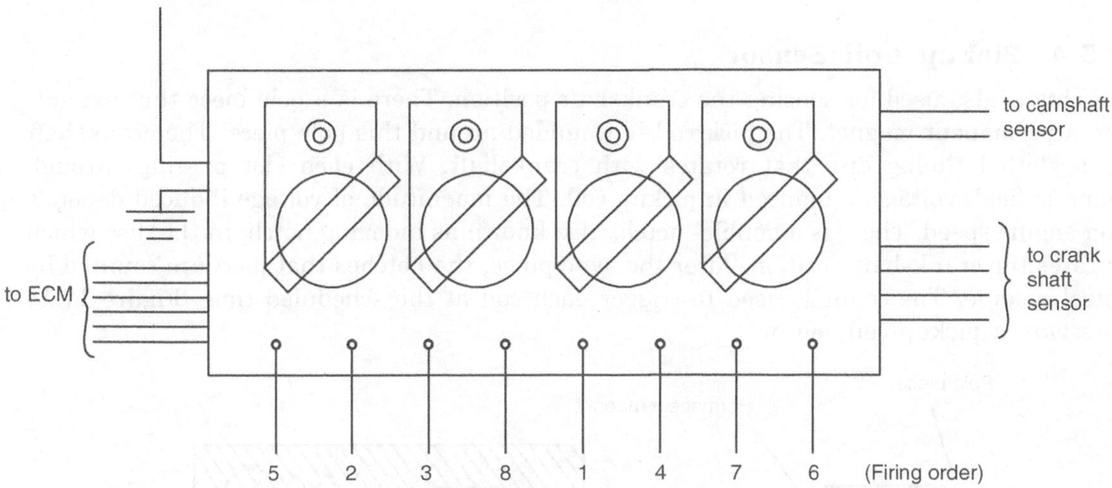

Fig. 17.12

It is possible to use one coil for two spark plugs. Each end of secondary coil is connected with two plugs in two cylinders whose pistons are moving together. The voltage is sent to both the plugs simultaneously. In one cylinder if there is compression stroke ending then in the other exhaust stroke is ending. Spark is wasted in cylinder where exhaust stroke is ending making the system not very suitable.

17.5.3 Hall-Effect Sensor

In some ignition systems, coil is replaced by hall-effect switch. This switch operates the supply voltage 'on' or 'off with 'presence' or 'absence' of magnetic field. The switch also senses the crankshaft position. Figure 17.13 represents sensor used for a six cylinder engine. The crankshaft speed is determined by control module. There are three vanes in the harmonic balancer of crankshaft. When these vanes pass through magnet and transducer, the magnetic field is interrupted. This turns the transistor 'on' or 'off' and culling the signal voltage. Module counts the voltage pulses and consequently the crankshaft speed.

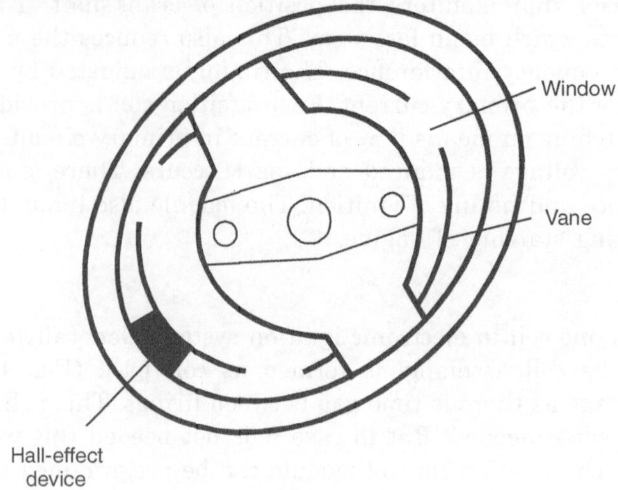

Fig. 17.13

17.5.4 Pickup Coil Sensor

This is also used for sensing the crankshaft position. There is a pole piece that extends from a permanent magnet. The pick coil is wounded around this pole piece. The crankshaft has a slotted timing disc that rotates with crankshaft. With each slot passing through magnetic field, voltage is induced in pickup coil. The magnitude of voltage induced depends upon engine speed. There is a double- notch, also known as indexing notch, in the disc which indicates the crankshaft position. After the sync pulse, the notches that pass are counted by control module. The count is used to trigger each coil at the scheduled time. Figure 17.14 represents a pickup coil sensor.

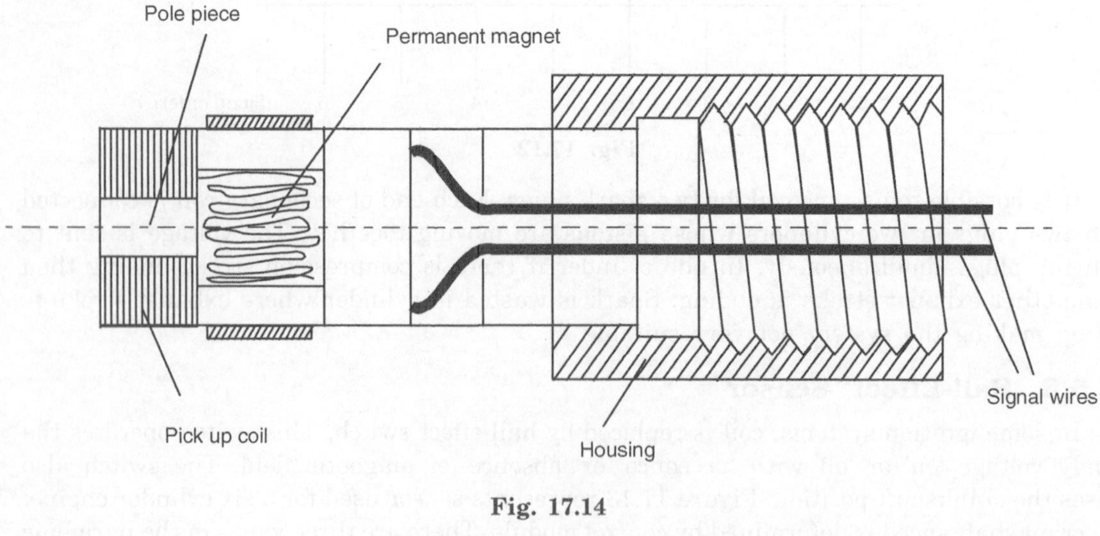

Fig. 17.14

17.5.5 Camshaft Sensor

This determines the camshaft position. The voltage pulse identifies the position of piston in cylinder number 1. The control module, at the start of each cycle, recognizes the signal from this sensor and fires spark plug in all the cylinders as per firing order and at the scheduled time (Fig. 17.15).

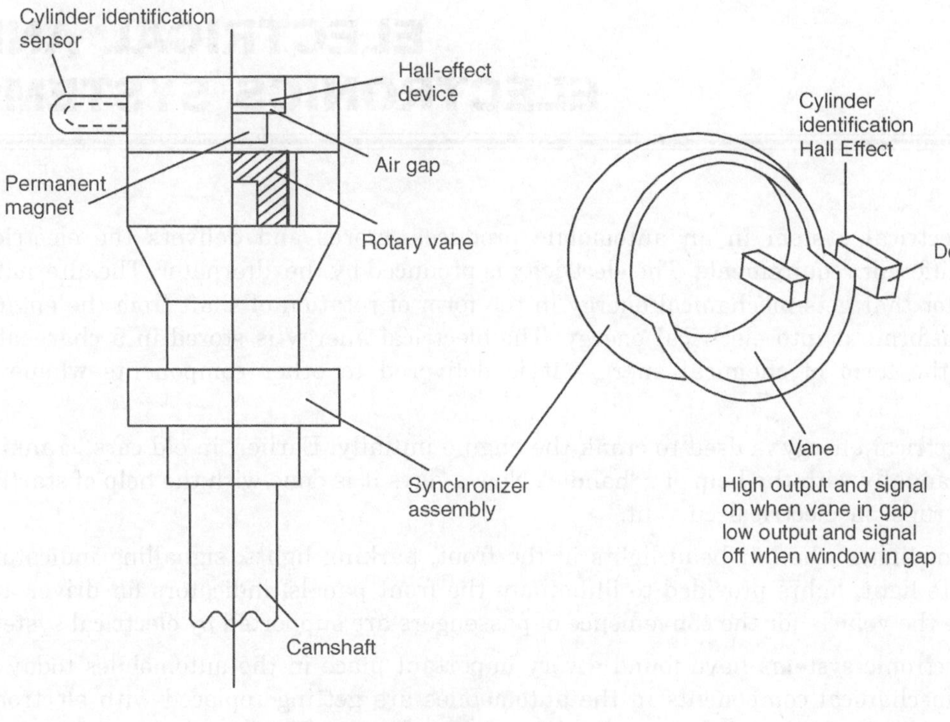

Fig. 17.15

The electronic ignition system is better in so many respects when compared to basic system with distributor. They are more reliable as they do not have components that wear out. They have better accuracy also. They are easy to maintain. Due to these reasons more and more manufacturers are adapting these systems.

QUESTIONS

1. What are the functions of ignition system?
2. Explain briefly different types of ignition systems.
3. Describe the ignition switch with the help of a diagram.
4. Explain the constructional details and working of spark plug.
5. What is centrifugal advance mechanism? Explain.
6. Explain Half-effect sensor.
7. Explain the benefits of electronic ignition system over ignition system with distributor.

18

ELECTRICAL AND ELECTRONICS SYSTEMS

The electrical system in an automobile produces, stores and delivers the electrical energy for different requirements. The electricity is produced by the alternator. The alternator is a generator that gets mechanical energy in the form of rotation of shaft from the engine. It then transforms it into electrical energy. The electrical energy is stored in a chargeable battery in the form of chemical energy. It is delivered to other components whenever required.

The electrical energy is used to crank the engine initially. Earlier, in old cars, cranking was done manually with the help of a 'handle'. Now-a-days it is done with the help of starting motor that runs on electrical current.

Different lights whether head lights in the front, parking lights, signalling indicators, number plate light, lights provided to illuminate the front panels, indicators for driver and lights inside the vehicle for the convenience of passengers are supported by electrical system.

The electronic systems have found a very important place in the automobiles today. A number of mechanical components in the automobiles are getting replaced with electronic components for they are accurate, reliable and last longer. There are other electronic components also that are being used in an automobile for its efficient performance. These components are provided instructions from Electronic Control Module (ECM). The fact is that different parameters of engine are monitored by engine control system and are available for the service personnel. Engine controlled system, automatic transmission, brakes, steering system and air-conditioning in modern vehicles is also governed by Electronic Control Module (ECM).

Other components that require electric current include signalling system, horn, instrument panel lights, indicators, audio and video system. Now-a-days seats windows, door locks, wipers etc., are also operated through electric current. These components are connected through wires. This makes essential use of wires, electrical circuits and other components.

18.1 WIRING CIRCUITS

Alternator acts as source of electrical energy. Battery acts its storage system. The wires or conductors are used to connect different components and battery and alternator. The connections are made through terminals at the end of wire. These terminals (Fig. 18.1) act as temporary joints and can be removed and connected again as the need may be.

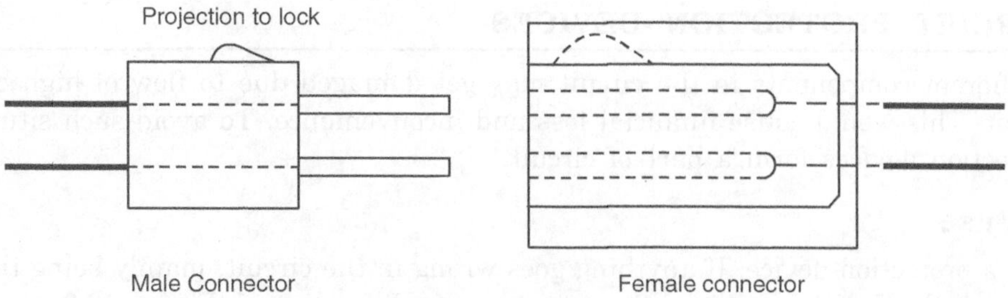

Fig. 18.1

If the current or voltage in the circuit exceeds the permissible limit there is a possibility that due to excessive heat the wire, joints or even components may burn and get damaged. To avoid this situation it is essential that safety devices such as fuse or circuit breakers are used in the circuit. The fuse gets burnt and disconnects the components whereas a circuit breaker opens and the flow of current stops. Switches allow and control the flow of current in a particular circuit when desired. An electric circuit has load or resistor. This is the component that uses the current supplied in the circuit. Resistor offers resistance to flow of current whereas load can be a motor or light or any other part that consumes electricity. There is needed a return path so that current flows back to source and circuit is completed.

The size or diameter of the wire depends upon the amount of current that flows through it. If due care is not taken and thin wire is used it may burn and damage the circuit and its components. In an automobile there are a number of electrical components and these are located in every nook and corner. This makes the length of the wires very large. Also the number of wires is very large. The wires meant for a particular location, say rear side of vehicle, are grouped together. These are wrapped with help of flexible conduit or even insulating tape. This process grouping the wires is wire harnessing. Different colour codes are used that make recognition of wire easy while servicing. Manuals are provided with circuit diagrams so that there is no difficulty while servicing. Figure 18.2 represents a part of harnessing.

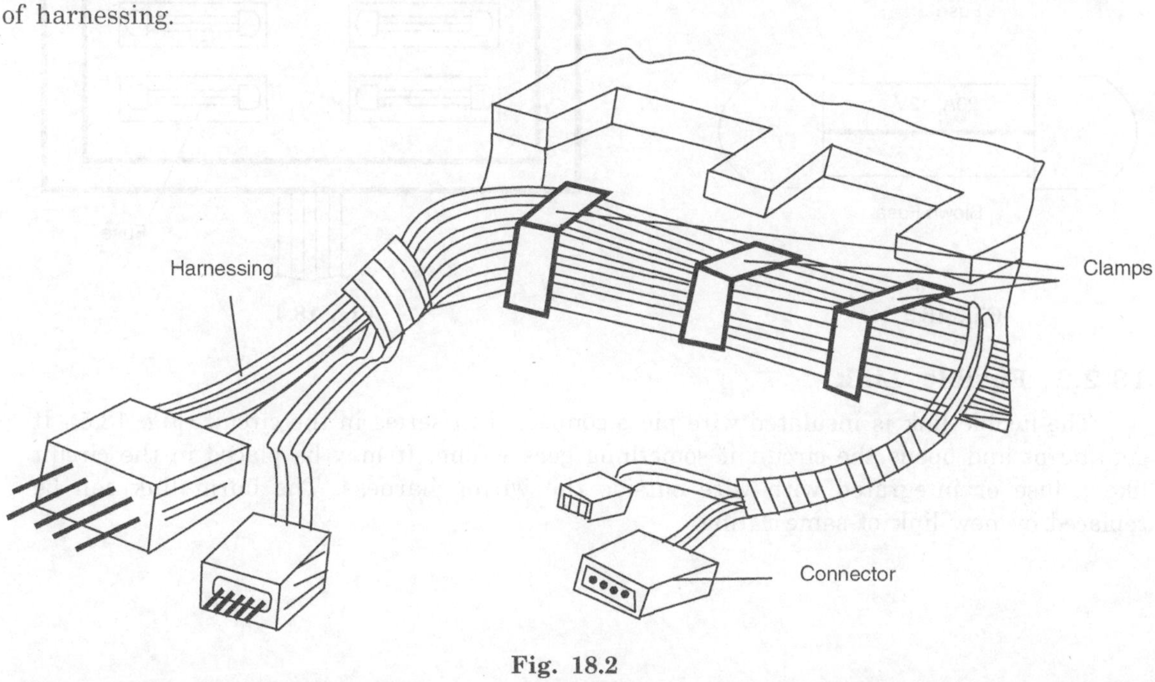

Fig. 18.2

18.2 CIRCUIT PROTECTION DEVICES

The different components in the circuit may get damaged due to flow of high current in the circuit. This would cause financial loss and inconvenience. To avoid such situations, circuit protection devices form a part of circuit.

18.2.1 Fuse

This is a protection device. If anything goes wrong in the circuit, mainly being the flow of high current, the device blows and flow of current is discontinued. Figure 18.3 represents 'cartridge fuse'. It contains a metallic wire connected to the ends of the fuse. There is a fuse panel, (as shown in Fig. 18.4) which accommodates all the fuses needed in the circuits. Placing them on a single panel makes convenient their handling. The location of panel is below instrument panel. The fuses are provided with 'rating'. Rating is the maximum current in amperes that the fuse can handle. It is very essential that while replacing, the fuse of same rating is used as replacement. If fuse of lower rating is used it will blow as soon as current flows through it. On the other side, if fuse of higher rating is used, it would not blow even if high current is flowing through the circuit and ultimately other components of the circuit would be damaged. This would forfeit the very purpose of fuse.

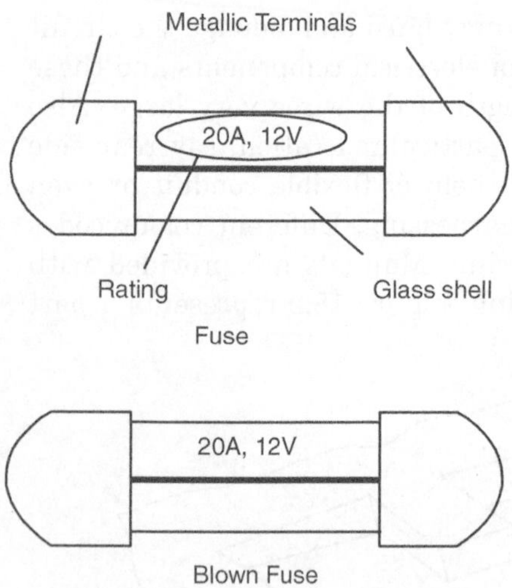

Fig. 18.3

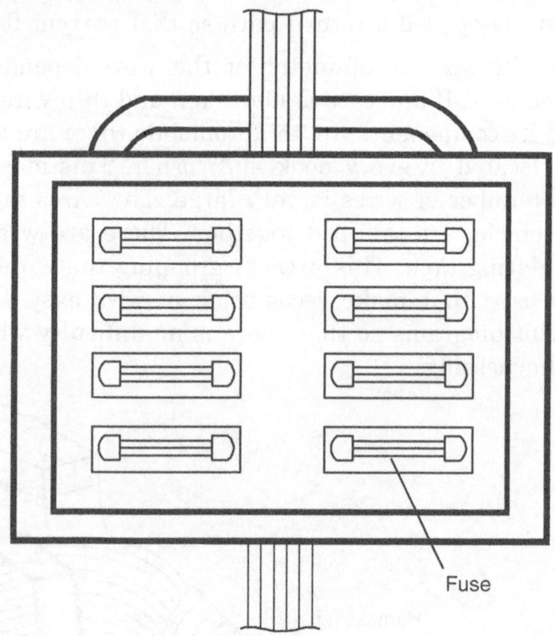

Fig. 18.4

18.2.2 Fusible Link

The fusible link is insulated wire piece connected in series in the circuit (Fig. 18.5). It gets burnt and opens the circuit if something goes wrong. It may be placed in the circuit like a fuse or integrated with wire outside the wiring harness. The burnt link can be replaced by new link of same rating.

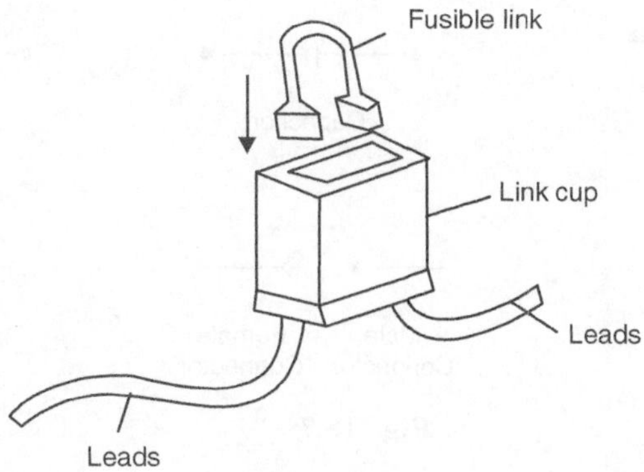

Fig. 18.5

18.2.3 Circuit Breakers

These also act in the same way as fuse or fusible link. Generally these are employed, for protection, where large amount of current flows such as headlights or wiper circuits. This is a strip of two metals joined together. The strip is placed across the terminals and circuit is closed (Fig. 18.6). When excessive current flows the strip tends to straighten and the contact is lost. This prevents the flow of current. When the flow of current becomes proper and strip cools down it again bends. This again closes the circuit. Thus corrective action is taken on its own though warning may be given to driver through flashing indicator.

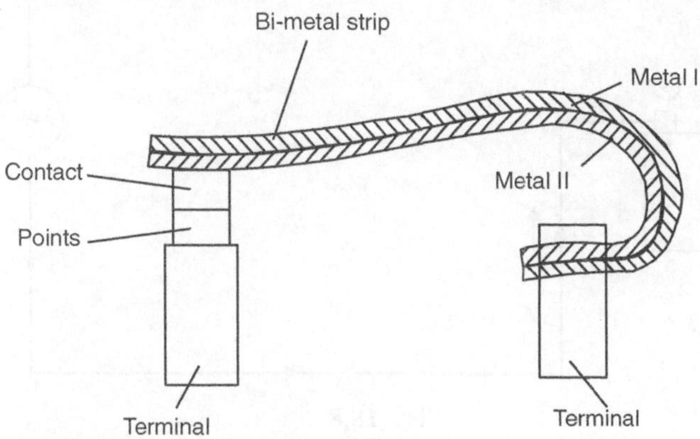

Fig. 18.6

18.2.4 Circuit Diagram

The circuits in an automobile are provided in the form of wiring diagram on the paper in operating manuals. This is done so that maintenance and repairing of circuit is possible. These diagrams provide information about the location of different components in the circuit and also the amount current and voltage across a particular component. This makes the repair of the circuit easy.

18.2.5 Electrical Symbols

Wiring diagrams use various symbols for different components. These symbols are standard symbols that are used throughout the world. This is essential otherwise maintenance engineers won't be able to read the wiring diagrams. Few symbols are shown in Fig. 18.7.

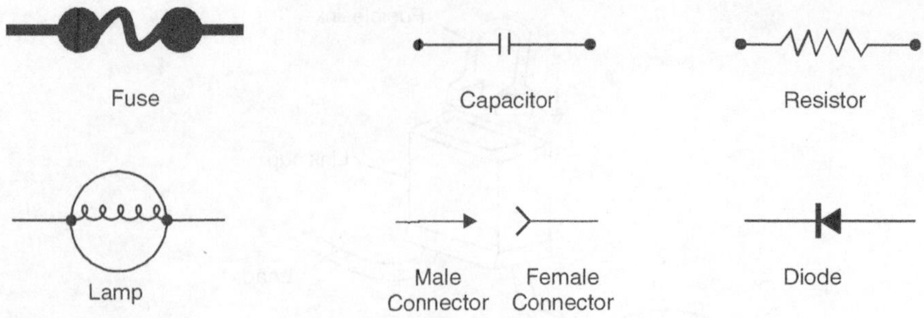

Fig. 18.7

18.3 SERIES CIRCUITS

In this type of circuit the components are connected end to end forming a single path. This is termed as components placed in 'series' and therefore circuit is series circuit. In this type of circuit every device needs a separate control which is disadvantageous (Fig. 18.8).

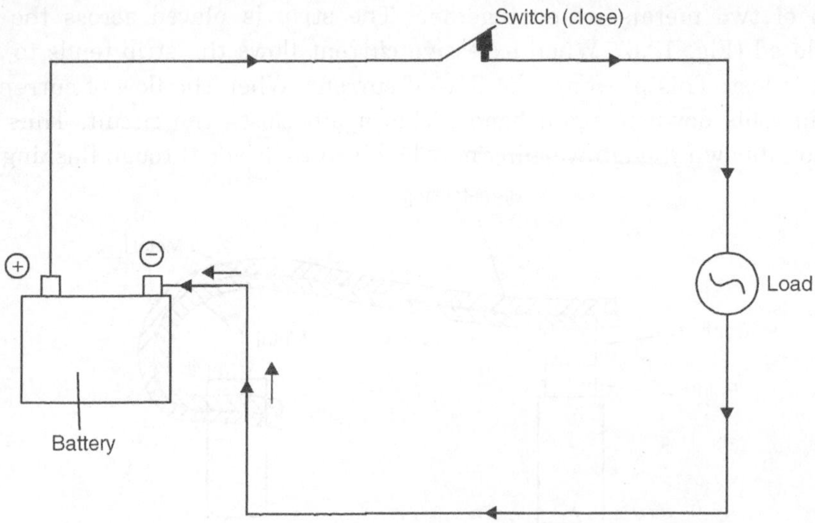

Fig. 18.8

18.4 PARALLEL CIRCUITS

In this type of circuit, the components are connected in separate branches. For example, two loads are connected in two branches as shown in the Fig. 18.9. They are connected to the same power source and controlled by a single switch. There can be more than two branches also. The current is divided in each branch but voltage across a particular component remains the same. One of the examples of parallel circuits is headlights. There are two headlights that have parallel circuit and controlled by single switch. Another common example of parallel circuit is tail lights, two in number controlled by single switch.

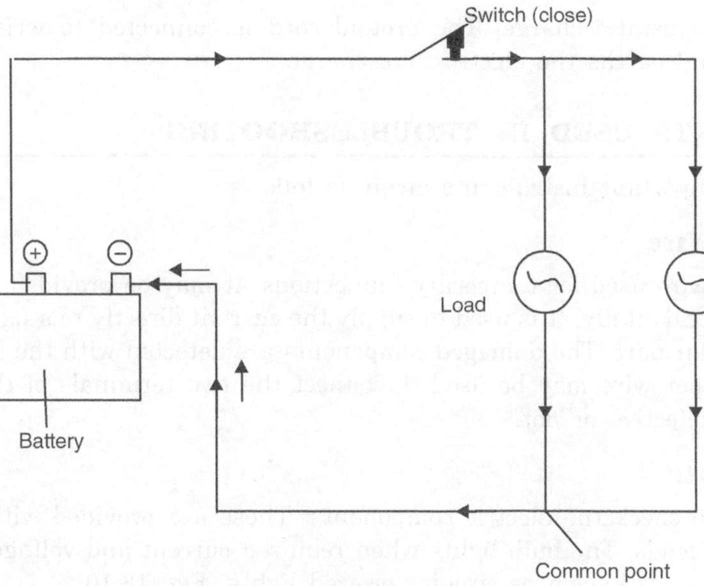

Fig. 18.9

18.5 ELECTRICAL TROUBLE DIAGNOSIS

To diagnose the trouble in an electric circuit, it is essential to know the symbols and read the circuit diagram. Once circuit diagram is understood the parts are located in the actual circuit. Current and voltage is measured across different components. If designed voltage/current does not exist, the part should be checked and replaced, if necessary. The manuals also give details of connectors and how to make connections.

18.5.1 Electric Circuit Problems

The circuit may have problem if it is open or it is short circuited or it is earthed. An open circuit may be due to supply problem and blown up fuse. In this case, the supply should be checked and corrected then fuse be replaced. The circuit may also open when some wire is broken or burnt. This broken/burnt wire may be replaced after checking and correcting the reason for break/burning of wire.

The circuit may have problem if it is short circuited. Short circuit means that the wires are in contact with each other. This may be due to removal of insulating layer from the wire. Otherwise also the joining between two terminals may occur due to heat. Fuse works in this case as safety device and disconnects the supply. The fuse should be replaced only after the short circuiting is removed otherwise fuse will blow again.

Circuit may be earthed if the wire comes in contact with earthed part of the automobile. This would discharge the battery. This may occur due to wearing of insulation with time.

Apart from the problems mentioned above, handling of electronic components requires special care. These components are sensitive to even a very small amount of current. The tools such as screw driver get charged when being used. This is electrostatic charge. Even the hand of a mechanic may become charged. This charge or current passes through electronic components and can damage them. Therefore, in such situation, the tool or hands must be discharged. For this, the pins or terminals of the components must not be touched. The use of elastic wrist band, flexible rubber mat and ground cord helps to protect the electronic components from electrostatic charge. The ground cord is connected to wrist band and flexible rubber mat and earths the electrostatic charge.

18.6 INSTRUMENTS USED IN TROUBLESHOOTING

Here are some important instruments given as follows:

18.6.1 Jumper Wire

This is a simple wire used for temporary connections. It may be provided with fuse so that it not damaged accidentally. It is used to supply the current directly to a particular part or to bypass a particular part. The damaged components are detected with the jumper wire. For example, the jumper wire may be used to connect the two terminals of the switch to detect whether it is defective or not.

18.6.2 Test Lights

These are used to check the electric components. These are provided with a suitable bulb between the test leads. The bulb lights when required current and voltage is present. This type of test lights are known as circuit powered lights (Fig. 18.10).

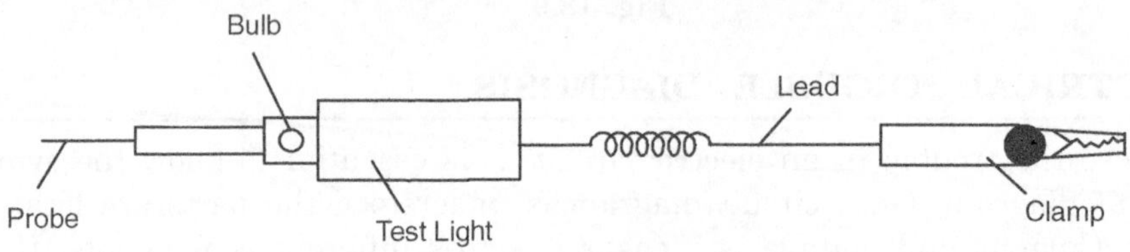

Fig. 18.10

Another variant of test lights is self-powered lights. It has a battery that provides the required voltage and current to light the bulb in the circuit. If the bulb lights it indicates the wire is not broken. It is also known as continuity tester.

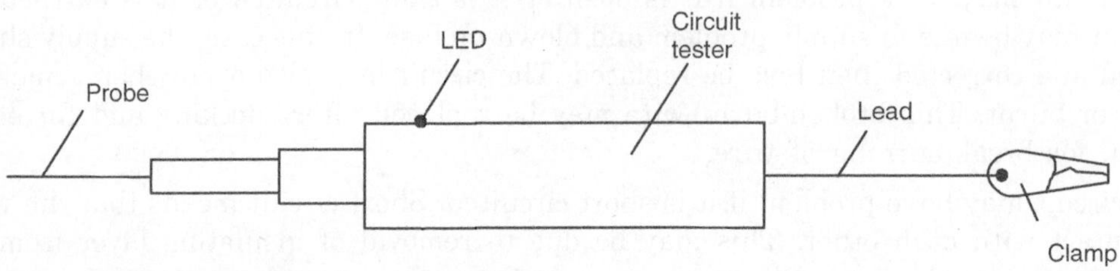

Fig. 18.11

18.6.3 Electric Meters

Various parameters such as current, voltage and resistance are required to be measured to verify whether these are existing in the circuit in the correct magnitude. These are measured with the help of electric meters. To measure current, there is ampere meter, to measure voltage there is voltmeter and to measure resistance there is ohm meter. Then these parameters may be in a.c. circuits or d.c. circuit. Also there magnitude may be in different ranges. To make the measurement convenient, a single instrument, known as multi-meter, is used. With this instrument, current, voltage, resistance can be measured in a.c. or d.c. circuit. Also it is provided with different ranges in which measurements can be made. By selecting a proper range accurate value can be read.

Initially these instruments were analogue (Fig. 18.12). But now-a-days digital instruments are available that are more accurate and reliable (Fig. 18.13). The two probes are provided with the instrument that can be used to connect the terminals.

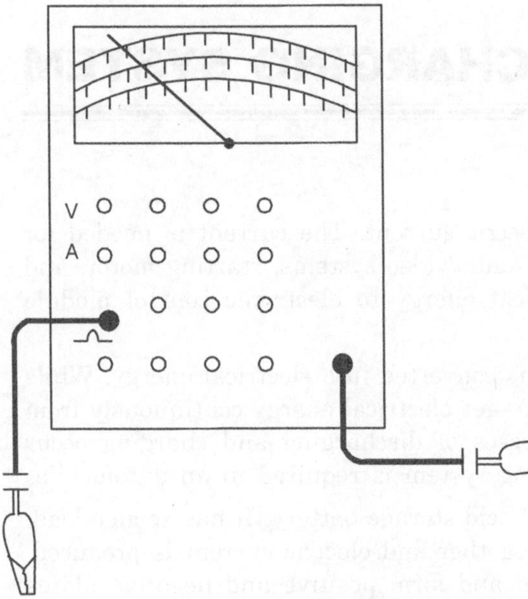

Fig. 18.12

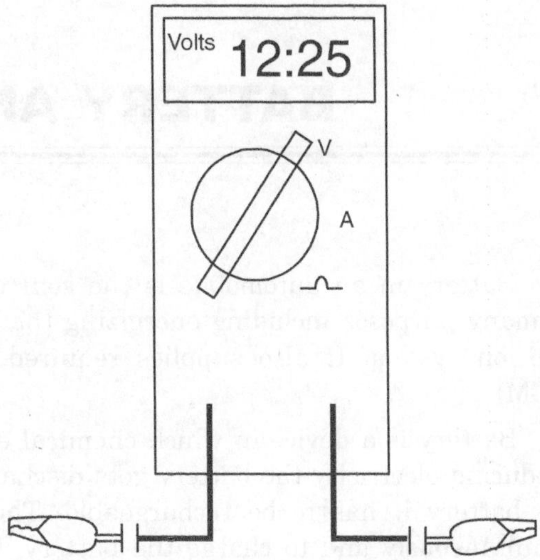

Fig. 18.13

QUESTIONS

1. Explain the importance of electrical systems in an automobile.
2. Draw a neat diagram of terminal and explain its necessity.
3. What is a circuit protection device?
4. Explain the working of fusible link with the help of suitable diagram.
5. Explain the importance of circuit diagram.
6. What is the difference between series and parallel circuits?
7. How the trouble is in the electric circuit is diagnosed?
8. What are the instruments used in trouble shooting of electric circuits? Explain the use of test lights and electric meters.

19

BATTERY AND CHARGING SYSTEM

Battery in an automobile is the source of electric current. The current is needed for so many purposes including energizing the lights, audio/video systems, starting motor, and ignition system. It also supplies required electrical energy to electronic control module (ECM).

Battery is a device in which chemical energy is converted into electrical energy. While producing electricity the battery gets discharged. To get electrical energy continuously from the battery it has to be rechargeable. The processes of discharging and charging occur simultaneously and to charge the battery. Charging system is required in an automobile.

The battery that is used in automobiles is lead-acid storage battery. It has sponge lead, lead oxide and sulphuric acid. These three react together and electric current is produced. Sponge lead and lead oxide are held in plate grid and form positive and negative plates (Fig. 19.1). Sulphuric acid is mixed with water to form electrolyte. The proportions in which water and sulphuric acid is mixed is 60:40. With discharge, the sulphuric acid is spent and water remains.

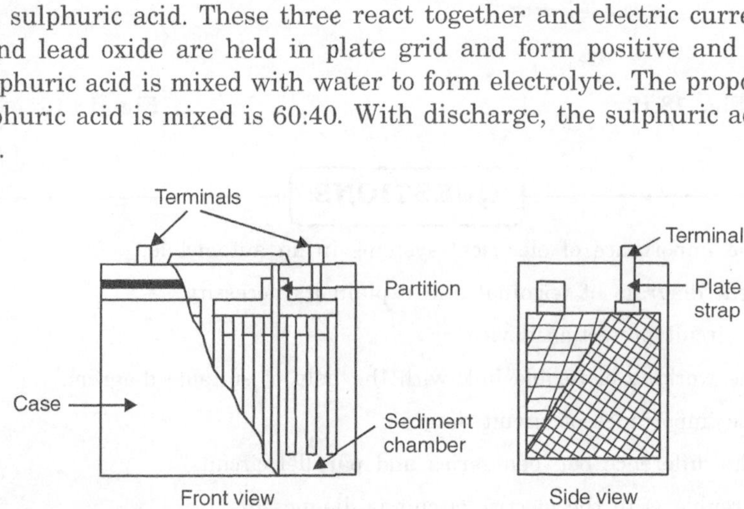

Fig. 19.1

19.1 CONSTRUCTIONAL DETAILS

The battery consists of a number of plates that are grouped together through strap. The groups are positive or negative and these are located adjacent to each other. These are required to be insulated from each other otherwise there would be short circuiting that would damage the battery. The insulators are termed as 'separators'. Each pair of positive and negative plates along with separator is termed as 'cell'. Combination of these cells forms the battery. The cells are arranged properly in plastic shell that forms the outer box of the battery. The box has a plastic cover at the top. The cover has openings for filter plugs. These are used for adding water to the battery from time to time.

Now-a-days the batteries do not require addition of water so frequently. These are termed as low-maintenance batteries. In another improved version of batteries, water is not at all required to be added up. There are no removable caps because of this. These batteries are maintenance free batteries.

Another variant of battery is hybrid battery. This battery can withstand six cycles and can maintain its full reserve capacity. Its grid consists of 2.75% of antimony alloy on the positive plate and calcium alloy on the negative plate. This prevents the grid corrosion. The grid plates have lug near the centre and vertical and horizontal grid bars are arranged in a radial design (Fig. 19.2). Due to shorter path up to lug the resistance is reduced.

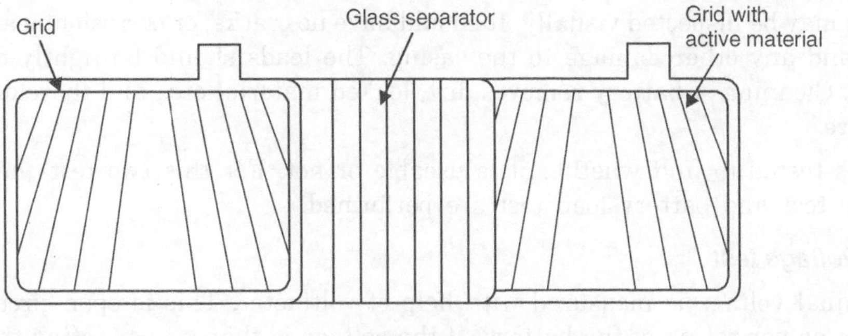

Fig. 19.2

The terminals are also provided at the top in most of the batteries. For easy identification the positive terminal is larger than negative terminal. Some batteries are also provided with charge indicator that conveys the status of the battery. Figure 19.3 shows the top of a battery.

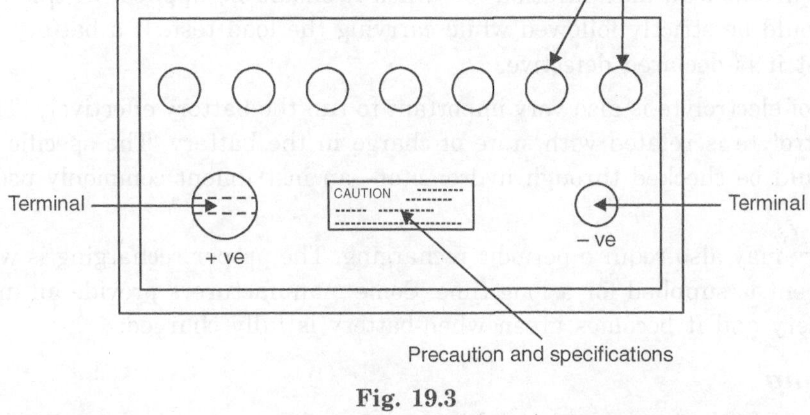

Fig. 19.3

Generally, a battery consists of six cells. These are connected in series. In series the voltage of each cell is added together. This means for a 12 volt battery each cell should have an electromotive force (EMF) of 2 volts.

Other parameters which indicate the status of a battery is its reserve capacity and cold-cranking rate. Reserve capacity of a battery is the duration for which, a fully charged battery, would deliver the rated current. Rated current is dependent upon battery rating. Battery rating is the amount of current a battery can deliver for 20 hours while voltage drop across each cell drops by less than 1.75 volts.

Cold-cranking rate of battery is its capability to crank engine when it is cold. It is the amount of current in Amperes that it delivers for 30 seconds at -18°C (0°F) and the voltage does not fall below 7.2 volts.

Maintenance of battery

The maintenance of battery is essential as it is a vital component in automobile. If something goes wrong with battery it may immobilize the automobile.

A battery may be inspected visually. It should have no cracks, or corrosion near terminals particularly, and any other damage to the casing. The leads should be tightly clamped on the terminals. Cleaning of battery removes dirt, leaked material etc., and therefore is a part of maintenance.

Battery is tested to find whether it is useable or not. For this two tests namely open circuit voltage test and battery load test are performed.

Open circuit voltage test

The terminal voltage is measured with help of voltmeter. This is open circuit voltage. This should be as per rating of the battery. If the voltage is than as per rating then battery should be charged.

Battery load test

In this test, terminal voltage is measured when battery is getting discharged at high rate. The load is applied through a variable resistor. Voltmeter and Ampere meter is also included in the circuit. Cold cranking amperes are mentioned by the manufacturer and accordingly load is placed in the circuit. Often, manufacturers also provide information regarding the current load and duration for which it should be applied. In this case, these parameters should be strictly followed while carrying the load test. If a battery fails twice in the load test it is declared defective.

The state of electrolyte is also very important to run the battery effectively. The specific gravity of electrolyte is related with state of charge in the battery. The specific gravity of electrolyte should be checked through hydrometer—an instrument commonly used for this purpose.

The battery may also require periodic recharging. The proper recharging is when small amount of current is supplied for a long time. Some manufacturers provide an indicator at the top of battery and it becomes green when-battery is fully charged.

Jump starting

When battery in the car is discharged it is not possible to start it through self. Pushing the car provides the required cranking necessary to start the engine. It is improper to push the cars with multi-point fuel injection system. In this case the working battery located in other vehicle can be used (Fig. 19.4). This can be done by following the proper method.

1. The two cars should not be in contact with each other otherwise large amount of current will flow through the wires and this would damage them.
2. Parking brakes are applied to make cars immobile.
3. Ignition switch and all accessories are turned off in both cars.
4. Connect positive terminals of both batteries (discharged and charged) with positive jumper cable.
5. Connect one end of negative jumper cable to negative terminal of charged battery and the other end to the engine ground of the car with discharged battery.
6. Start the car with discharged battery. If it does not start, start the other car and run it at fast idle. This is done to prevent the drawing of excessive current from battery.
7. Try to start the car with discharged battery. When the car starts remove the negative jumper cable end from the ground of the engine first and then from the other side.
8. Positive jumper cable should be removed from charged battery first and then from discharged battery.

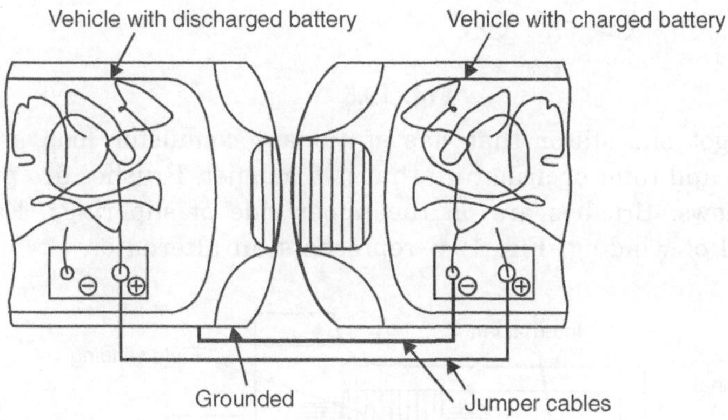

Fig. 19.4

19.2 CHARGING SYSTEM

The battery is consumed when it cranks the engine. It also undertakes the load of lights, audio/video systems and other electrical and electronics equipments. During these operations the battery gets discharged. It is essential to recharge the battery so that it can perform properly. The charging system does the job of charging the battery. The system consists of alternator with regulator and connecting wires.

19.2.1 Alternator

An alternator is a generator producing alternate current (AC). It gets mechanical energy, as input, in the form of rotation of shaft from the internal combustion engine in an automobile. This energy is converted into electrical energy. The regulator, an essential part, controls the voltage and does not allow it to go beyond permissible limit.

When a bar magnet is moved through a conductor the flow of electrons occurs in the conductor. It is shown in figure 19.5. This shows a rotating bar magnet. The magnetic field, thus produced, passes through the two sides of stationary wire loop. When South Pole moves past the upper side of loop current is induced in one direction. At the same time, as the North Pole moves past the lower side of the loop, current is induced in the opposite direction. The current is induced but in opposite directions when similar thing happens and North Pole moves past the upper side of loop and South Pole moves past the lower side of loop. Thus, alternate current (AC) flows in the loop. The current can be increased by enhancing the strength of magnetic field, speed of rotation of magnetic field and by increasing the number of loops.

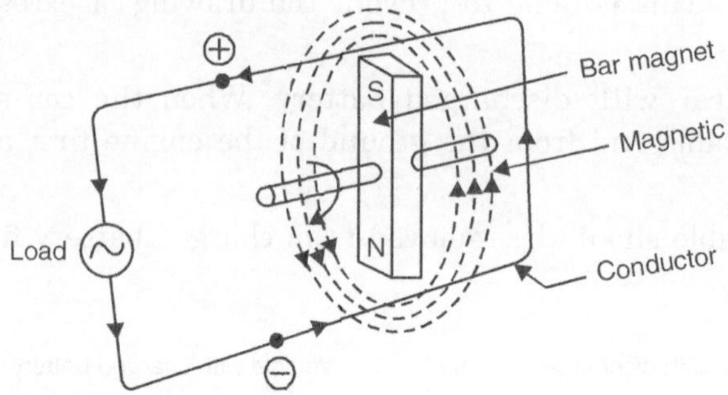

Fig. 19.5

Alternator has got one stator that has stationary conductor loop assembled into a laminated iron frame and rotor or armature that is a magnet. Brushes are provided through which the current flows. Brushes are on the upper side of slip rings. Each slip ring is connected to one end of winding. Fig. 19.6 represents an alternator.

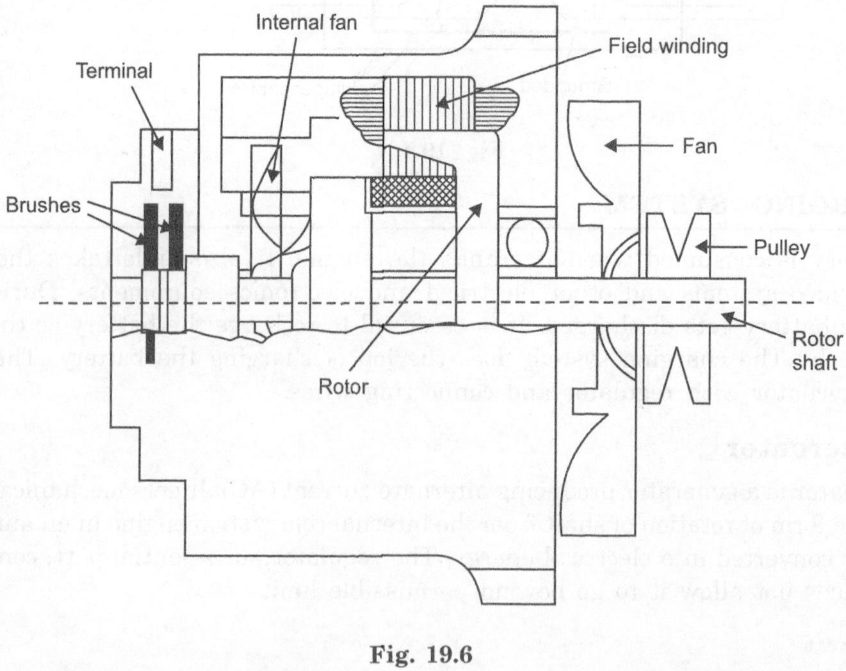

Fig. 19.6

The alternate current (AC) is produced by alternator whereas most of the equipment in an automobile require direct current (DC). This is done by providing the rectifiers that convert the alternate current (AC) in direct current (DC). Figure 19.7 shows the rectifier. This is a six diode rectifier. The diode allows current flow only in one direction. The loops in the stator are divided into three groups and these form a delta-connected stator or Y-connector stator. The six diodes are connected to three legs as shown.

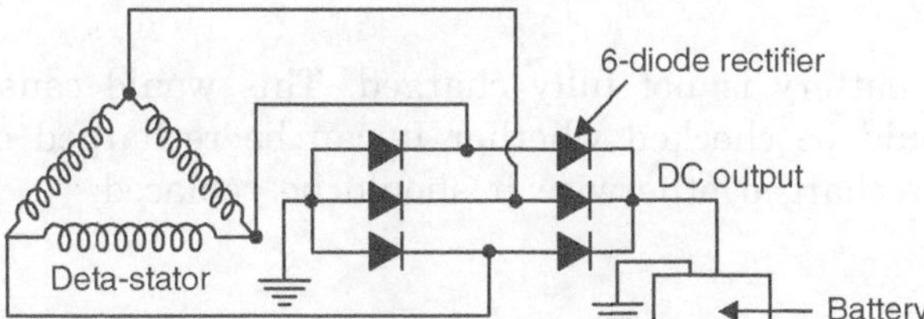

Fig. 19.7

Alternator regulator

It is a voltage regulator that keeps the voltage under the permissible limit. It can be an electromechanical type regulator using relays or it can be an electronic regulator built into the alternator. In modern cars, the voltage is regulated by electronic control module. An alternator regulator is shown in Fig. 19.8.

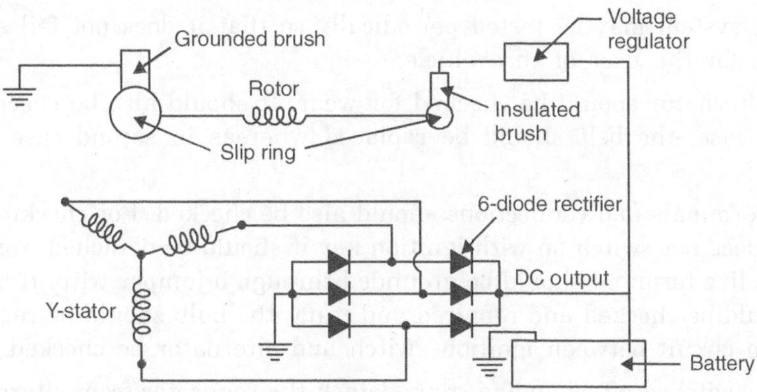

Fig. 19.8

An alternator while producing electricity, generates some heat also as whole of the input cannot be converted into useful electric current and a part of it is wasted in term of heat. This loss is mainly due to friction between different moving components. The heat so generated would accumulate in the alternator and would ultimately damage its components. Therefore, it is essential to dissipate this heat. For this purpose, a fan is provided behind the pulley (Fig. 19.6). There are vents provided so that cool atmospheric air can enter. This helps to keep cool the rectifiers and regulator.

Charge indicator

This is provided on the instrument panel so that driver can monitor the status of charging system. Ordinarily, it is in the form of light indicator. The light is switched on initially when ignition is turned on. As the engine runs and battery is charged, the light goes off. If there is no charging, the light is again on and an alarm for the driver to take corrective measures.

Troubles in charging system

Following are the major troubles that can occur in a charging system:

Undercharged battery

This indicates that battery is not fully charged. This would cause slow cranking of engine. The battery should be checked whether it can be recharged or not. If it can be recharged, it should be recharged otherwise it should be replaced.

Overcharged battery

If battery is overcharged, electrolyte comes out of vents provided in the casing of battery. This lowers the level of electrolyte. Overcharging is undesirable as it damages electrical parts and even battery. High alternator voltage may cause overcharging hence regulator in the alternator should be checked and repaired.

These troubles are indicated by the charging indicator. The light indicator may not switch off after the engine has started running. In some automobiles, where electronic control module monitors all the components, check engine light switches on in case of trouble.

Charging system testing

The charging system may be tested periodically so that it does not fail suddenly and causes emergency for the user of the vehicle.

The belt of alternator should be checked for wear. It should also be checked that it is not lose. In first case, the belt should be replaced whereas in second case it should be tightened.

The wiring, terminals and connections should also be checked. For checking the charge indicator, if bulb does not switch on with ignition key it should be detached from alternator. The lead from the live terminal should be grounded through a jumper wire. If light switches on alternator should be checked and repaired and if not the bulb should be replaced. If still the bulb is not on circuit between ignition switch and alternator be checked.

If light is not switched off as engine runs, detach the connector from alternator. If light goes off repair the alternator. If light continues, wire from live terminal might have grounded.

Charging system output test

The maximum output from the alternator is measured at a specified voltage. To do this an ampere meter is connected at battery terminal of alternator. A voltmeter is connected between battery terminal of alternator and ground. This test can also be performed by computerized engine analyzer.

The engine is run with lights on at a speed of about 2300 rpm. Adjust the load control till the maximum current is indicated from the alternator. The voltage should not be allowed to drop below specified value indicated by manufacturer. If the output from charging system is within specified limit, it is considered to be working well. If not so it should be checked for stator winding, regulator and connections.

Full field test

When the output from charging system is not as specified, this test is performed. Here, full battery voltage is applied to alternator bypassing the regulator. The voltmeter reading should not surpass the specified voltage. If that happens the alternator needs repair.

Key off current drain

When ignition key is off there may be small current Flow from battery particularly in engines with electronic controls. This is key off current drain. In this situation if short circuiting occurs there may be enhanced drainage of current discharging the battery. The key off drain current is specified by the manufacturers and it should be seen that it does not go beyond this value. If so, corrective measures should be taken.

QUESTIONS

1. Explain the constructional details of battery with the help of diagram.
2. What are different tests carried on a battery?
3. Explain battery load test.
4. Explain the procedure of jump starting.
5. Explain the constructional details and working of alternator.
6. What are the effects of under-charged and over-charged battery?
7. Explain different charging system tests.

20

STARTING SYSTEM

In an automobile, starting system cranks the engine initially. It has replaced manual effort to crank the engine with the help of cranking rod that was used in ancient days. Initially, the engine requires cranking but once the cycle is completed it starts and runs on its own. In two wheelers, it is common to 'kick start' the engine but in recent times a number of manufacturers have introduced 'button start'. For initial cranking an electric motor is provided that gets electric current as input from battery. The mechanical energy, in the form of rotation of shaft, is transmitted to engine. This provides initial movement of crankshaft, connecting rod and piston. As soon as spark occurs the fuel is ignited and output becomes available from engine. No more cranking is needed and starting system stops working and engine runs on its own. The starting system makes starting of vehicle convenient.

A starting system consists of starting motor, magnetic switch, safety switch, battery, cables and ignition switch. These components are connected with each other through two circuits. One is starting circuit, in which high current flows which is used to start the engine. Second is control circuit, in which low current flows. The ignition switch acts as switch for starting circuits also. In starting circuit, the current flows from battery to starter motor through solenoid or magnetic switch. The control circuit connects magnetic switch with, battery through ignition switch (Fig. 20.1).

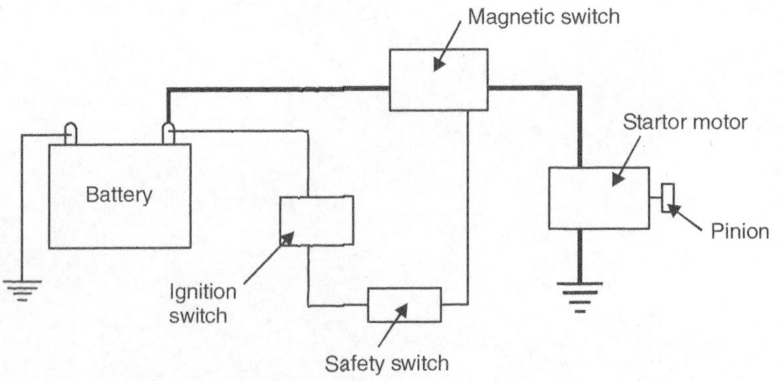

Fig. 20.1

20.1 STARTER MOTOR

The starter motor is like any other electrical motor but it is designed to work under high electrical overloads and produces very high power. Due to this, the motor can operate for short durations. High current is needed to operate it that generates heat. Time is also required to dissipate this heat. Therefore, it is advisable that the motor be given enough gaps between more than one starting attempts.

The motor has got field coils with pole shoes, armature and a housing that encloses them. Apart from these it has brushes, bushings that make its operation efficient.

The field coils and pole shoes produce strong stationary electromagnetic fields as current is passed through them. Magnetic polarity (N or S) depends upon the direction in which the current flows. The magnetic fields produced are opposite in nature.

The armature is located between drive and end frames. It has windings and the commutator mounted on the armature shaft. The windings are made of a number of coils of a single loop each. These are insulated from each other and fit into slots in the armature shaft. The commutator has heavy copper segments surrounding the shaft but are insulated from each other and the shaft. The armature is surrounded by field coils. Current is supplied to armature and it produces magnetic field in each conductor. The magnetic fields are also produced by field coils. The reaction between these magnetic fields causes the rotation of armature. The rotation is transferred to crankshaft of the engine through armature shaft. This causes cranking of engine.

The current from field coils to the armature is transferred through brushes. These brushes are held with the help of springs against the commutator. The brushes can be from two to six in number for smooth motion and constant torque delivery.

Figure 20.2 represents starting system with all its components.

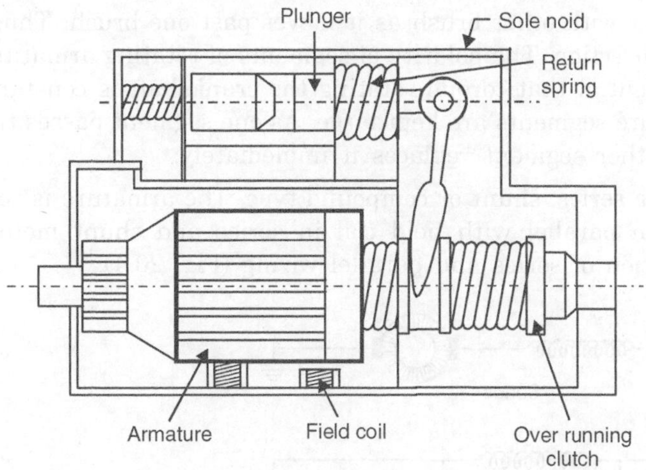

Fig. 20.2

The field coils produce stationary magnetic field. The armature windings are placed in this stationary magnetic field and current is passed through it. A secondary magnetic field is generated. The lines of forces of stationary magnetic field move across the winding. They combine on one side and enhance the strength of magnetic field. On the other side they are opposed and, therefore, weaken the magnetic field. There is unbalanced magnetic force that causes push towards weaker magnetic field. Armature windings are in the shape of coils. The current flows in and out in opposite directions. This makes orientation of magnetic forces in opposite direction in each segment of winding. When it is placed in stationary magnetic field one part of armature winding is pushed in one direction and the other part in opposite direction. This causes the rotation of armature winding. The winding coil being mounted on shaft causes the rotation of shaft (Fig. 20.3).

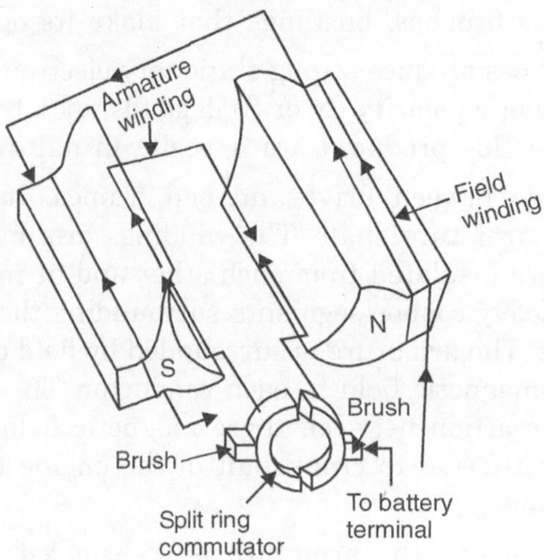

Fig. 20.3

As the armature rotates through half revolution, the flow of current is reversed due to contact between brushes and commutator. The commutator segment is attached to each coil and it comes in contact with other brush as it moves past one brush. Thus, flow of current is maintained in one direction. The polarity of segments of rotating armature coil is reversed as it rotates. It is essential that torque rotating the crankshaft is constant and to achieve this number of armature segments are kept large. As one segment passes through secondary magnetic field pole, other segment replaces it immediately.

The motors can be series, shunt or compound type. The armature is connected in series with field coils and in parallel with field coil in series and shunt motors. In compound motors, it is combination of series and parallel wiring (Fig. 20.4).

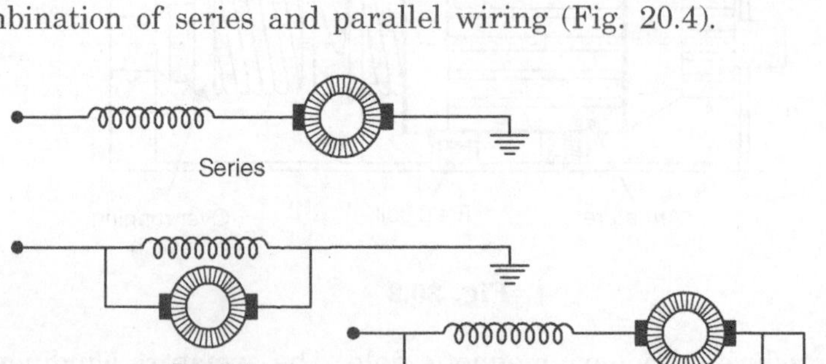

Fig. 20.4

The magnitude of torque from a motor depends upon the current it draws. The motor draws higher current when runs slow. As more torque is needed to crank the engine shaft starting motor requires higher current.

20.2 STARTER DRIVE

Starter drive transfers motion from starter motor shaft to the crankshaft of the engine. It has got a pinion that meshes with flywheel mounted on the crankshaft (Fig. 20.5). The flywheel is provided with teeth to mesh with pinion. The meshing of the pinion and flywheel occurs prior to starting of motor. This is done to avoid any damage to teeth either on pinion or on flywheel.

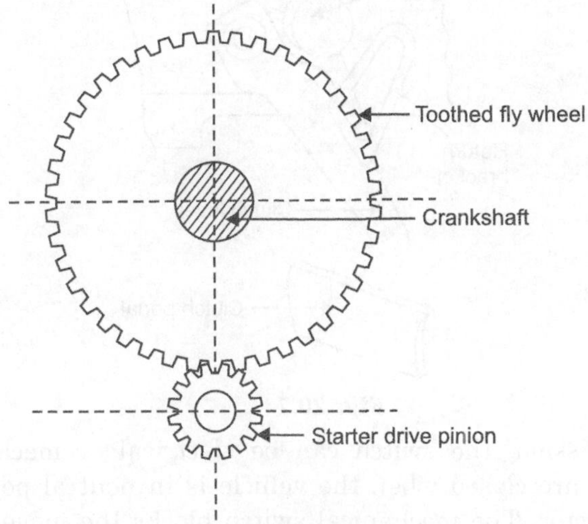

Fig. 20.5

An overrunning clutch is included to protect starter motor. After the engine has started, and crankshaft starts rotating at speed higher than that of starting motor, the armature is disengaged from the flywheel with the help of overrunning clutch. If not disengaged, the armature would rotate with very high speed (engine speed) that may destroy its winding.

The overrun clutch has housing, fixed on armature shaft through internal splines. Spring loaded rollers are provided and these wedge tightly against the pinion barrel when forced into their tapered slots. Pinion and clutch housing is locked together and this causes transfer of motion from armature shaft to crankshaft. When the crankshaft speed goes beyond armature shaft speed, the rollers are released and pinion gear and armature shaft are unlocked (Fig. 20.6). At this stage, the pinion overruns the armature shaft till it is withdrawn by starting drive linkage. Starting drive linkage also operates the overrun clutch.

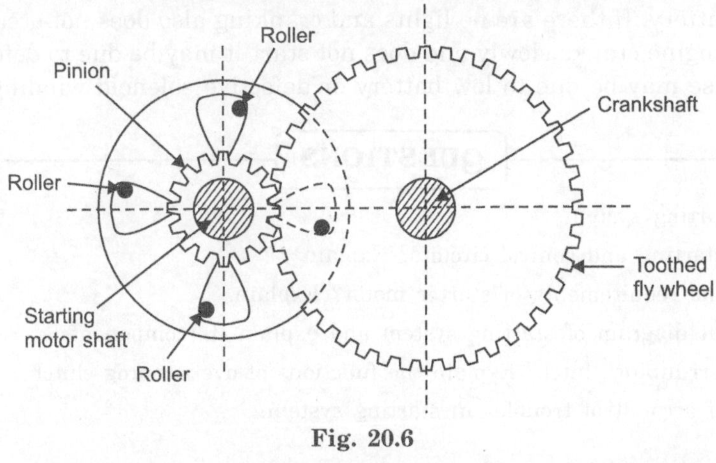

Fig. 20.6

The control circuit has a safety switch that is also known as neutral safety switch. It prevents the operation of starting system when transmission is in gear. Different switches are used for manual and automatic transmissions. For manual transmission, it is an electrical switch located on the floor (Fig. 20.7). It is operated and its contacts are closed when clutch pedal is pressed.

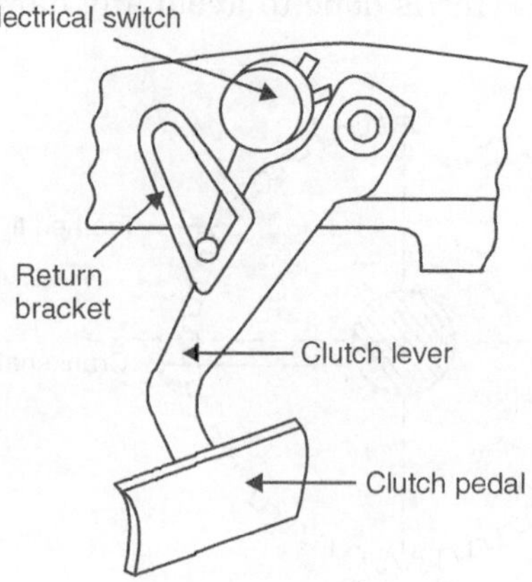

Fig. 20.7

In automatic transmission, the switch can be electrical or mechanical. In electrical switch, the contact points are closed when the vehicle is in neutral position. The switch is located near the gear selector. The mechanical switch blocks the movement of ignition key when gears are engaged.

20.3 TROUBLE DIAGNOSIS

The starting system may have troubles such as the engine does not crank or the engine cranks but does not start. Apart from these troubles, the solenoid may have some noise; the pinion may not disengage properly.

To diagnose the trouble, the headlights be switched on and observed. If lights do not dim and there is no cranking and check the whether there is voltage at ignition switch and starting motor terminals with ignition key on 'start'. If lights dim heavily and there is no cranking the possibility of battery is discharged. If lights dim slightly and no cranking occurs, the pinion may not be engaging properly with crankshaft. Also there may be an open circuit in the starting motor. If lights go out completely and cranking does not occur, there may be improper connection in the battery. If there are no lights and cranking also does not occur, the battery is open or dead. If engine cranks slowly and does not start it may be due to defective starting motor. Solenoid noise may be due to low battery or defective solenoid winding.

QUESTIONS

1. What is starting system?
2. What are starting and control circuits? Explain.
3. What are the requirements of starter motor? Explain.
4. Draw a neat diagram of starting system and explain its components.
5. What is overrunning clutch? Explain the functions of overrunning clutch.
6. Give a brief account of troubles in starting system.

21

LIGHTING SYSTEM, ELECTRICAL INSTRUMENTS AND ACCESSORIES

Lighting system in an automobile plays a vital role. It is not possible to drive a vehicle in dark hours during night without headlamps, tail lamps and other lights. Apart from head lights and tail lights, indicator lights are required to indicate when vehicle is taking a turn. In case of emergency, flashers are provided. All the indicators flash and convey emergency. Horn, safety devices like air bags, power windows, and adjustable seats are other accessories provided in the modern car. Lights are also provided in engine compartment, glove box and luggage compartment.

21.1 HEADLIGHTS

Headlights provide illumination of front area of road. These help driver to see the road clearly. Generally, these lights are two in number one on the left side and other on the right side. Sometimes two lights may be used on each left and right side making total number of lights four. These lights have the following variants:

21.1.1 Sealed Beam Light

These were used in old times up to seventies of 20th century. These had circular shape and sealed beam construction. The filament of light has no glass cover. The glass lens forms a part of parabolic reflector (Fig. 21.1). This reflector has a coating of vaporized aluminium. The lamp is filled with argon gas. The oxygen is removed to prevent oxidation of filament. Reflector intensifies the light whereas the lens forms a beam of the light from bulb. There are two filaments provided—one for low beam and other for high beam. The lower bulb is meant for high beam and upper bulb is meant for low beam (Fig. 21.2).

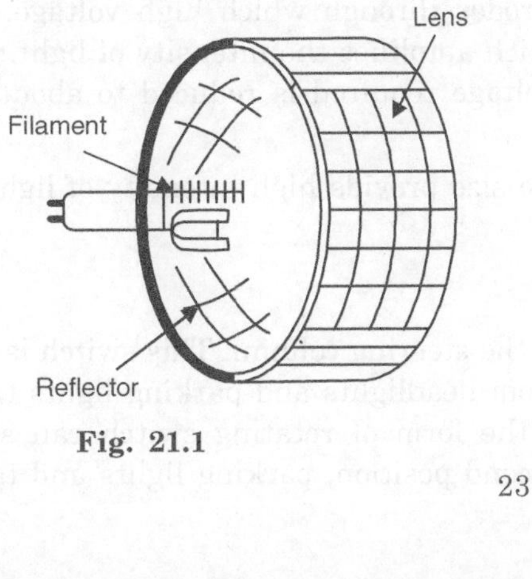

Fig. 21.1

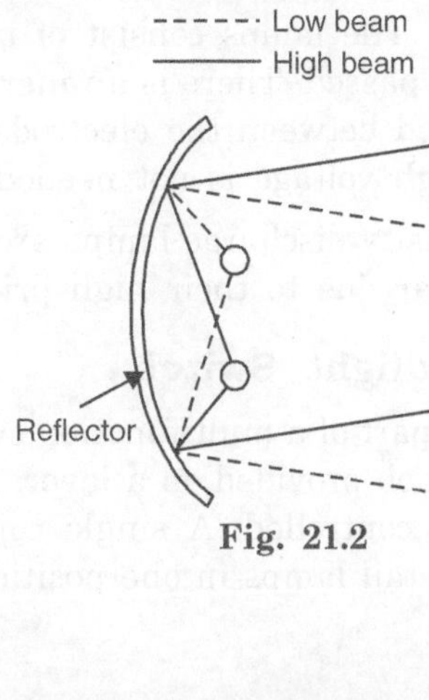

Fig. 21.2

21.1.2 Halogen Light

The bulb in halogen lamp is filled with iodine vapours. It has high temperature resistance glass shell surrounding the tungsten filament. The bulb is placed in sealed glass housing (Fig. 21.3). Presently, halogen is filled in the bulb which is a mixture of chlorine, fluorine and iodine. The halogen, in the bulb, helps the tungsten filament to attain higher temperature and burn brighter. The iodine vapours that are generated due to high temperature deposit back on the filament. In sealed beam headlight, these vapours deposit on the reflector and cause black spots. These black spots on reflector reduce the light output.

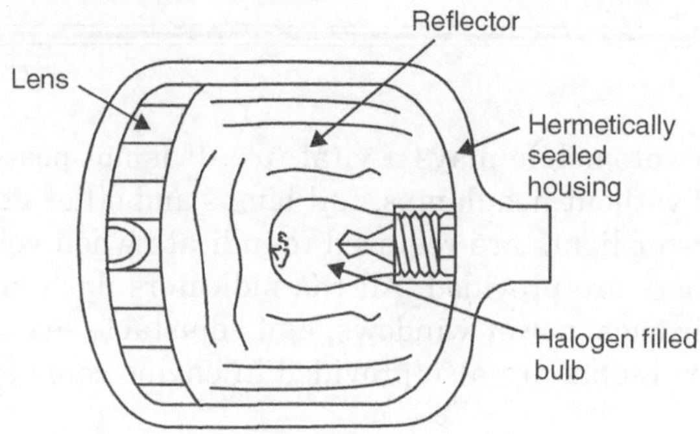

Fig. 21.3

21.1.3 Composite Light

In some automobiles, the halogen bulbs in head light system can be replaced. These are known as composite headlights. This allows the manufacturers to shape headlight lens in any shape and attain aerodynamic shape. In headlights, vents are also provided to counter the heat produced inside. Due to vents, condensation may occur in the headlight but as soon as the lights are switched on the condensate is evaporated and does not cause any adverse effect.

21.1.4 High Intensity Discharge (HID) Lamps

These are lately introduced lamps. These are stronger and intensity of light produced by them is three times that produced by halogen lamps. They provide better spreading of light on the road. Their life is longer than normal lamps. These are capable of producing light in ultraviolet as well as visible wavelength. This reflects light in a better manner from the road signs. The lamps consist of two electrodes through which high voltage, of about 15,000 volts, is passed. There is an inert gas which amplifies the intensity of light. Once the bridge is created between the electrodes the voltage required is reduced to about 80 volts as then the high voltage is not needed.

High intensity discharge lamps are small in size provide high intensity of light but are not very popular due to their high price.

21.1.5 Headlight Switch

These are part of a multifunction switch on the steering column. This switch is operated by turning a knob provided on a lever. Apart from headlights and parking lights tail lamps should also be controlled. A single control in the form of rotating switch can switch on headlights and tail lamps in one position. In second position, parking lights and tail lamps are switched on. In third position all the lights are off (Fig. 21.4). The light on the instrument panel is also switched on as soon as head lights or parking lights are switched on.

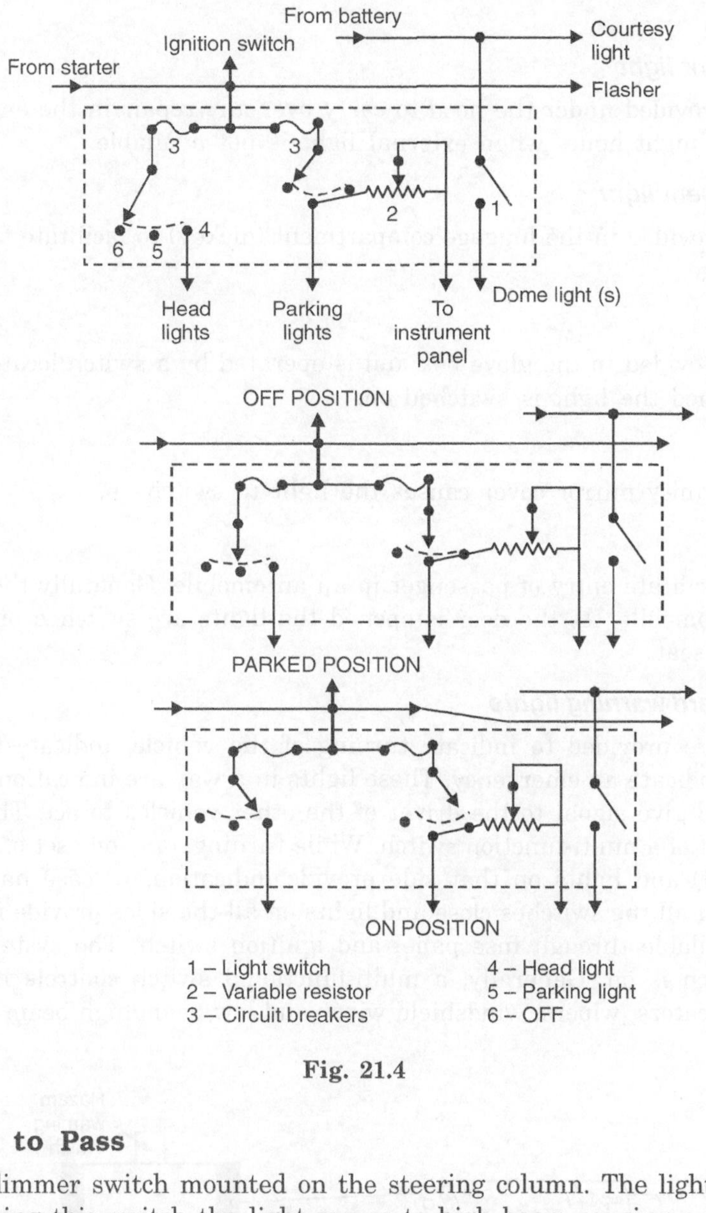

Fig. 21.4

21.1.6 Flash to Pass

There is a dimmer switch mounted on the steering column. The light is off or on low beam. By activating this switch the light moves to high beam causing a flash effect. This signal is used to overtake the vehicle during night and thus termed as flash to pass.

In some modern automobiles, automatic light system is provided. It eliminates the manual operation of switches. It consists of photo-cell sensors or amplifiers along with a relay. The ambient light is sensed by these sensors and as it is reduced to a particular level the sensors or amplifier supplies power and lights are illuminated. The feature is executed by switching the system on. The lights can be switched off manually, if needed, but are switched off automatically also as the intensity of ambient light increases to a particular level.

21.1.7 Interior Light Assemblies

The interior light assemblies are provided for the convenience of the user of automobile. These can differ from vehicle to vehicle. Some commonly provided assemblies are described.

Engine compartment light

This light is provided under the hood to carry over any repair in the engine and other components during night hours when external light is not available.

Luggage compartment light

This light is provided in the luggage compartment (dickey) to facilitate the loading and unloading of luggage.

Glove box light

This light is provided in the glove box and is operated by a switch located in the door. As the door is opened the light is switched on.

Vanity light

Opening the vanity mirror cover causes the light to switch on.

Courtesy lights

These lights facilitate entry of passenger in an automobile. Generally these are located in doors of the automobile. As the door is opened the lights are switched on throwing the light on entry and seat.

Turn, stop and hazard warning lights

These lights are provided to indicate turning of the vehicle, indicate that vehicle is about to stop and indicate an emergency. These lights in a way are indications to other cars moving on road and give signal to the driver of the other vehicles to act. The switches for these lights are part of a multi-function switch. While turning, only one set of switch contact closed (left or right) and lights on that side provide indication. In case hazard switch is activated contacts of all the switches close and lights on all the sides provide indication. The power is made available through fuse panel and ignition switch. The system is activated when ignition switch is on. Generally, a multi-functional switch controls head lights/tail lamps, turning indicators, wipers, windshield wiper and low beam/high beam switch (dipper switch) (Fig. 21.5).

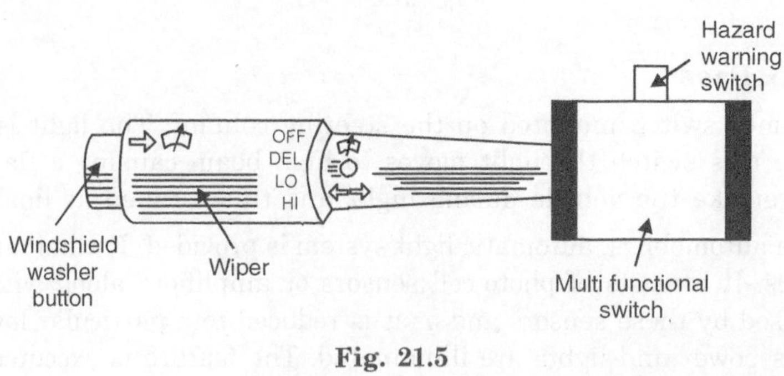

Fig. 21.5

Flasher

These are used as turning and hazard indicators. These have two metallic strips joined together and are sensitive to temperature (Fig. 21.6). There is heating element. The strip is connected to one side of contacts. As vehicle takes turn the current flows through circuit and light is switched on. Due to current, heating of strip occurs, it bends outward and contact is lost. This causes the switching off of light. As the current flow stops, the strip becomes cool; bends inwards and contact is established switching on the light. This process continues till vehicle continues to take turn and current flows through the circuit.

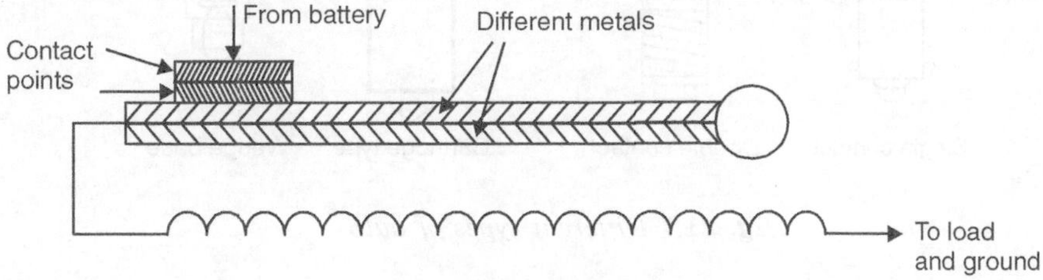

Fig. 21.6

Fog lights

These lights help in safe driving in foggy weather. Ordinary lights are reflected back as glare. In these lights, the bulb tries to pierce through the darkness. They attempt to deliver a flat, wide beam of light. The lights are kept low and parallel to road. Therefore, these lights are kept in the lower most part sometimes even below the bumper. The lights are provided yellow or amber colour through reflector. This is done to reduce the glare as fog tends to reflect the light.

Stop light

As the vehicle stops it is very essential to indicate it to the driver of the vehicle just behind. The driver of the vehicle just behind becomes aware and takes corrective actions to avoid collision. This becomes more essential when a vehicle is required to stop suddenly. There are stop lights with tail lamps, red in colour that is switched on. For this purpose, beneath the brake pedal a switch is provided. This switch is activated when brake pedal is pressed. As soon as brake pedal is released the switch is deactivated and lights are off. In recent past, stop lights are also provided in the rear window. This makes them high and is better visible to the driver of vehicle coming from behind. This is also known as hi-mounted stop light.

Reverse lights

These lights are switched on when reverse gear is engaged. The purpose is to make cautious the persons who may be behind the car when it is moving backwards. Now-a-days these lights are accompanied by loud audible sounds to alert the people.

Earlier tail lamps, turning indicators, stop lights and reverse lights used to be located at bumper level but in modern vehicles, to improve the visibility, and gaining better attention; these are being placed at higher level.

Light bulbs

There are several bulbs needed in an automobile other than head lamps. These have different wattage rating and are meant for uses at different locations in an automobile.

These are fitted into sockets. The sockets are connected through a single wire and car body provides the ground connection back to battery. Double filament bulbs require two live wires and one ground connection. These have three terminals to connect three wires. Figure 21.7 shows some of bulbs used in automobiles.

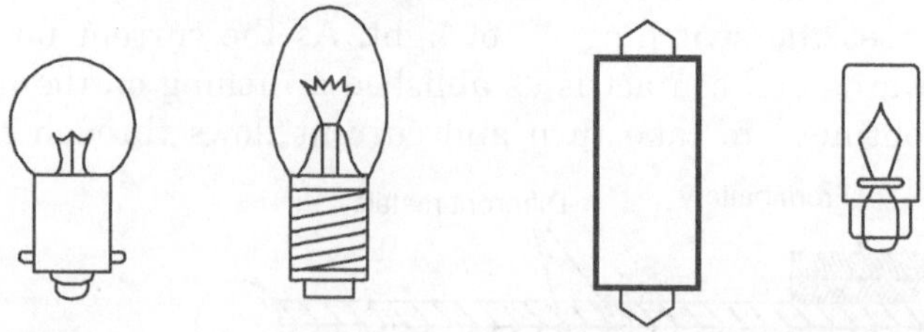

Fig. 21.7 *Different types of bulbs*

21.2 ELECTRICAL INSTRUMENTS

An automobile is equipped with a number of electrical instruments. Generally, it depends upon the price of vehicle that what instruments are provided. The fact is that these instruments are not essential. These act as accessories and helps driver to get better performance from the vehicle.

21.2.1 Electronic Control Module (ECM)

This device has microprocessor similar to that used in computers. The microprocessor can be programmed once. It is so programmed that it can monitor a number of parameters such as lighting system, antilock braking, suspension system, and automatic transmission, climate control inside the automobile and different parameters of engine performance. Even light indicators, rear window fogger, and wipers are sometimes controlled by ECM.

The operation of microprocessor is divided into four parts: Input, processing, storage and output. The different parameters are monitored and fed to the processor where these are stored and can be used for diagnosis. If a parameter works on abnormal value, outside the permissible range, the microprocessor detects the malfunction and stores it. The malfunction this way can be detected later by the service engineer. To minimize the effect of malfunction, the microprocessor goes into failsafe mode of operation and controls the system based on programmed value instead of signals from the inputs. This prevents the system from completely shutting down and it works on a limited basis. That is why modern cars, due to malfunction, do not halt on the road. They can be kept moving till they reach the service centre.

21.2.2 Instrument Panels

These are mounted on the dash board in front of driver. These have gauges, switches and controls which the driver can see and monitor. They are made quite attractive and can draw attention easily. These can have analogue or digital display. In first case, there may be a needle indicator while in second case there is direct display in the form of digits. Analogue systems are simple whereas digital systems are costly, accurate and last long. The digital display has light emitting diodes (LEDs) as indicators. These can be arranged to provide display in the form of letters. Then there are liquid crystal display (LCD) and vacuum fluorescent. The liquid crystal display is quite delicate and requires proper care but is capable of displaying information in an attractive manner.

Vacuum fluorescent are glass tubes filled with argon or neon gas. When current flows through the tubes these glow brightly.

21.2.3 Instruments Gauges

These gauges provide the scaled information to the driver such as temperature or fuel availability in the fuel tank. The values in the analogue gauges may not be very accurate but digital gauges display an accurate value. The various gauges require particular voltage to operate and this is provided by instrument voltage regulator. The regulator is accompanied by a sensor unit that changes the electrical resistance in the circuit in response to change or movement done by the external component. The regulator also limits and stabilizes the voltage being supplied to a particular gauge. Figure 21.8 represents an instrument voltage regulator.

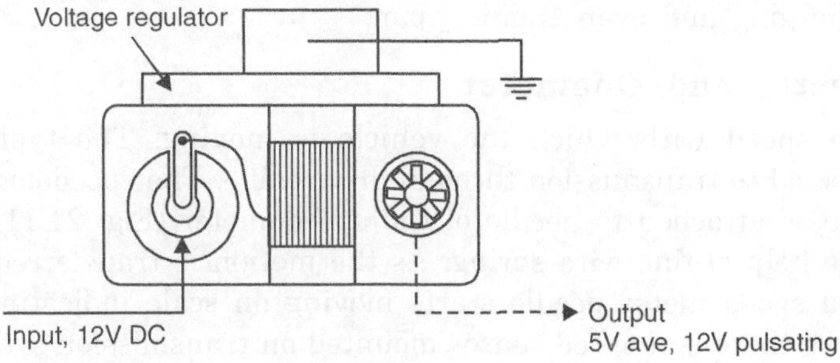

Fig. 21.8

21.2.4 Magnetic Gauges

These gauges have magnet, an armature and a needle. When current flows through conductor, magnetic field is produced around it. This induces magnetism in the armature that opposes the magnetism due to permanent magnet. The magnetic force causes the swinging of needle and required indication becomes available on the dial of the gauge (Fig. 21.9).

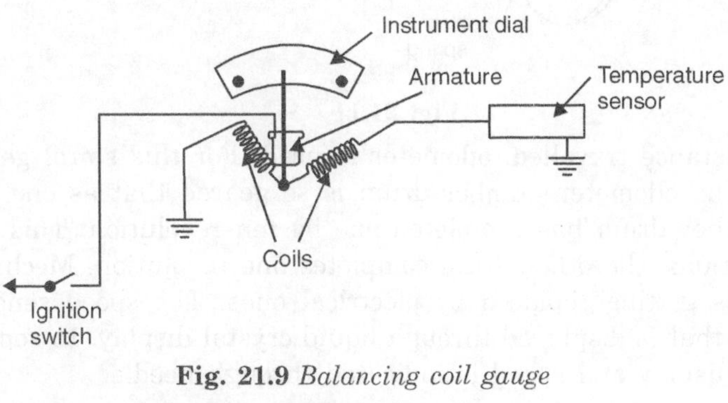

Fig. 21.9 *Balancing coil gauge*

21.2.5 Thermal or Bimetal Gauges

When current flows the heat is produced. Thermal gauges are made of two different metals and are sensitive to heat. The amount of heat is controlled through a suitable resistor. The amount of current determines the deflection of bimetal strip (*i.e.*, gauge). The needle is also deflected accordingly and information is displayed on the dial (Fig. 21.10).

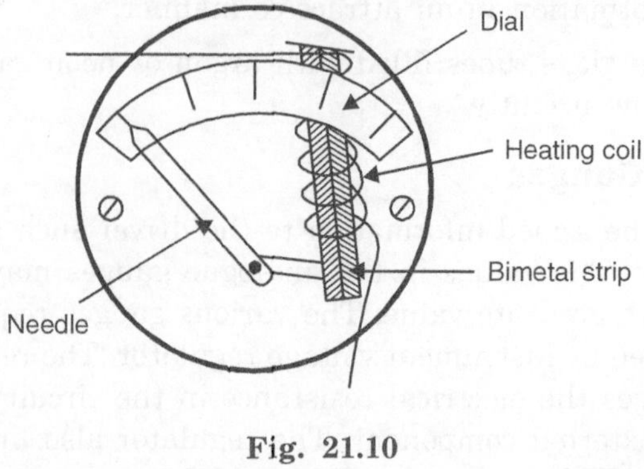

Fig. 21.10

Besides gauges explained above there are some gauges providing basic information such as speed of the vehicle, distance travelled, engine temperature, oil pressure in the engine, fuel level, battery charging and even engine r.p.m.

21.2.6 Speedometer and Odometer

It indicates the speed with which the vehicle is moving. The typical mechanical speedometer is connected to transmission through drive cable. The speedometer has magnet inside metal cup that is attached to needle of the speedometer (Fig. 21.11). The needle is kept at zero with the help of fine wire spring. As the motion is transferred to drive cable, it starts rotating and speedometer needle starts moving on scale indicating the speed. In electrical speedometers there is a speed sensor mounted on transmission system. The signal generated by the sensor is fed to electronic control module and it is used to measure the distance covered and control the speed in addition to indicating the vehicle speed.

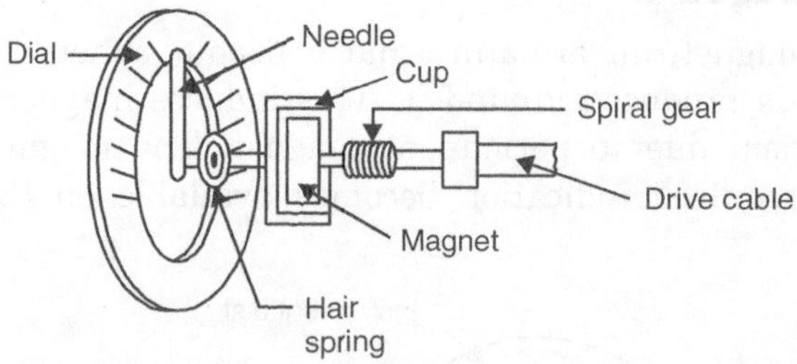

Fig. 21.11

To record the distance travelled, odometer is used. For this spiral gear is cut on the speedometer shaft. The odometer number drum is so geared that as one drum completes one revolution the other drum has completed one by ten revolution. This when one drum completes ten revolutions, the other drum completes one revolution. Mechanical odometer, as explained above is getting replaced by electrical ones. The speed sensor provides the required information that is displayed through liquid crystal display. In some cars, distance covered is on liquid display and speed is indicated through needle.

21.2.7 Oil Pressure Gauge

The oil pressure in the engine should be kept at particular level. This level depends upon the engine running conditions. If the engine is at low speed the pressure should also be low but if engine is running at high speed the oil pressure should also be high. The flow of current through the gauge depends upon the pressure exerted by oil. The pressure is sensed by piezo-electric sensor placed in the oil delivery passage (Fig. 21.12). There is a flexible diaphragm that moves due pressure exerted by oil. The movement is transferred to arm sliding on a resistor. The position of sliding arm determines the resistance and amount of current that flows. The current sent to winding attached to needle. The needle moves on a scale indicating the pressure. When pressure is normal the needle is placed in the middle of scale.

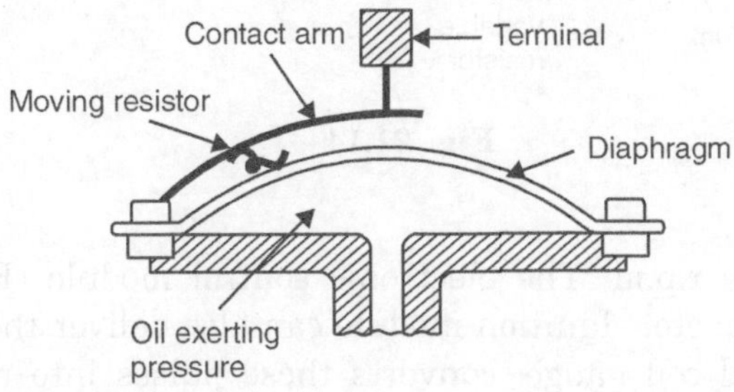

Fig. 21.12

21.2.8 Coolant Temperature Gauge

This is to indicate the coolant temperature which is an important parameter to keep the temperature of engine within permissible limit. It has a thermistor (Fig. 21.13) which is a resistor that varies with temperature. This senses the temperature and accordingly offers electrical resistance in the circuit. If the temperature approaches the limiting value warning the indicator is switched on and driver can take corrective action.

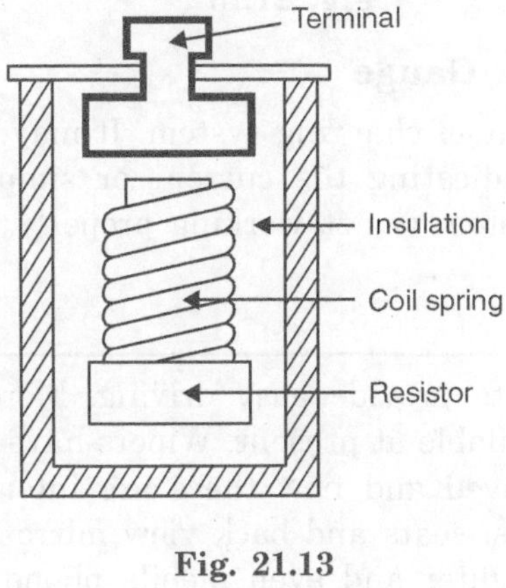

Fig. 21.13

21.2.9 Fuel Level Gauge

It is very important to know the remaining fuel in the tank for which this gauge is needed. It has magnetic indication system. There is float at the surface of oil in the tank. This float operates a variable resistor. The resistance in the resistor is low when the level of oil is low and high when it is high. This information is transferred to needle indicator or digital display through electric current. Figure 21.14 represents circuit diagram of a fuel gauge.

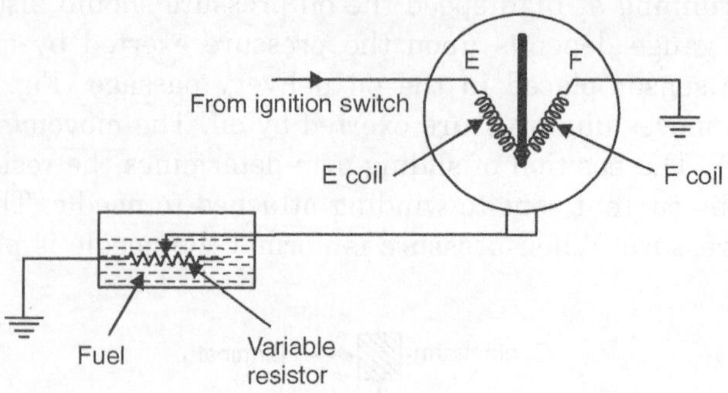

Fig. 21.14

21.2.10 Tachometer

It indicates the engine r.p.m. The electronic control module (ECM) provides electric pulses that are fed to tachometer. Ignition module can also deliver these pulses (Fig. 21.15). The tachometer, a balanced coil gauge, converts these pulses into revolutions per minute (r.p.m.) that can be read on a dial.

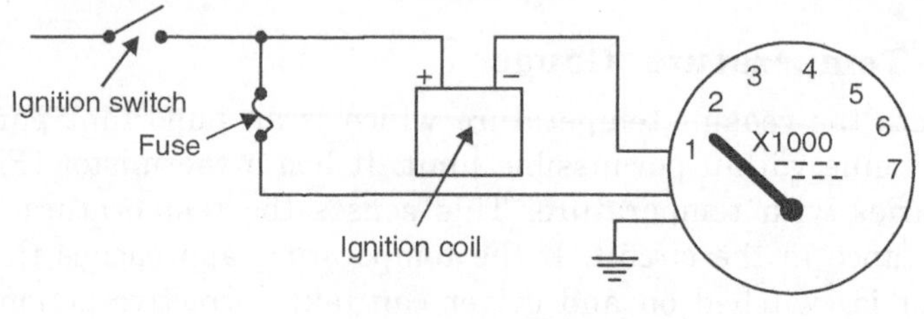

Fig. 21.15

21.2.11 Charging System Gauge

This gauge indicates the state-of charging system. It may have a voltmeter, indicating the voltage or ampere meter indicating the current or simply a light indicator that is switched on in case charging system is not working properly.

21.3 ACCESSORIES

The accessories are meant to provide easy driving. More and more accessories are finding place in cars that are available at present. Wipers have been there in cars for a long time but these have been improved and now there are intermittent wiper systems. The modern cars have power windows, seats and back view mirror, defrosters, sun roof, audio/video system, clocks, cigarette lighter and even mobile phone charger.

1. Wipers

At present, the wiper systems are two/three speed systems and intermittent two/three speed systems. The motor used in both the system are same. However, the motor may have permanent magnetic field or electromagnetic field. In case of motor with permanent magnetic field, the placement of brushes on the commuter controls the speed of motor. Three brushes are provided that are common, high and low. The speed of the motor is controlled by the number of armature winding being supplied the current. If fewer windings are supplied current the counter electromotive force is less and high speed can be attained. Similarly, by supplying current to large number of windings, due to high counter electromotive force, the speed attained by the motor is less.

Intermittent wiper system can operate at variable interval of wiper sweep. The system has module containing an electronic switch. Internal timer triggers the switch to close the circuit for the governor relay which closes the circuit to low-speed brush. The delay between the wiper sweeps is due to resistance chosen by driver through a potentiometer control. There is a capacitor charged through potentiometer and when it becomes fully charged the current flows to the wiper motor. The gap between wiper sweeps depends upon the time taken by the capacitor to get fully charged that can be adjusted by driver through potentiometer control.

2. Windshield Washers

These are used to spray fluid (often water) on the screen and then sweeping the wiper to clean the wind shield. There is a water pump in the fluid reservoir that pumps the fluid upwards before it comes out in the form of spray through nozzle placed in the bonnet. The washer is operated through switch.

3. Power Windows

The windows can be raised or lowered by pushing a button. The system consists of master control switch, individual control switches for windows and drive motors for each window (Fig. 21.16). The master control switch controls the whole system that gets power through ignition switch. The control switch has independent segments to control each window. A child safety device is included so that children cannot operate the windows. The electric motors raise or lower glass in the windows. The direction of movement of glass depends upon the movement of motor which in turn depends upon the polarity of supply voltage. The switch has two contact points (two way switch) one up and other down. When 'up' contact is activated the 'down' contact acts as earth for the motor and raises the glass. When 'down' contact is activated 'up' contact acts as earth for motor and glass is lowered. Each motor has internal circuit breakers acting as safety device.

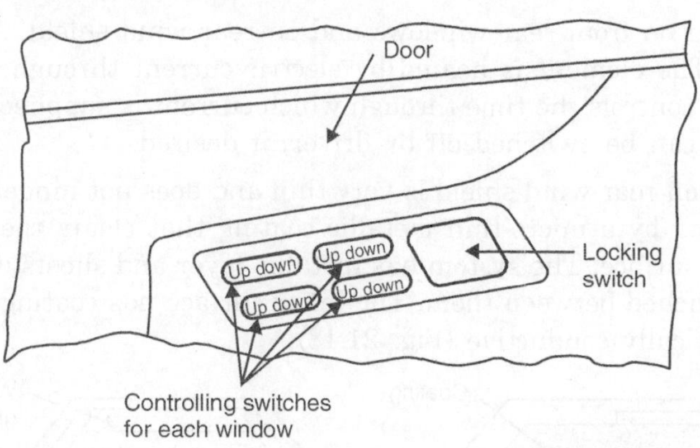

Fig. 21.16

4. Power Seats

A power seat is adjustable to provide most comfortable position to the driver or passenger. There is a seat control switch and a motor. These can be four way seat or six way seats.

In four way system, the whole seat moves in four directions namely forward, backward, up and down. This system is generally used in bench seats. In six way systems, in addition to these four directions of adjustments, height of the front and/or rear of the seat can also be adjusted. This system is suitable for bucket seats.

5. Power Back View Mirror

Both back view mirrors, on left and right side, are controlled electrically through joystick switch. There is a dual motor drive assembly to provide the motion to the mirrors (Fig. 21.17). By moving the joysticks up and down the direction of the mirror can be adjusted.

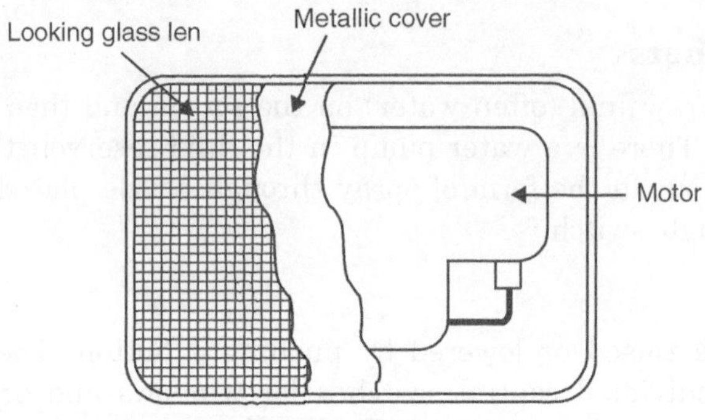

Fig. 21.17

6. Photo Chromatic Rear View Mirror

This is provided inside the car. The photo chromatic mirror operates on the intensity of the glare. The glare when reflected may disturb the vision of driver which is undesirable. When glare is more the photo chromatic mirror darkens and glares is not reflected and driver is not disturbed. When glare is gone it becomes clear and reflects the light normally.

7. Defroster

These are provided on front/rear windows and on rear wind shield. A heating element is fused in the glass. The element is heated by electric current through switch. The delay circuit is provided that controls the time through which current is supplied and defroster are activated. The system can be switched off by driver, if desired.

The grid provided on rear wind shield is very thin and does not hinder the driver's view but it has been replaced by a micro-thin metallic coating that clears the wind shield. The coating is on the inner surface. The system has metallic layer and sheets of glass those have plastic laminate sandwiched between them. The inner surface has coating of silver and zinc oxide to make it electrically conductive (Fig. 21.18).

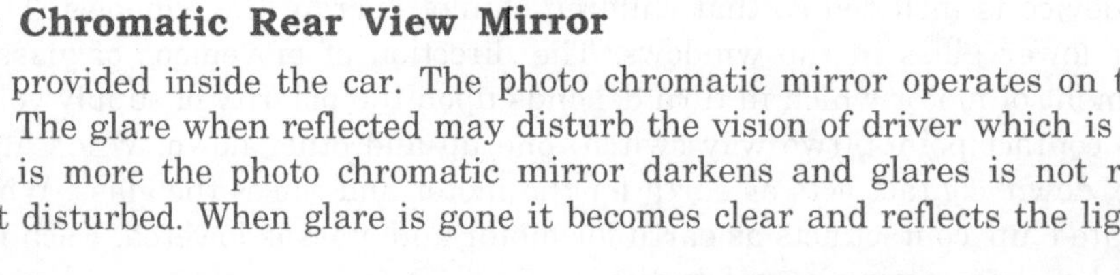

Fig. 21.18

8. Sun Roof

The sun roof can be tilted upwards or slides open to provide natural sunlight and fresh air. A motor, relay, switch and sliding roof panel comprise the sun roof system. The system is activated with ignition switch. The motor activated through switch causes the movement of panel either to create opening or to close the opening. There is a in-line circuit breaker as safety device to system.

9. Audio/Video System

These systems are the source of entertainment for the passengers of the car. These are operated through car battery as source of electrical energy. The flow of electric current is controlled through switch and safe guarded by circuit breaker. The audio system may consist of radio and CD player/cassette player. The video system includes a liquid crystal display panel of suitable size. The CD player, in this case, can play video CDs and DVDs also.

10. Clock

The clock is operated through fuse panel. The clocks have liquid crystal display that shows time, sometimes date and even temperature. The power is supplied from battery.

11. Cigarette Lighter

It has got a heating element energized by battery. To start the system the element is pushed in. The element is automatically released from pushed in after a particular temperature is attained (sufficient to ignite the cigarette). In some countries, where smoking is offence in public place this accessory is not provided.

12. Mobile Phone Charger

These days mobile phone charger has become almost an essential accessory. Only socket is provided that supplies desirable current to the charger. The charger is specially designed for use in car.

21.4 WARNING INDICATORS

These indicators are to warn the driver and take corrective measures. These are mostly related with safety devices in an automobile such as seat belts, air bags, lights and hand brake warning.

1. Seat Belt 'not Fastened' Indicator

There may be a chime or indicator that illuminates when ignition is turned on and seat belt is not fastened.

2. Air Bag Light

The light indicates that air bag system is working. The light briefly becomes 'on' when ignition is switched on and engine is started. If the light remains 'off' the driver can take suitable corrective measures.

3. Lights 'on' Indicator

This indicator is in the form of chime which is activated when the head lights are on and door is opened. This is an indication that head lights are still 'on' and driver is preparing to leave the vehicle.

4. Hand Brake Warning

There is an indicator light on the dash board that indicates that hand brake has been applied and can damage the brake shoes if car is run in this condition.

5. Oil Indicator

This light on the dash board indicates that oil level is not proper and there may be some problem in lubrication of different parts. The light becomes 'on' with ignition switch but it should go off as the engine is started. If it remains on corrective measures are required.

6. Charge Indicator Light

This light should go off when engine is started. It is switched on with ignition key. If it remains 'on', corrective measures need to be applied.

7. Door Ajar Warning

If door remains open on any side of the vehicle there is a chime and driver can properly close the doors before starting the journey.

QUESTIONS

1. What is the function of head lights? Explain.
2. Explain different types of headlights briefly.
3. Why interior light assemblies are required in a car? Explain briefly any two such assemblies.
4. What are the functions of turn, stop and hazard warning lights?
5. How does a flasher work? Explain with the help of a diagram.
6. What are different electrical instruments provided in an automobile?
7. Explain the working of magnetic gauges.
8. What is the necessity of speedometer and odometer in an automobile?
9. How does an oil pressure gauge work?
10. What are different accessories provided in the car? Name them.
11. How does a power window work?
12. What is the function of defroster? Where it is located?
13. What is the relevance of different warning indicators in an automobile?
14. Explain any three warning systems used in an automobile.

22

THE CARRIAGE UNIT

The carriage unit is required in an automobile to carry all the components including its body. The body is needed to accommodate the passengers. In case of goods carrying vehicle body is used to store the goods to be transported. This is the basic structure that should have enough strength to take up the load of different components and the passengers or goods.

22.1 THE FRAME

In modern automobiles, there is no separate frame. But in old times, they were used to have a separate frame similar to that is used in commercial vehicles of today. The frame is used to carry all the major components and sub assemblies such as engine, transmission system, suspension system etc.

There were different types of frames used in automobiles. These were designed according to weight, size and shape of the automobiles. A simple frame used in old automobiles is shown in (Fig. 22.1). It is a ladder type frame. The main longitudinal member may be straight as shown or may have an inward turn as illustrated in figure 22.2. The frame is provided with transverse members to take torsional and other loadings. Their number is according to the weight of the vehicle. These also support various components of the automobile. Transverse members are fitted between mounting points so that torsional load applied to each side member is converted into bending load.

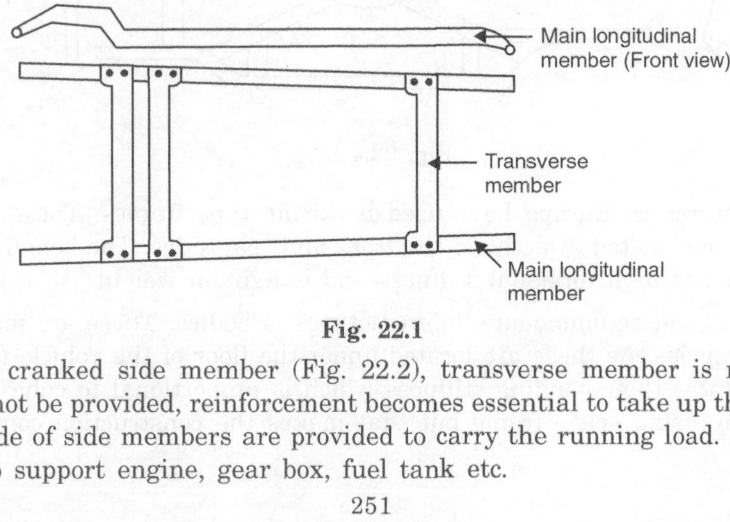

Fig. 22.1

In case of cranked side member (Fig. 22.2), transverse member is needed near the crank. If it cannot be provided, reinforcement becomes essential to take up the load. Brackets on the lower side of side members are provided to carry the running load. Similar brackets are provided to support engine, gear box, fuel tank etc.

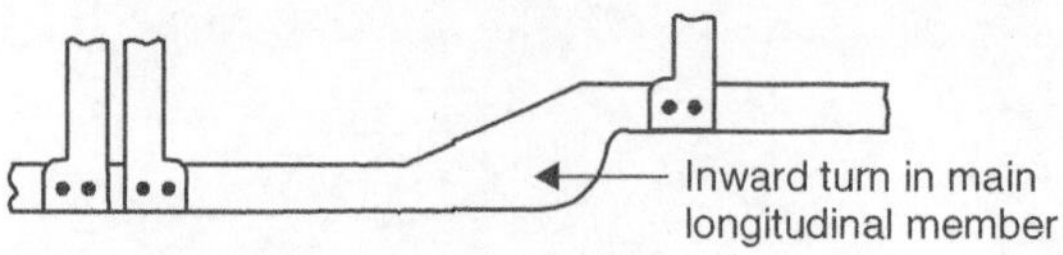

Fig. 22.2

Different sections are used for longitudinal as well as transverse members to make frame. Most common being channel section shown at number one in Figure 22.3. Figure shows some other commonly used sections also.

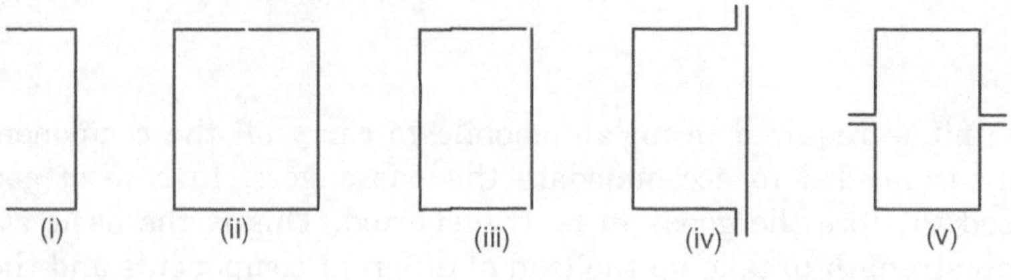

Fig. 22.3

One of the frames used by an automobile manufacturer in U.K. has been illustrated in Figure 22.4. It is provided with box section members. The front cross member has large proportions with a plate welded to its bottom flanges. This helps in providing brackets to carry suspension links, springs and dampers. There is a cruciform bracing member provided in the frame. At the centre of member, suitable hole is provided for the propeller shaft to pass through. The length of the side member between ends of cruciform and cranked portion is provided with reinforcement in the form of gusseting.

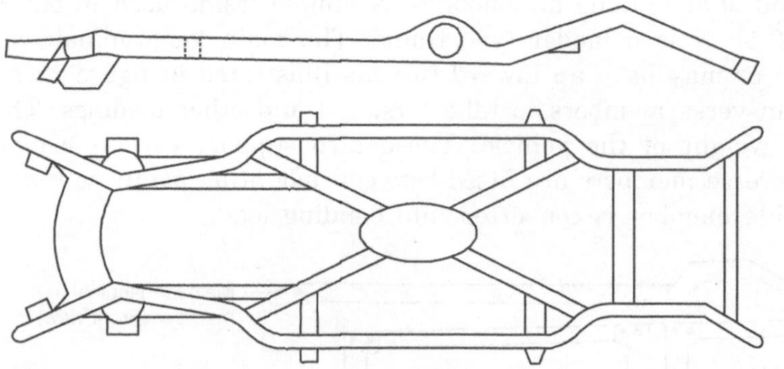

Fig. 22.4

Some manufactures in Europe have used backbone type frames. These are provided with tubular backbones bolted directly on the front and rear sides. The benefit of backbone frame being that it has high torsional stiffness and is light in weight.

Separate frames can accommodate different types of bodies. These are mainly used in light commercial vehicles. As these are located under the floor of the vehicle their depth is restricted. This reduces their bending stiffness which is proportional to cube of the depth. This is overcome by using space frame but that makes the construction complicated and costly.

22.2 SUB-FRAMES

These are used to isolate the vibrations of high frequency. This includes vibrations from an engine and suspension system. In this case, rubber mountings are put between the sub-frame and main frame to take up vibrations. A sub-frame provides additional support to the main frame by interposing three-point mounting system. One mounting is on longitudinal axis about which the main frame may twist and the other two are on its sides. The sub-frame also carries the sub-assemblies such as front or rear suspension systems. This helps in attaining a simple assembly and reduction in cost.

22.3 INTEGRATED CONSTRUCTION

In this type of construction, the chassis frame is welded the body of the automobile. It does not have separate chassis. The first automobile that was manufactured in this manner was Austin A30 in early fifties of twentieth century.

This type of construction has several variants. The fact is that different designs were improved and adopted by numerous manufacturers. The beams are formed by the body panels that have more depth and chassis frame that has less depth. The area enclosed by complete body is much bigger than that enclosed by the cross-section of a frame side or transverse member. The strength of beam is proportional to the square of its depth and stiffness proportional to cube of depth. For a box section, the torsional stress and stiffness is proportional to the area enclosed by it. This means that strength and stiffness of a shell is much greater than strength and stiffness of a chassis frame. If not buckled or distorted, a simple flange made of sheet metal can carry a large load. A flange can roughly take up stress to its yield strength if it has width less than sixteen times its thickness. This can prevail till the deflection of the beam is negligible. If not, it would fail. In such beams, the loading induces compressive stresses at the top and tensile stress in the lower parts. As the yield stress approaches, the beam would bend downwards. The tensile loading in the flanges would tend to straighten. This would reduce the moment of inertia of the section. The stresses in the flanges would be higher than permissible limit and the section would buckle.

In integrated construction, spot welding is used extensively. It has become possible to have improvement in the design with the advent of better welding technology available today.

QUESTIONS

1. Explain the functions of frame in automobile.
2. What is sub-frame? Explain briefly.
3. Explain integrated construction and its benefits.

23

PASSENGER COMFORT

Even a few years back, air-conditioned cars were considered to be luxury and only few could afford them. But as the time passes, more and more cars are equipped with air-conditioning and heating systems. These systems make journey very comfortable in any season.

In an air-conditioning system, refrigerants carry heat from inside to outside the vehicle. The boiling point of refrigerant depends upon the pressure. If the pressure is high the boiling point is also high. R-134a and R-12 are the commonly used refrigerants. Refrigerant R-134a evaporates at 18°C at a pressure of 476kPa and at 49°C at a pressure of 1214kPa. Similarly, refrigerant R-12 evaporates at 18°C at a pressure of 510 k Pa and at 47°C at a pressure of 1078 k Pa. The temperature and pressure of refrigerant is kept low for absorption of heat and high for dissipation of heat. The heat absorption means conversion of phase from liquid to vapor and heat dissipation means conversion of state from vapor to liquid. Heat absorption means cooling. The absorption process occurs inside the vehicle and dissipation occurs outside the vehicle. Both the processes occur continuously and refrigerant is used in cyclic manner.

23.1 THE AIR-CONDITIONING SYSTEM

In an automobile air-conditioning system is closed one. It has major components as following:

23.1.1 Compressor

The low-pressure and low-temperature refrigerant in the vapor form is compressed into high temperature, high pressure vapor. The compressors are piston type, rotary vane type and scroll type.

Piston type compressor

There is an intake and compression stroke as the piston moves. In intake stroke the refrigerant goes inside the cylinder and during compression stroke its pressure (and thus temperature) is raised. The refrigerant is drawn into compressor from evaporator. There are two valves provided to control the flow of refrigerant in the cylinder (Fig. 23.1). The valves are reed type. A variant in piston type compressor is variable displacement type that can also control the amount of refrigerant. The pistons are connected to wobble plate and stroke of piston depends upon the angle of wobble plate. The angle of wobble plate depends upon the pressure difference between outlet and inlet of the compressor. Increase in stroke means more quantity of refrigerants entering the compressor. The compressed refrigerant goes to the condenser.

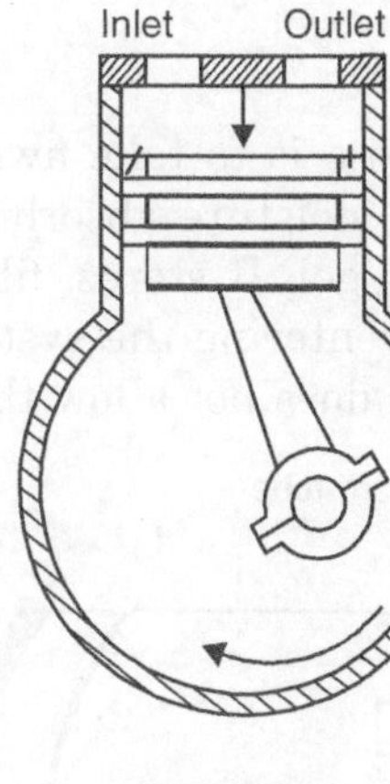

Fig. 23.1

Rotary vane type compressor

This type of compressor consists of rotor with a number of vanes instead of piston. The rotor is kept in housing and as it rotates the refrigerant is drawn into chamber formed between vanes and housing. The discharge port takes away the compressed refrigerant. The oil sump is located on discharge side and supplies oil for lubrication.

Scroll type compressor

It has a movable and a fixed scroll. The refrigerant is forced against fixed scroll towards the centre of compressor by movable scroll. This increases the pressure of the refrigerant. The discharge port is located in the centre of the compressor. The operation of this type of compressor is smooth.

23.1.2 Condenser

The purpose of condenser is to reduce the temperature of refrigerant through heat dissipation (Fig. 23.2). It receives refrigerant at a temperature of around 180°C. It enters at the top, moves through the condenser, and finally comes out from the bottom. From here it is sent to dryer. Space to accommodate condenser can be a constraint in the automobile.

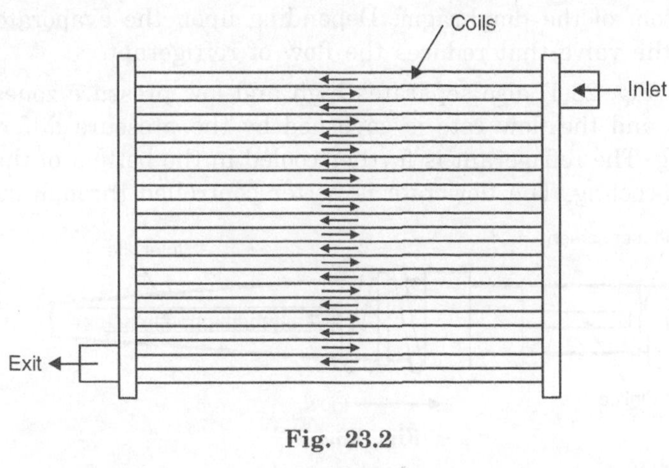

Fig. 23.2

Hence, it is required to be of small size and efficient. It has coils mounted in series and provided with thin fins. Fins enhance the heat dissipation. The condenser is located in front of the radiator.

23.1.3 Dryer

Also known as accumulator, its purpose is to take away moisture from the refrigerant (Fig. 23.3). For this, it is provided with a moisture absorbing material known as desiccant. One of material used as desiccant is silica gel. It stores, filters and dries the refrigerant. It is necessary to prevent the moisture from entering the system as it can cause corrosion and damage the system. Also, the accumulator does not allow the liquid to enter the compressor.

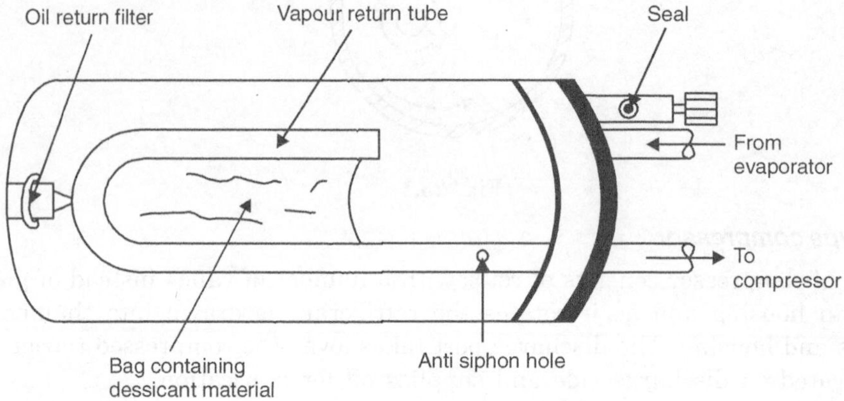

Fig. 23.3

23.1.4 Expansion Valve

The liquid refrigerant in the evaporator must evaporate completely to get maximum cooling effect. Also, the quantity of refrigerant entering the evaporator should be under control. The expansion valve that performs these functions is a thermostat. The valve is mounted at the inlet of evaporator. It allows only a proper quantity of refrigerant in the evaporator.

Initially, the pressures on both the sides of diaphragm are equal. The valve is closed through the spring. As the system starts, the pressure below diaphragm is reduced and valve opens. The refrigerant enters the evaporator and starts evaporating. The compressor sucks the refrigerant and it moves to the upper part of the evaporator. The evaporator pressure is sensed at the bottom of the diaphragm. Depending upon the evaporator pressure, the diaphragm actuates the valve that reduces the flow of refrigerant.

The orifice tube (Fig. 23.4) also separates high and low pressure zones of the system. It has a fixed orifice and the flow rate is governed by the pressure difference across the orifice and sub-cooling. The refrigerant is further cooled in the bottom of the condenser and this is known as sub-cooling. The flow rate is better controlled through sub-cooling.

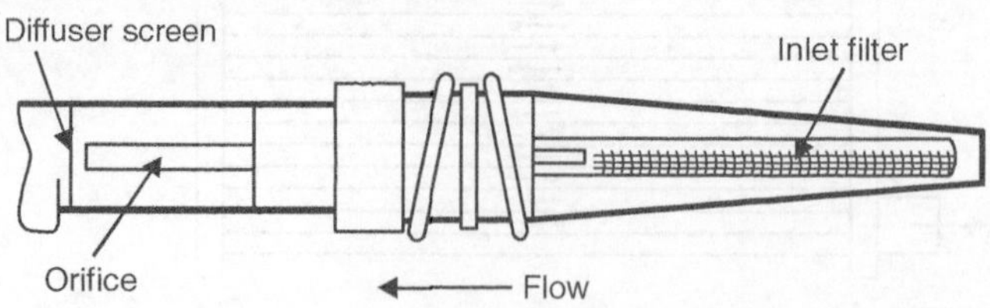

Fig. 23.4

23.1.5 Evaporator

It is similar to condenser and has coils mounted in a series (Fig. 23.5). The reason being that here also maximum amount of heat is to be transferred. The difference being that heat would be extracted from the interior space of car. The evaporator is located beneath the dash board. The refrigerant in the form spray enters the evaporator after coming out from the expansion valve or orifice tube. The refrigerant is immediately converted into vapor. The process continues as refrigerant is supplied continuously. The moisture contained in the air (in the interior space of car) is condensed outside the evaporator core and is removed as water outside the vehicle. The removal of moisture becomes necessary in the hot and humid weather conditions. The flow of refrigerant to the evaporator is controlled as per cooling needs of the automobile. More than required refrigerant means higher temperature and pressure and reduced cooling. Less than required refrigerant again means reduced cooling as evaporation is reduced. It is to be ensured that refrigerant entering the condenser does not contain any droplet as that would damage the condenser.

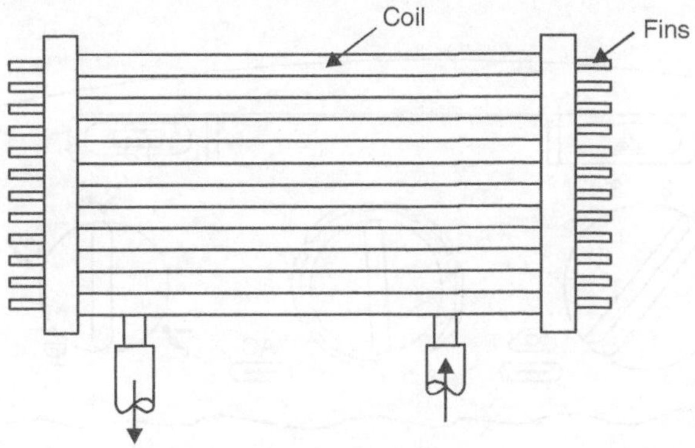

Fig. 23.5

23.1.6 Blower Motor and Fan Assembly

There is a fan and blower motor present in the evaporator. This is to enhance the airflow in the interior space of car. The blower draws stale air from the interior space and sends in fresh cool air. The fan switch controls the blower motor. In winters, this assembly can provide air heated by heating coil.

Apart from these main components, air conditioning systems are provided with refrigerant lines that are meant to transport refrigerant between the components mentioned above. These are hose or tubing of appropriate length. The tubing meant to carry refrigerant at low temperature and low pressure is on suction side whereas that capable of carrying refrigerant at high temperature and pressure are on discharge side. The liquid lines connect the condenser to dryer and dryer to inlet of expansion valve. Refrigerant such as R-134a requires special types of tubing to prevent its escape from the pores of the tube.

23.1.7 Air-Conditioning and Temperature Control System

The air-conditioning system is provided with controls for different purposes. There is evaporator pressure control to maintain the back pressure in the evaporator. This regulates the evaporator temperature also and prevents the freezing of moisture. The thermostatic switch senses the outlet air temperature and when it attains the higher permissible value it closes and energizes the compressor. This starts the cooling. When desired low temperature is attained the switch acts in the reverse manner and compressor is shut down. The operation of compressor is also controlled through pressure cycling switch that is connected in series with electromagnetic clutch. Controls are also provided to monitor the low pressure and high pressure conditions. Prevention from excessive pressure is achieved through high-pressure relief valve installed on dryer. Another control may be provided to regulate the crankcase pressure in some compressors. This modulates the displacement of compressor according to pressure or temperature.

Temperature controls may be manual or automatic. In manual controls, the amount of cooling is controlled manually through blower speed (Fig. 23.6). In automatic temperature control system, there are heat sensors that send the signal to electronic control module (ECM) that operates the compressor, valve, blower etc. The system is supported by coolant temperature sensor, in-car temperature sensor, outside temperature sensor, high and low side temperature control switches and even vehicle speed sensor, sun load sensor and throttle position sensor.

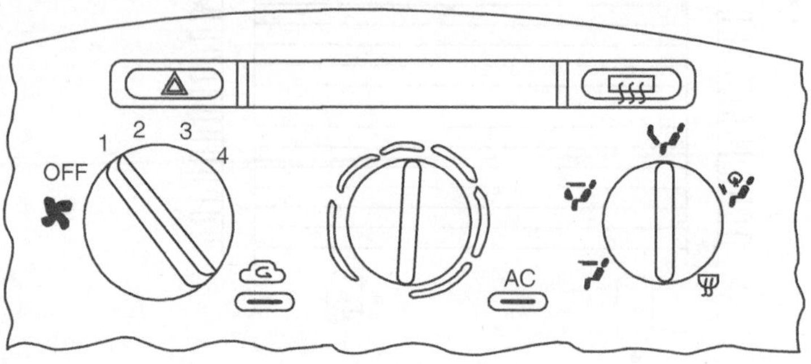

Fig. 23.6

The control panel gives the information regarding the in-car temperature, outside temperature and other parameters to the user of the car. This panel may be manual, push button or touch pad type. There is a microprocessor that can store all the input data including the temperature to be maintained. The user of the vehicle may not be required to feed data every time and it remains stored till the battery is disconnected.

23.1.8 Ventilation and Heating System

For the comfort of passengers in a car it is very essential to provide ventilation and fresh air. There are vents provided in a car for this purpose. One of the systems is shown in the Figure 23.7. The outside air flows inside the car as it moves. When car is stationary fan may provide the supply of air. The air flows inside through the grill that cause circulation of air in different directions according to the need of passengers and is adjustable. The air is taken out through exhaust area.

The heating system in an automobile is required to provide comfort conditions during winter season. It supplies hot air inside the car to keep it warm and comfortable. It also keeps car window glasses and wind shields free from fog or frost.

It is known that all the heat produced through the combustion of fuel is not converted into useful work in the engine. This heat is dissipated by using coolant. This heat, that is wasted otherwise, can be utilized to heat the passenger compartment. The heat contained in the coolant is transferred to heater core. Air is blown over this core and heat is transferred to air from core. This hot air is circulated in the passenger compartment. The coolant, after it loses its heat to heater core, is circulated back around the engine to take heat again.

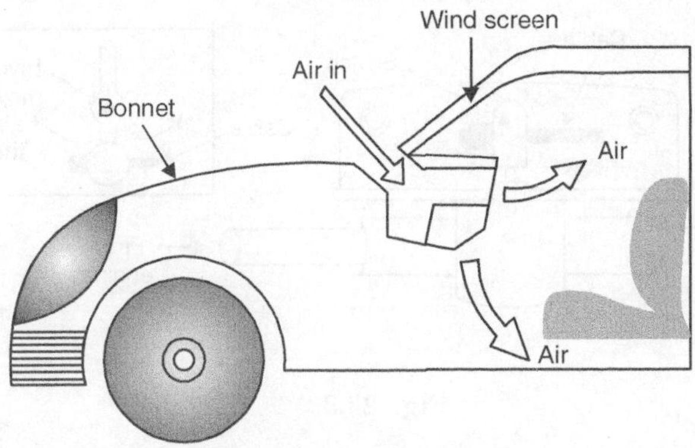

Fig. 23.7

The heating system in an automobile consists of heater, control valve, blower motor and fan (Fig. 23.8). Heater and defrost ducts are provided to take heated air to different parts of the passenger compartment.

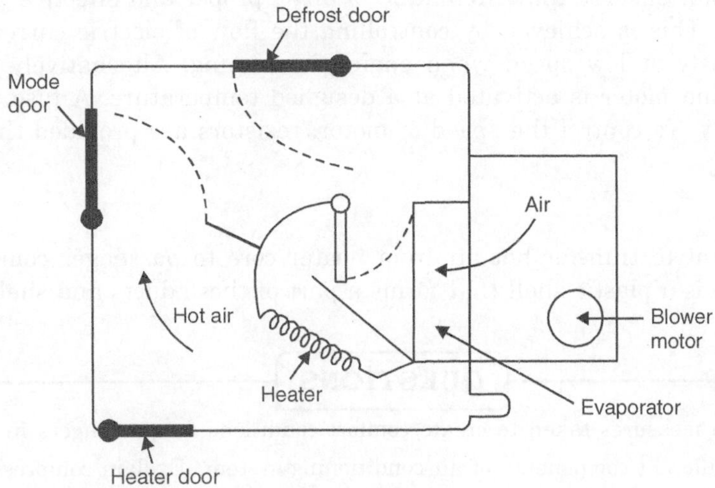

Fig. 23.8

Heater core

It consists of inlet, outlet and tubes with fins. The heater core is designed by individual car manufacturer depending up specific needs of the automobile. The maintenance requirements of heater core are cleaning of tubes as scale formation may be there. Anti corrosion measures are also needed to avoid corrosion.

Control valve

This valve controls the flow of coolant. The flow occurs from engine to heater core when valve is open (Fig. 23.9). When heater core is heated to the desired level the valve is closed to stop the flow. The air flowing over core takes away heat to passenger compartment. When the core becomes cool, the valve is opened again to heat the core. The valve may be thermostatically controlled with a liquid filled capillary tube located in discharge air-stream that senses the temperature and modulates the flow of coolant.

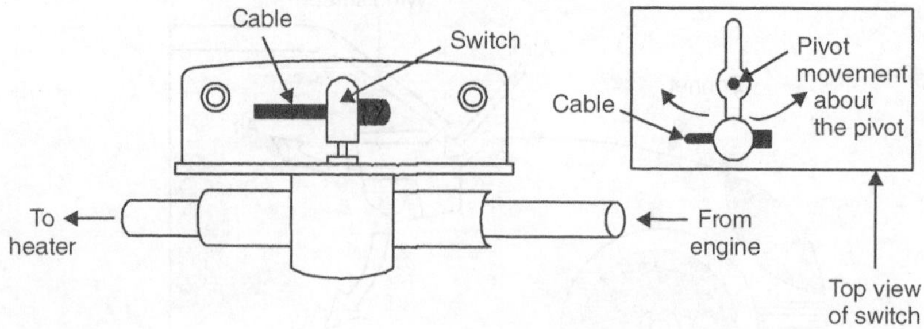

Fig. 23.9

The heater valve is vacuum operated. There is a diaphragm provided inside the valve. This diaphragm is raised, opening the valve, against the force of spring. The valve is closed when diaphragm is lowered due to force exerted by spring. To avoid corrosion, the valves are made of plastic. Another advantage is that use of plastic makes them light.

Blower motor

The blower motor ensures that circulation of air is proper and effective. It can operate at different speeds. This is achieved by controlling the flow of electric current. The motor may run continuously at low speed when engine is running. Alternatively, there can be automatic control and motor is activated at a designed temperature. A fuse is provided in the circuit for safety. To control the speed of motor, resistors are provided that control the flow of current.

Duct hoses

These are meant to transfer hot air from heater core to passenger compartment and wind shields. There is a plastic shell that forms a part of these ducts and shell connects the duct to vents.

QUESTIONS

1. Explain the measures taken to create comfort conditions for passengers in an automobile.
2. What are different components of air-conditioning system? Explain compressor used in air-conditioning system.
3. What are the functions of a condenser?
4. How does an expansion valve work?
5. What is evaporator? Explain with the help of a diagram.
6. Explain the utility of air-conditioning and temperature control systems.
7. Explain briefly different components of heating system.

24

SAFETY AND SECURITY

24.1 INTRODUCTION

Safety in the title indicates the safety of passenger while security indicates the security of automobile. Safety of passenger is most important. Nothing is lost if passenger comes out safe in an accident. The emphasis while designing, particularly the body of the vehicle, is that body should take the maximum impact of collision. For the extra safety of passenger, the modern automobiles are provided with seat belts, air bags, antilock brakes and traction control. Some of the features are available as standard features and these include seat belts. Other features are available only in high end cars.

24.1.1 Seat Belts

Generally, the passengers are required to fasten seat belts on them when they occupy their seat. But in some cars, the system works automatically and becomes operative as soon as the passengers occupy the seat. These are passive restraint systems. Electric motors are provided that automatically put shoulder belts across the driver and passenger occupying the front seat. There is a carrier moving at the top of door frame. The upper end of belt is attached to it. The other end, on the lower side, is attached to inertia lock retractor. The retractor draws the belt back when the passenger removes it. It is essential that seat belts are inspected and serviced periodically. The belts should not have any damaged part including buckle and latch plate. Figure 24.1 represents some of the manners in which a belt may be damaged, belt should be released when button on the buckle is pressed with a pressure of about 1 kg. If pressure required to release the belt is more seat belt assembly should be replaced. A cracked buckle cover should be replaced immediately. Retractor should also be inspected periodically. When the belt is fully out the retractor should lock automatically. The retractor in automatic seat belt system may be webbing sensitive or vehicle sensitive. If it does not lock, it should be replaced. When retractors are attached to car body seat belt anchors are used. In case of collision, these must be checked and if found damaged these should be replaced. Some vehicles are provided with seat belt tensioners that contain small explosive charge. These detonate during collision and tighten the seat belt in case of severe impact on front side. Rear seat belt system should also be checked periodically. The position of lap belt and single loop belt anchor may be tightened in same position and torque as specified in the user's manual. The centre seat belt, if provided, may be checked for anchors and the locking side along with webbing. Check the slide lock and if it does not function properly, replace it with new one.

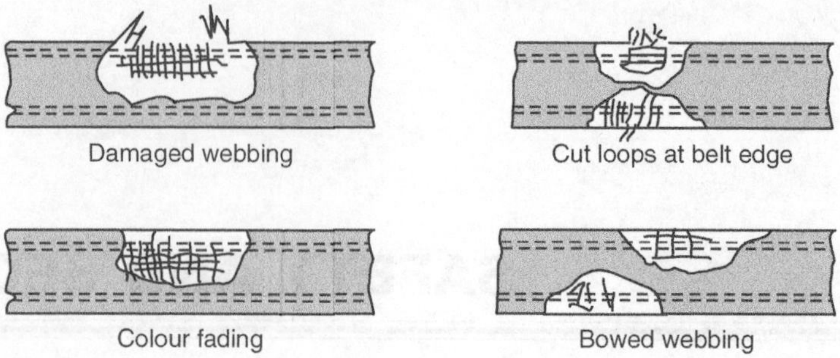

Fig. 24.1

24.1.2 Child Seat

Child seats can be rear facing, front facing or a combination of two. Child safety seats are provided with tether strap. This strap end attaches to anchor behind the rear seat. There may be integral child seats provided in some cars. These are behind the back cushion of normal seat and can be used by removing the back cushion. The child seats are provided in the rear side of the vehicle and to use these seats instructions given by the manufacturer should be followed strictly.

24.1.3 Warning Lights

Some vehicle provide warning lights or sound alarms to indicate whether seat belts have been fastened or whether ignition key is left or whether lights are left 'on' or whether door is properly locked or not. These systems warn the user of the vehicle of any malfunction mentioned above.

24.1.4 Air Bag

Air bags are meant to protect the occupants of the front seats from impact due to collision. Earlier air bags simply inflated at the time of collision and protected the passengers from hitting the front. But it was found that inflated air bag covered the passenger's face in such a way that it did not allow the passenger to breathe and caused death. The design was further improved and air bags deflated immediately. The whole process takes seventy milliseconds. Twenty milliseconds after impact, the sensors send signal to air bag module. The air bag takes three milliseconds to inflate and covers the chest of the passenger. In seventeen milliseconds, the air bag is fully inflated. At this juncture the body of the passenger starts moving forward due to impact. In the next thirty milliseconds, the bag absorbs the forward movement of the passenger and begins to deflate.

An air bag system has an independent electrical circuit, air bag module and knee diverter. The bags are located underneath the steering column in front of driver. The other bag is located above the glove box to protect the second passenger (Fig. 24.2). The knee diverter is in front of driver. This cushions the knees from impact. It also does not allow the driver to slide down under the air bag.

The electrical circuit is one of the components of air bag system. It is an independent circuit. It conducts the self checks and indicates to the driver that it is functioning normally. It has sensors, diagnostic assembly, clock spring and readiness light.

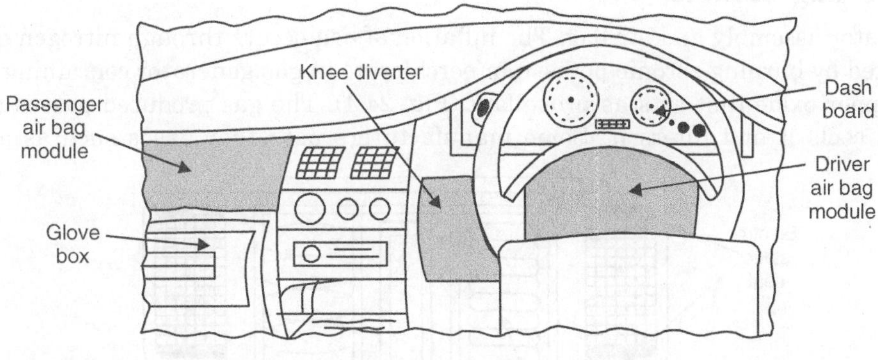

Fig. 24.2

Sensors

It has two sensors, one to sense impact and other to operate the system. The sensor has a gold plated ball held in place by magnet (Fig. 24.3). When sufficient force is applied, the ball moves away from magnet, makes contact, and completes the circuit. The number of sensors depends upon the design of the system. Another electrical circuit when completed activates the igniter. This starts chemical reaction producing nitrogen gas that fills the air bag.

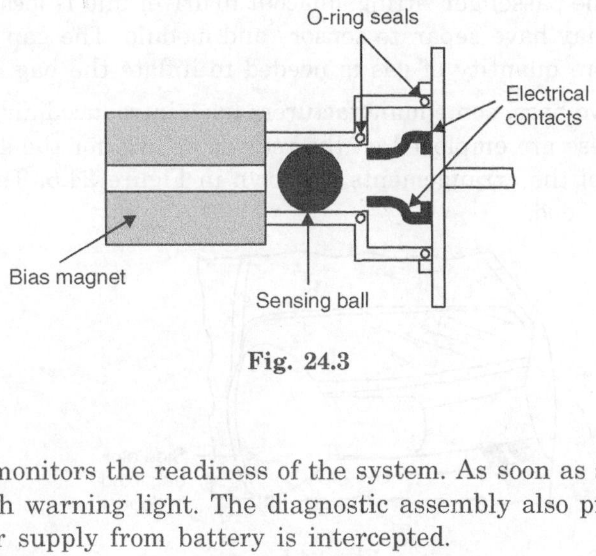

Fig. 24.3

Diagnostic assembly

This constantly monitors the readiness of the system. As soon as some fault is detected it is indicated through warning light. The diagnostic assembly also provides backup power in case normal power supply from battery is intercepted.

Lock spring

It keeps the electrical contact to the air bag system all the time. It is located between steering wheel and steering column.

Readiness light

This indicates whether the air bag system is ready or not. It becomes 'on' as ignition is switched on. If it remains 'on' it means that air bag system is not ready. Sometimes warning is also indicated through sound alarm.

21.1.5 Air Bag Module

It is inflator assembly and air bag. The inflation of bag occurs through nitrogen gas. The gas is produced by burning zeronic potassium perchlorate in gas generator containing sodium azide and copper oxide that acts as propellant (Fig. 24.4). The gas produced passes through diffuser that cools it and filters it. Some manufacturers use other gases such as argon.

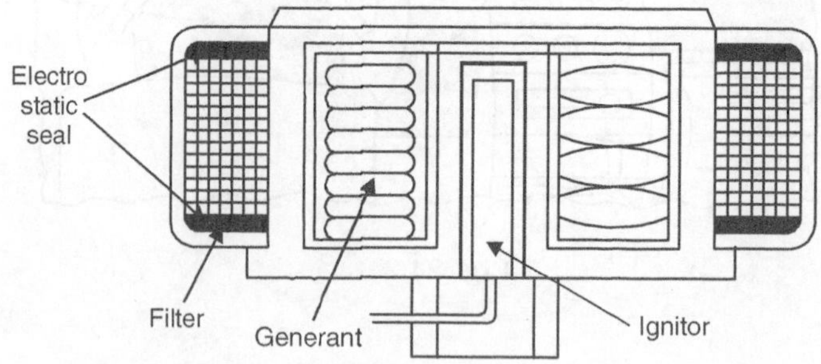

Fig. 24.4

The air bag is made of nylon coated with neoprene from inside. A mounting plate and retainer ring attaches the air bag with inflator.

21.1.6 Side Air Bag

This air bag is for the passenger sitting adjacent to driver and is located under the hand glove compartment. It may have separate sensors and module. The gap between passenger and bag being large more quantity of gas is needed to inflate the bag.

Apart from these two bags, some manufacturers have introduced air bags located in the doors of the vehicle. These are employed on the rear doors also for the safety of passengers on the rear seats. One of the arrangements is shown in Figure 24.5. The bag is located in the interior trim on the door.

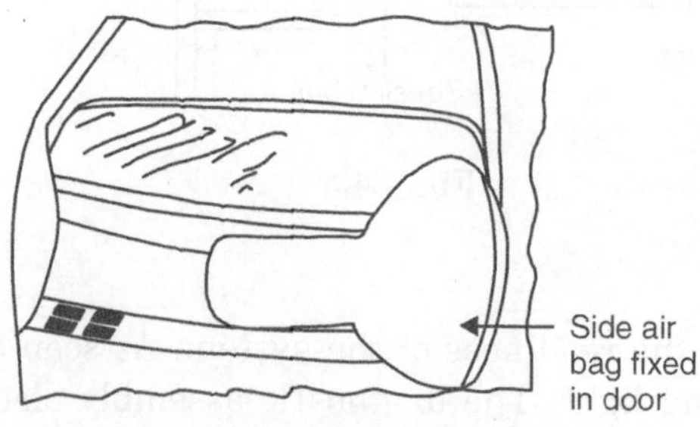

Fig. 24.5

Another very important aspect in an automobile is its safety from theft. For this purpose, in automobile antitheft devices are incorporated. These are locking devices, disabling devices and alarming devices. Some of the devices may be installed by the manufacturer and others may be installed by the customer.

Lock

The locks are provided so that doors or trunk compartment cannot be opened to prevent unauthorized entry in the vehicle. The locks are provided to the fuel tank or ignition or steering wheel. This again helps in preventing the different types of thefts. Some locks are automatically actuated as the vehicle moves. Rear doors are locked so that these are not accidentally opened by child.

Keyless entry

This allows the driver to operate the locks from outside the vehicle. It has an electronic control module and coded button keypad. The keypad is outside the driver's door. By pressing the code on keypad door locks and other locks can be operated. Now-a-days as an alternative, the driver can operate the locks of the vehicle through remote control. A small transmitter located in the key ring. There are separate switches that can open or lock the vehicle. These devices have reasonably good range of operation. This makes the locking and unlocking of the vehicle very convenient. The system may operate all the doors and trunk of the vehicle.

Disabling devices

These devices can be operated through a key or a code or a touch pad or even a hidden button. These devices disable the vehicle and make it non-operative. These act as safety devices against theft.

Pass-key system

The key is equipped with electronic circuitry and is programmed to respond to only one vehicle. In this case, though the key may enter the key hole but it would respond to only one vehicle for which it is designed.

In some cars, the key contains a radio frequency unit for locking and unlocking. Key is touched with ignition switch to unlock steering column and engine through infra red data exchange.

Alarm system

There are some alarm systems that are activated automatically as soon as doors are locked while others are activated manually through a key. The alarm can be triggered through mechanical switch. Whenever doors, trunk or hood are opened the switch is closed and there is a sound alarm. It is switched off when intruder stops trying to open the vehicle. Automatically, the alarm is activated again to prevent second such attempt. There may be current sensitive or ultrasonic sensors in place of mechanical switches to operate the alarm system.

Apart from safety and security systems, described above, the vehicles are also provided with some other devices. These are vehicle tracking system to locate the vehicle, navigation system to help the driver to find the way and mobile phone to help him in emergency.

Vehicle tracking system

These are operated with the help of satellites. There is continuous communication with the vehicle through satellite. These are useful in locating the vehicle during emergency or during storms or floods and other natural calamities. Some vehicles are capable of indicating whether air bag system has been operated. This indicates the possibility of some accident and helps in taking the suitable measures.

Navigation system

These systems operate from the signals received from satellite. They use global positioning satellite that helps the driver to find the way. There is a display of the map on a monitor that displays the paths for the driver to follow. It may be difficult for the driver to see the map and drive the vehicle. To overcome this, the system provides audible instructions to driver to move on the road and by following these instructions he can reach his destination.

Mobile phone

A mobile phone may be installed in the car. This provides facility for the driver to contact for help from anywhere any time. Now-a-days with availability of mobile phones with the users of the car the cars may not be having their mobile phones. The purpose of contacting for help in emergency is served by the mobile phone possessed by the user of the car.

QUESTIONS

1. Explain the importance of security of passengers in an automobile.
2. What are means to provide safety to passengers in the car?
3. How air bags operate in a car? Explain.
4. What are different components of air bag system? Explain briefly.
5. What is keyless entry? Explain.
6. What is vehicle tracking system? What purposes it serves?

25

AUTOMOBILE EMISSION CONTROL

25.1 INTRODUCTION

As the number of automobiles is increasing day by day the problem of controlling the emission from them is becoming more and more prominent. In metropolitan cities the problem was realized few years back and may be the situation has improved since then but presently it is influencing the environment of small cities and unfortunately no body seems to be much concerned.

The pollutants emitted by vehicles mix with air and water. Both the things are essential for human beings and as these are consumed they show adverse affect on the health of human beings.

The main pollutants coming out with emission from vehicles are hydrocarbons (HC), carbon monoxide (CO) and oxides of nitrogen (NO_X). In addition to these three the emission also contains particulate material.

25.1.1 Hydrocarbons (HC)

Incomplete combustion of fuel causes emission of hydrocarbons (HC). As the flame propagates through the combustion chamber, its front comes in contact with combustion chamber walls that are 'cooler' that is at a relatively low temperature. This cause some unburnt hydrocarbons. This situation may arise with too lean or too rich fuel air mixture. Hydrocarbon emission is minimum when stoichiometrically correct air fuel mixture is used. Other cause of hydrocarbon emission is evaporative emission in fuel tank.

25.1.2 Carbon Monoxide (CO)

Carbon monoxide (CO) is due to improper fuel air combination and comes out as the product of combustion. It is produced when enough oxygen is not available during combustion. Enough oxygen means formation of carbon dioxide (CO_2). Carbon dioxide (CO_2) is otherwise a harmless gas but its presence causes what is known as 'greenhouse effect'. It is one of the reasons of global warming. Carbon monoxide (CO) is toxic and harmful for human beings. When air fuel mixture is lean or when it is stoichiometrically correct the carbon monoxide (CO) emission is low but with rich mixture its emission becomes high.

25.1.3 Oxides of Nitrogen (NOx)

Oxides of nitrogen (NO_x) are the product of combination of nitrogen and oxygen at high temperatures. The two gases together form different compounds. The 'x' in the formula represents the proportion of oxygen mixed with nitrogen. If x is 1 *i.e.,* compound is NO, it nitrogen monoxide, when it is 2 the compound is NO_2 *i.e.,* nitrogen dioxide and so on.

Mainly, nitrogen monoxide (NO) is formed among all the oxides of nitrogen (NO_x). It is highly toxic gas causing the formation of ozone and acid rain in addition to smog.

As the number of vehicles is increasing the emission from these vehicles is a matter of concern for the scientists and engineers. In this direction in late fifties of twentieth century, standards of emission were established in California. The technical improvements in the engine design came into existence in early sixties of previous century, when crankcase was provided with ventilation. Evaporation control system was incorporated in 1970. Almost during the same time, attempts were made to reduce oxides of nitrogen. Another modification was done in fuel by removing lead from petrol. In mid seventies of the twentieth century catalytic converter was used in exhaust system that oxidized the monoxides present in emission. All these modifications occurred in developed countries like United States of America.

Euro standards for vehicle emission were introduced in 1992. These were known as Euro 1 standards. These standards specified the amount of CO, HC+NO_x and particulate matters in vehicular emission. These standards were set for all kinds of vehicles whether with petrol engine or diesel engine whether cars or light commercial or heavy commercial vehicles. Euro 2 were launched between 1996 and 1999. Euro 3 came into existence in 2000 and Euro 4 came into existence in 2005. As the standards were improved these permitted less and less amount of CO. HC, NO_X and particulate materials in vehicle emission. The idea was to prevent ambient air getting affected from these harmful constituents present in vehicular emission. The Table 25.1 gives the details of permitted materials in cars in gram/ kilometer of travel.

Table 25.1

Cars with Diesel engine (in grams per kilometer of travel)

Standard	*Year*	*CO*	*HC + NO_X*	*Particulate Materials*
Euro 1	1992	2.72	0.97	0.14
Euro 2 IDl	1996	1.0	0.7	0.08
Euro 2 DI	1999	1.0	0.9	0.10

In the year 2000, Euro 3 was introduced. It specified amount of HC + NOx and NO_X separately.

Standard	*Year*	*CO*	*HC + NO_x*	*NO_x*	*Particulate Materials*
Euro 3	2000	0.64	0.56	0.5	0.05
Euro 4	2005	0.50	0.30	0.25	0.025

Cars with Petrol engine (in grams per kilometer of travel)

Standard	*Year*	*CO*	*HC*	*NO_x*	
Euro 3	2000	2.30	0.20	0.15	
Euro 4	2005	1.00	0.10	0.08	

In India, the problem was felt in late eighties of twentieth century. The number of cars on Indian roads has increased very fast since then. Unfortunately, the matters have moved at slow pace in our country. It was as per judgment of Supreme Court that Euro 3 standards (Known as Bharat stage III) were made mandatory with effect from 1st April 2005 but only in metropolitan cities. There were some other measures also taken like selling only lead free petrol since 1998 or replacing diesel vehicles with CNG vehicles. Unfortunately, all of these measures were not adopted in small cities and these are facing the problem of increasing pollution day by day. Another reason of enhanced pollution and deterioration of environment is the ignorance and illiteracy among the people.

25.2 EMISSION CONTROL SYSTEMS

As soon as the problem of pollutants in emission of the vehicle was realized, scientist and engineers started working on means to control them. Initially emission of fuel vapours in air was realized. To control it, several measures were adopted. The design of fuel tank was modified. It limited the amount of fuel that could be filled. The vented cap of the tank was replaced by pressure relief cap. A vapour separator was provided in some tanks that collected the vapours and sent these back to fuel tank.

25.2.1 Charcoal Canister

Apart from these measures the vapours from fuel were controlled by charcoal canister. It is located in fuel tank's vapour line. When vehicle is stationary, the vapours are absorbed by the charcoal granules. When the vehicle starts the vapours are sent to carburetor where these are sent to engine cylinder. There are different means to purge the canister. These vary from Electronic Control Module (ECM) to purge valve to thermal delay valves. The function of all the devices is to allow the vapours to go to carburetor when vehicle starts moving. The canister is provided with liquid fuel trap that collects the liquid fuel. The vapours condensed and converted into liquid are returned from canister to tank when vacuum occurs in the tank (Fig. 25.1).

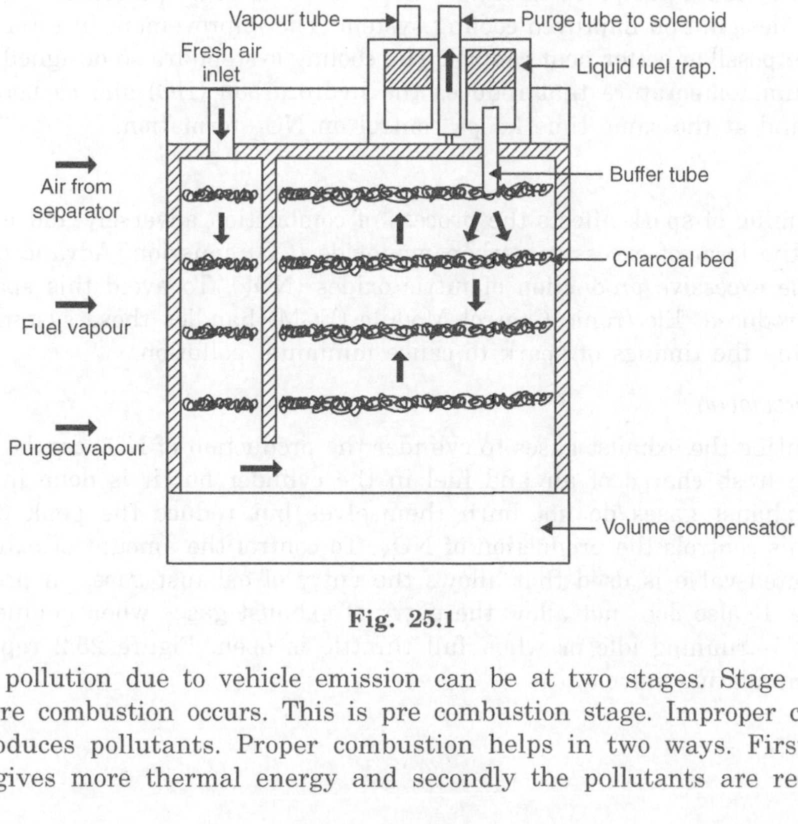

Fig. 25.1

The air pollution due to vehicle emission can be at two stages. Stage one is inside cylinder where combustion occurs. This is pre combustion stage. Improper combustion of fuel also produces pollutants. Proper combustion helps in two ways. Firstly, complete combustion gives more thermal energy and secondly the pollutants are reduced. When combustion gives more thermal energy, it also enhances the efficiency of the engine. To attain complete combustion the engine design has been modified. Stage two is post combustion stage where pollution is caused due to the products of combustion.

Pollution occurring in pre combustion stage

Blowby gases. Some unburnt fuel and products of combustion leak through the piston rings and go to crankcase. These are termed as blowby gases. These gases from crankcase go to atmosphere and cause pollution. Initially, these are reduced by having piston rings that can provide better sealing. For this the cylinder walls have also been improved to provide better and friction less contact between them and piston rings. The gases which still go to crankcase are dealt through positive crankcase ventilation system. This system prevents the emission of blowby gases from crankcase to atmosphere. The fresh air mixes with blowby gases and the mixture is sent back to cylinder where it burns. This way blowby gases are utilized.

Combustion chamber design

The modification in combustion chamber design has been done to reduce the 'quench area' which is 'relatively cooler' and causes formation of unburnt hydrocarbons. The location of spark plug is important as that determines the propagation of flame. Another important parameter is mixing of fuel and air that is achieved by creating turbulence.

Intake manifold design

To control the formation of NO_X, it is essential to keep combustion temperature low. This is achieved by lower compression ratio. There are some other parameters also such as intake manifold design, and improved cooling system. The improvement in intake manifold design has made possible better heat control. The cooling system are so designed that these allow an optimum temperature that reduces the hydrocarbon (HC) and carbon monoxide (CO) emission and at the same time keeps control on NO_X formation.

Spark timing

Incorrect timing of spark affects the process of combustion adversely and is the cause of pollution in the form of excessive carbon monoxide (CO) emission. Advanced timing of spark may cause excessive production of nitric oxides (NO_X). To avoid this spark control systems are introduced. Electronic Control Module (ECM) handles these systems and are capable of varying the timings of spark to cause minimum pollution.

Exhaust gas recirculation

By recirculating the exhaust gases to cylinder the production of NO_X can be controlled. This dilutes the fresh charge of air and fuel in the cylinder but it is done in controlled manner. The exhaust gases do not burn themselves but reduce the peak combustion temperature. This controls the production of NO_X. To control the amount of exhaust gases getting recirculated valve is used that allows the entry of exhaust gases in proportion to throttle opening. It also does not allow the entry of exhaust gases when engine starts up cold or when it is running idle or when full throttle is open. Figure 25.2 represents an exhaust gas recirculating valve.

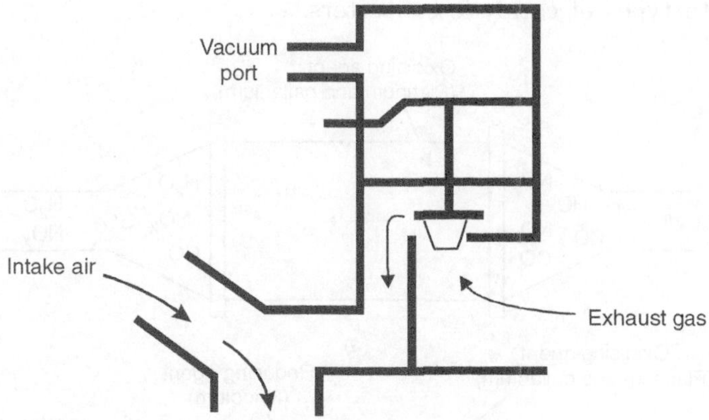

Fig. 25.2

Intake heat control system

When the engine is started in cold condition, the hydrocarbon (HC) and carbon monoxide (CO) contents are maximum in emission. To reduce it warm air is provided for the combustion. To achieve this intake manifolds are provided with heating devices. These devices also prevent the condensation of fuel particles in the intake manifold. Alternatively, the exhaust gases are routed through intake manifold and these cause warming up of air.

Some modern cars have computer controlled mixture heater. These are controlled through Electronic Control Module (ECM) provided in the car. Basically, fuel evaporation heater is a resistance grid that is located in the venturi of the carburetor.

Pollution occurring in post combustion stage

Considering the harmful affects of pollutants, devices are employed in the automobiles at the exhaust outlet to tackle the pollutants going along with the emission. Catalytic converter is the most effective device used in post combustion stage. It helps in reducing hydrocarbons (HC) and oxides of nitrogen and carbon in the emission.

Catalytic converter

Though prior to use of catalytic converter other means—such as auxiliary air injection system, exhaust gas recirculation system—were used but these affected the engine performance adversely.

Catalytic converter consists of ceramic element coated with catalyst. The catalyst can enhance the chemical reaction but is not a part of reaction. The catalyst mainly converts monoxides into dioxides that are not harmful. The catalysts used are platinum, palladium and rhodium. Platinum and Palladium are oxidizing agents. These oxidize hydrocarbons (HC) and carbon monoxide (CO) in the emission into water (H_2O) and carbon dioxide (CO_2). Rhodium acts as reducing agent. It reduces oxides of nitrogen (NO_x) into Nitrogen (N) by removing oxygen. The catalytic converters are of two types. These could be of three way catalyst type that can tackle hydrocarbons (HC), carbon monoxide (CO) and oxides of nitrogen (NO_x). These have platinum, palladium and rhodium. Second types of catalytic converters are those having platinum and palladium only. These are only oxidizing the hydrocarbons

(HC) and carbon monoxide (CO) and do not reduce oxides of nitrogen (NO_x). Figure 25.3 (*a*) and (*b*) represent both types of catalytic converters.

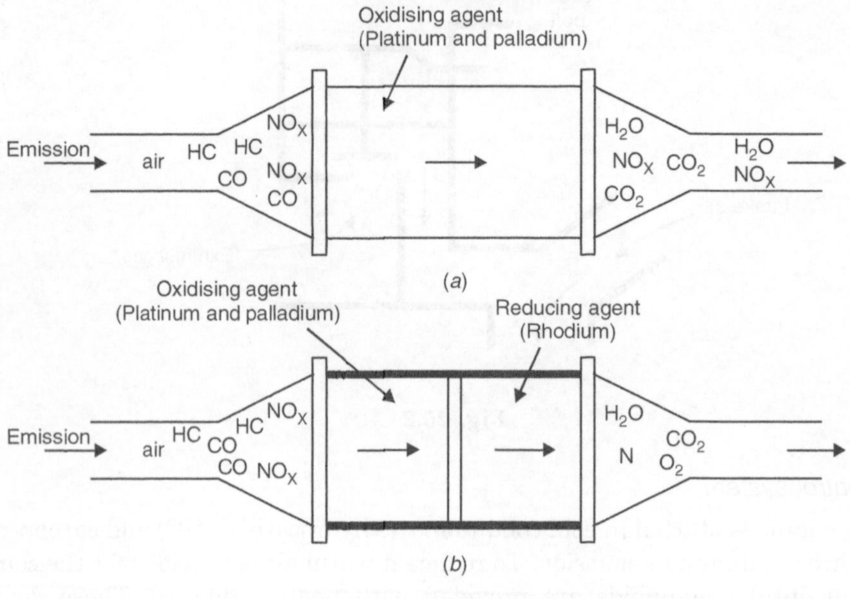

Fig. 25.3

Prior to catalytic converters air injection systems were used. These provided fresh air. The oxygen in the air provided the oxidation of hydrocarbons (HC) and carbon monoxide (CO). These were pump type or pulse type. In pump type air injectors, air pump provided pressurized air that was sent to exhaust manifold. The air pump was driven by engine through crank shaft. In pulse type injectors, utilized the natural exhaust pressure pulses to suck the air from air cleaner and force it into the exhaust manifold.

QUESTIONS

1. What are different pollutants in automobile emission?
2. What are different standards of emission adopted in the world? Describe the brief history.
3. What is the status of emission standards in India?
4. How the pollutants are controlled in emission? Describe briefly.
5. Explain the constructional details and working of catalytic converter.

26

HYBRID CARS

The realization that conventional fossil fuel is available in limited quantity and its rising price has produced so many innovations. Small cars were produced to use the available fuel as economically as possible. Similarly, work was undertaken to replace the conventional fuel by non-conventional fuel. The attempts were also made to run the cars on two or more sources of power. These cars that had two or more sources of power were termed as 'hybrid'.

In nineteenth century before internal combustion engine came into existence the attempts were to run a self propelled vehicle through steam engine or an electric motor. The results were encouraging but the vehicles had their own limitations. The steam engines were very heavy and cumbersome. The electric vehicles were not that efficient as battery technology was not that developed. First successful attempt is reported in 1897 when Walter Bersey designed a car that could run 50 miles between the charges. This was known as Bersey cab. The first hybrid car using an internal combustion engine and an electric motor is reported in 1898 built by Porsche. Similarly, in 1900 a Belgian car maker, Pieper, introduced a 3.5 horse power car with petrol engine mated to an electric motor. When car was 'cruising' its electric motor worked as generator and charged the batteries. But when it was required to move upward over an inclined surface, extra effort was provided by electric motor.

Internal combustion engine replaced electric motor in 1904 after engineers were able to overcome the problems of noise, vibrations and odor related with the engine. The attempts continued by few engineers and we find that H. Piper came out with car having electric motor and internal combustion engine which could attain a speed of 25 miles per hour. Similarly, in 1910 a truck was produced that used a four cylinder petrol engine to run a generator that in turn operated an electric motor. But in 1913, when conventional vehicles using an internal combustion engine, were provided with self-starter it was an end to all other options. It became very easy to use the vehicle.

In seventies of twentieth century, with Arab oil embargo, the price of petrol soared to new heights. This made scientists and engineers to sit together and think of alternative to fossil fuel and means to use the fossil fuel as economically as possible. This gave rise to use solar energy, electrical energy and making cars using small engines. Again the attempts began to support the conventional internal combustion engine with other source of energy. In this regard an electric motor supporting the conventional engine seems to be most feasible and we have development of hybrid vehicles using both as source of energy.

During past twenty years, several manufacturers, such as Audi, Honda, General Motors, Toyota, and Ford in United State of America and Europe have undertaken the manufacturing of hybrid cars. In 2004, Toyota manufactured 47,000 cars for its US market and sold them successfully.

26.1 HYBRID TERMS

Hybrid system of automobiles can be subdivided into Full hybrid and Mild hybrid.

26.1.1 Full Hybrid

If a vehicle can launch forward at low speed without consuming petrol it is termed as full hybrid. Presently, Lexus, Toyota and Ford hybrids are full hybrid.

26.1.2 Mild Hybrid

Output from internal combustion engine only can make possible their movement from rest. Electric motor can only supplement the power provided by internal combustion engine. Mild hybrid can further be classified as Stop/Start hybrid systems. Integrated Starter Alternator with Damping hybrid systems and Integrated Motor Assist Hybrid systems.

Start/stop hybrid system

When the engine is running idle, it is switched off and can be restarted when needed.

Integrated starter alternator with damping hybrid system

Here the electric motor provides starting and stopping capacity and also supplements the power when the vehicle is moving.

Integrated motor assist hybrid system

Here the electric motor is bigger and provides more amount of electricity for the vehicle to move.

By another criterion, the hybrid systems could be divided into parallel and series hybrid systems.

26.1.3 Parallel Hybrid System

Here simultaneously, the petrol is fed to engine and batteries provide power to electric motor. Both of these provide propulsion power.

26.1.4 Series Hybrid System

In this case, petrol engine runs a generator. This generator runs an electric motor. It also charges the batteries. Here, the petrol engine does not provide to propel the vehicle.

26.2 HOW HYBRID CARS WORK?

The hybrid cars are provided with smaller internal combustion engine. The reduced output from engine is compensated by electric motor that runs on chargeable battery. Generally, when driver begins journey, up to a speed of say 25 kilometer per hour, the electric motor provides the power and engine is not working so no consumption of fuel and no emission. Next beyond this speed the engine also starts and provides power to run the car. The computer decides, according to the requirement of driver, what should be the contribution of engine or electric motor. In some cars, the engines continue to contribute power till the car is decelerated. Upon deceleration, the engine stops and whole load is taken up by electric motor. The fact is that problems regarding emission are more prominent when the car is running at slow speed. An indictor on the dashboard provides information regarding use of electric motor or engine or their combination to provide required power to the car.

The application of computer has made the operation of the hybrid car very simple and efficient. It can sense the requirements of the vehicle and acts accordingly. This makes drive absolutely noiseless, emission free from pollutants and economical.

26.3 MAIN COMPONENTS OF HYBRID SYSTEM

26.3.1 Battery

It is an important part of the hybrid system as it is solely responsible for providing energy to it in large amount. The battery in a hybrid car is similar to that in ordinary car except that it can store large amount energy. This battery is known as deep cycle battery. The battery can be fully discharged and charged again. As amount of energy stored in the battery is large, it becomes very heavy. The attempts are being made to reduce its weight. Nickel-Metal-Hybrid (NiMH) is being used in batteries instead of lead-acid. This makes them light and small.

Toyota is working on Lithium batteries for their hybrid cars. Lithium is the lightest element with chemical weight of three. Its very light, available as solid, and is cheap. Therefore, it is the choice of scientists for use as energy carrier. Honda is using NiMH prismatic design battery for their Prius and Civic hybrid cars whereas Ford Escape hybrids have NiMH battery pack supplied by Sanyo.

The disposal of used batteries used in hybrid requires expert handling. Lithium is toxic and creates environmental problems if left in open. The lead and nickel used in batteries cause similar problems but with greater magnitudes. The car manufacturers are keen that used batteries of hybrid cars are returned to them. The battery is dissembled, its plastic material is shredded, the metal is recovered and processed and alkaline material is neutralized. Proper disposal of these batteries is essential to save environment.

26.3.2 Regenerative Braking

Braking in cars causes generation of heat that is dissipated and goes to waste. The force is applied on wheels that cause braking action. This force can be converted into torque at the electric motor shaft where it is converted into electrical energy. This energy can be used to recharge the batteries. The conversion into torque can be done conveniently through motor technology and motor controller. This is termed as 'regenerative braking' as through regeneration the energy can be reused and is not wasted.

26.3.3 Ultracapacitors

The hybrid power train can recapture energy from the wheels when brakes are applied. This energy can be used when vehicle accelerates. The energy can be recovered in electric form, electrochemical form or compressed fluid form. Capacitors can store energy in electrical form. It is stored as charge on metal plates separated by air or a non-conductive media. In case of hybrid system, the electric energy is to be stored and discharged in large quantity. This is possible with an Ultracapacitor. These capacitors are capable of absorbing and releasing large amount of electric charge.

As hybrid, simply, means combination of two, we have different types of hybrid such as diesel hybrid, ethanol hybrid, flywheel hybrid or hydraulic hybrid. These have been developed and are being used in automobiles with their own advantages and disadvantages.

Diesel hybrid

Heavy vehicles, that are meant to carry goods from one place to another run on diesel engine. The unhealthy emission from diesel engine is being controlled by using improvised fuels such as low Sulphur diesel. The engines of these vehicles have been combined with electric motor to give diesel hybrid. The diesel hybrids cost more initially but that is compensated with low running cost.

Flywheel hybrid

The hybrids use combination of two sources of power. One is energy supply unit that acts while the vehicle is running normally and another surge power unit that acts when it is accelerated. Normally surge power unit is electric motor. The battery is used to run the motor. The attempts have been made to utilize the kinetic energy that is being wasted while braking. This energy is stored and utilized when the vehicle accelerates. One method is that it is converted into electrical energy through motor or generator. This electrical energy is converted into chemical energy as battery charge up. Further, the energy is converted into electrical energy from where it is converted back to kinetic energy. This repeated conversion of energy causes losses at every stage. Alternatively, this energy can be stored in flywheel and can be used without any transformation. This would make the system more efficient. Using flywheel has been attempted and there are flywheel hybrid systems. The limitation with systems is that losses occur due to bearing friction, windage etc. These losses are so huge that they make the system even less efficient than battery based system. However, the flywheel hybrid systems can be used where storage required is for short periods. For example, in traffic when the vehicle is required to stop and accelerate immediately these systems can be used successfully. To make them more efficient surge power units are combination of flywheel and battery system. This way the battery is saved from shock loads of acceleration and braking. This enhances the battery life and improves the system.

Hydraulic hybrid

This system has pump and an accumulator. The accumulator is a pressure tank that stores compressed gas or liquid. When vehicle decelerates the fluid from low-pressure tank is transferred to high pressure tank by the pump. When vehicle accelerates reverse movement of fluid from high pressure tank to low pressure tank occurs through the pump. This movement of fluid causes application of torque to the wheels. This system is 80% efficient.

26.3.4 Safety Aspect of Hybrid Vehicles

As far as aspect of safety is concerned there is no difference whether vehicle is hybrid or otherwise. The safety features in cars whether they are hybrid or not are the same. It has been found in a recent survey in United States of America that hybrid vehicles are at a disadvantage in multiple vehicle crashes. In case of pickups and sports utility vehicles (SUVs), these are at disadvantage in single vehicle crash.

26.3.5 Hybrid Hazards

As it is there are specific hazards in hybrid cars. It is important to know the location of battery compartment. Also it is essential to have the knowledge of different cables and location of fuses. First of all in case of emergency the electrical system is to be disconnected. This can be done by simply switching off the ignition and removing the ignition key.

In case of fire, water should be available at hand. Extinguishing the fire with water is enough (after the electrical system has been disconnected). Water would also cool down the metal battery box and the plastic cells inside the battery pack.

QUESTIONS

1. Give brief history of hybrid cars.
2. What are the common hybrid terms? Explain briefly.
3. What are the main components of hybrid system?
4. Explain different types of hybrid systems used in cars.

27

THE MOTOR VEHICLE ACT

This chapter does not provide any technical information rather it provides the legal aspects of automobiles in India. The chapter finds its place in the book because legal information regarding the automobiles in India is quite relevant. The Motor Vehicle Act has been reproduced here as it is without any comments from the author for the benefit of readers. The attempt has been made to include all the amendments made in the act from time to time but still readers can update their information from the different sites of Government of India available on net.

THE MOTOR VEHICLE ACT, 1988

No. 59 OF 1988 [14th *October,* 1988.]

An Act to consolidate and amend the law relating to motor vehicles.

BE it enacted by Parliament in the Thirty-ninth Year of the Republic of India as follows:

CHAPTER 1

PRELIMINARY

1. Short Title, Extend and Commencement

(1) This Act may be called the Motor Vehicles Act, 1988.

(2) It extends to the whole of India.

(3) It shall come into force on such date *{1-7-1989; vide Notification No.S.O.368(E), dated 22-5-1989, Gazette of India, Extraordinary, 1989, Pt.ll; Sec.3(ii).}* as the Central Government may, by notification in the Official Gazette, appoint; and different dates may be appointed for different States and any reference in this Act to the commencement of this Act shall, in relation to a State, be construed as a reference to the coming into force of this Act in that State.

2. Definitions

In this Act, unless the context otherwise requires:

(1) "area", in relation to any provision of this Act, means such area as the State Government may, having regard to the requirements of that provision, specify by notification in the Official Gazette;

(2) "articulated vehicle" means a motor vehicle to which a semi-trailer is attached;

(3) "axle weight" means in relation to an axle of a vehicle the total weight transmitted by the several wheels attached to that axle to the surface on which the vehicle rests;

(4) "certificate to registration" means the certificate issued by a competent authority to the effect that a motor vehicle has been duly registered in accordance with the provisions of Chapter IV;

(5) "conductor", in relation to a stage carriage, means a person engaged in collecting fares from passengers, regulating their entrance into, or exit from, the stage carriage and performing such other functions as may be prescribed;

(6) "conductor's licence" means the licence issued by a competent authority under Chapter III authorising the person specified therein to act as a conductors;

(7) "contract carriage" means a motor vehicle which carries a passenger or passengers for hire or reward and is engaged under a contract, whether expressed or implied, for the use of such vehicle as a whole for the carriage of passengers mentioned therein and entered into by a person which a holder of a permit in relation to such vehicle or any person authorised by him in this behalf on a fixed or an agreed rate or sum:

(a) on a time basis, whether or not with reference to any route or distance; or

(b) from one point to another,

and in either case, without stopping to pick up or set down passengers not included in the contract anywhere during the journey, and includes:

(i) a maxicab; and

(ii) a motor cab notwithstanding that separate fares and charged for its passengers;

(8) "dealer" includes a person who is engaged:

(a) in the manufacture of motor vehicles; or

(b) in building bodies for attachment to chassis; or

(c) in the repair of motor vehicles; or

(d) in the business of hypothecation, leasing or hire-purchase of motor vehicle;

(9) "driver" includes, in relation to a motor vehicle which is drawn by another motor vehicle, the person who acts as a steersman of the drawn vehicle;

(10) "driving licence" means the licence issued by a competent authority under Chapter II authorising the person specified therein to drive, otherwise than as a learner, a motor vehicle or a motor vehicle of any specified class or description;

(11) "educational institution bus" means an omnibus, which is owned by a college, school or other educational institution and used solely for the purpose of transporting students or staff of the educational institution in connection with any of its activities;

(12) "fares" includes sums payable for a season ticket or in respect of the hire of a contract carriage;

(13) "goods" includes live-stock, and anything (other than equipment ordinarily used with the vehicle) carried by a vehicle except living persons, but does not include

luggage or personal effects carried in a motor car or in a trailer attached to a motor car or the personal luggage of passengers travelling in the vehicle;

(14) "goods carriage" means any motor vehicle constructed or adapted for use solely for the carriage of goods, or any motor vehicle not so constructed or adapted when used for the carriage of goods;

(15) "gross vehicle weight" means in respect of any vehicle the total weight of the vehicle and load certified and registered by the registering authority as permissible for that vehicle;

(16) "heavy goods vehicle" means any goods carriage the gross vehicle weight of which, or a tractor or a road-roller the unladen weight of either of which, exceeds 12,000 kilograms;

(17) "heavy passenger motor vehicle" means any public service vehicle or private service vehicle or educational institution bus or omnibus the gross vehicle weight of any of which, or a motor car the unladen weight of which, exceeds 12,000 kilograms;

(18) "invalid carriage" means a motor vehicle specially designed and constructed, and not merely, adapted, for the use of a person suffering from some physical defect or disability, and used solely by or for such a person;

(19) "learner's licence" means the licence issued by a competent authority under Chapter II authorising the person specified therein to drive as a learner, a motor vehicle or a motor vehicle of any specified class or description;

(20) "licensing authority" means an authority empowered to issue licences under Chapter II or, as the case may be, Chapter III;

(21) "light motor vehicle" means a transport vehicle or omnibus the gross vehicle weight of either of which or a motor car or tractor or road-roller the unladen weight of any of which, does not exceed 6,000 kilograms;

(22) "maxicab" means any motor vehicle constructed or adapted to carry more than six passengers, but not more than twelve passengers, excluding the driver, for hire or reward;

(23) "medium goods vehicle" means any goods carriage other than a light motor vehicle or a heavy goods vehicle;

(24) "medium passenger motor vehicle" means any public service vehicle or private service vehicle, or educational institution bus other than a motor cycle, invalid carriage, light motor vehicle or heavy passenger motor vehicle;

(25) "motorcab" means any motor vehicle constructed or adapted to carry not more than six passengers excluding the driving for hire or reward;

(26) "motor car" means any motor vehicle other than a transport vehicle, omnibus, road-roller, tractor, motor cycle or invalid carriage;

(27) "motor cycle" means two-wheeled motor vehicle, inclusive of any detachable side-car having an extra wheel, attached to the motor vehicle;

(28) "motor vehicle" or "vehicle" means any mechanically propelled vehicle adapted for use upon roads whether the power of propulsion is transmitted thereto from an

external or internal source and includes a chassis to which a body has not been attached and a trailer; but does not include a vehicle running upon fixed rails or a vehicle of a special type adapted for use only in a factory or in any other enclosed premises or a vehicle having less than four wheels fitted with engine capacity of not exceeding thirty five cubic centimetres;

(29) "omnibus" means any motor vehicle constructed or adapted to carry more than six persons excluding the driving;

(30) "owner" means a person in whose name a motor vehicle stands registered, and where such person is a minor, the guardian of such minor, and in relation to a motor vehicle which is the subject of a hire-purchase, agreement, or an agreement of lease or an agreement of hypothecation, the person in possession of the vehicle under that agreement;

(31) "permit" means a permit issued by a State or Regional Transport Authority or an authority prescribed in this behalf under this Act authorising the use of a motor vehicle as a transport vehicle;

(32) "prescribed" means prescribed by rules made under this Act;

(33) "private service vehicle" means a motor vehicle constructed or adapted to carry more than six persons excluding the driving and ordinarily used by or on behalf of the owner of such vehicle for the purpose of carrying persons for, or in connection with, his trade or business otherwise than for hire or reward but does not include a motor vehicle used for public purposes;

(34) "public place" means a road, street, way or other place, whether a thoroughfare or not, to which the public have a right of access, and includes any place or stand at which passengers are picked up or set down by a stage carriage;

(35) "public service vehicle" means any motor vehicle used or adapted to be used for the carriage of passengers for hire or reward, and includes a maxicab, a motorcab, contract carriage, and stage carriage;

(36) "registered axle weight" means in respect of the axle of any vehicle, the axle weight certified and registered by the registering authority as permissible for that axle;

(37) "registering authority" means an authority empowered to register motor vehicles under Chapter IV;

(38) "route" means a line of travel which specifies the highway which way be traversed by a motor vehicle between one terminus and another;

(39) "semi-trailer" means a trailer drawn by a motor vehicle and to constructed that a part of it is super-imposed on, and a part of its weight is borne by, the drawing vehicle;

(40) "stage carriage" means a motor vehicle constructed or adapted to carry more than six passengers excluding the driving for hire or reward at separate fares paid by or for individual passengers, either for the whole journey or for stages of the journey;

(41) "State Government" in relation to a Union territory means the Administrator thereof appointed under article 239 of the Constitution;

(42) "State transport undertaking" means any undertaking providing road transport service, where such undertaking is carried on by :

(*i*) the Central Government or a State Government;

(ii) any Road Transport Corporation established under section 3 of the Road Transport Corporations Act, 1950 (64 of 1950.);

(iii) any municipality or any corporation or company owned or controlled by the Central Government or one or more State Governments, or by the Central Government and one or more State Governments.

Explanation — For the purposes of this clause, "road transport vice" means a service of motor vehicles carrying passengers or goods or both by road for hire or reward;

(43) "tourist vehicle" means a contract carriage constructed or adapted and equipped and maintained in accordance with such specifications as may be prescribed in this behalf;

(44) "tractor" means a motor vehicle which is not itself constructed to carry any load (other than equipment used for the purpose of propulsion); but excludes a road-roller;

(45) "traffic signs" includes all signals, warning sign posts, direction posts, markings on the road or other devices for the information, guidance or direction of driving of motor vehicles;

(46) "trailer" means any vehicle, other than a semi-trailer and a side-car, drawn or intended to be drawn by a motor vehicle;

(47) "transport vehicle" means a public service vehicle, a goods carriage, an educational institution bus or a private service vehicle;

(48) "unladen weight" means the weight of a vehicle or trailer including all equipment ordinarily used with the vehicle or trailer when working, but excluding the weight of a driving or attendant; and where alternative parts or bodies are used the unladen weight of the vehicle means the weight of the vehicle with the heaving such alternative part or body;

(49) "weight" means the total weigh transmitted for the time being by a wheels of a vehicle to the surface on which the vehicle rests.

CHAPTER II

LICENSING OF DRIVING OF MOTOR VEHICLES

3. Necessity for Driving Licence

(1) No person shall drive a motor vehicle in any public place unless he holds an effective driving licence issued to him authorising him to drive the vehicle; and no person shall so drive a transport vehicle [other than a motor cab hired for his own use or rented under any scheme made under sub-section (2) of section 75] unless his driving licence specifically entitles him so to do.

(2) The conditions subject to which sub-section *(1)* shall not apply to a person receiving instructions in driving a motor vehicle shall be such as may be prescribed by the Central Government.

4. Age Limit in Connection with Driving of Motor Vehicles

(1) No person under the age of eighteen years shall drive a motor vehicle in any public place :

Provided that a motor cycle without gear may be driven in a public place by a person after attaining the age of sixteen years.

(2) Subject to the provisions of section 18, no person under the age of twenty years shall drive a transport vehicle in any public place.

(3) No learner's licence or driving licence shall be issued to any person to drive a vehicle of the class to which he has made an application unless he is eligible to drive that class of vehicle under this section.

5. Responsibility of Owners of Motor Vehicles for Contravention of Sections 3 and 4

No owner or person in charge of a motor vehicle shall cause or permit any person who does not satisfy the provisions of section 3 or section 4 to drive the vehicle.

6. Restriction on the Holding of Driving Licences

(1) No person shall, while he holds any driving licence for the time being in force, hold any other driving licence except a learner's licence or a driving licence issued in accordance with the provisions of section 18 or a document authorising, in accordance with the rules made under section 139, the person specified therein to drive a motor vehicle.

(2) No holder of a driving licence or a learner's licence shall permit it to be used by any other person.

(3) Nothing in this section shall prevent a licensing authority having the jurisdiction referred to in sub-section (1) of section 9 from adding to the classes of vehicles which the driving licence authorities the holder to drive.

7. Restrictions on the Granting of Learners Licences for Certain Vehicles

(1) No person shall be granted a learner's licence to drive a transport vehicle unless he has held a driving licence to drive a light motor vehicle for at least one year."

(a) to drive a heavy passenger motor vehicle unless he has held a driving licence for at least two years to drive a light motor vehicle or for at least one year to drive a medium passenger motor vehicle;

(b) to drive a medium goods vehicle or a medium passenger motor vehicle unless he has held a driving licence for at least one year to drive a light motor vehicle.

(2) No person under the age of eighteen years shall be granted a learner's licence to drive a motor cycle without gear except with the consent in writing of the person having the care of the person desiring the learner's licence.

8. Grant of Learners Licence

(1) Any person who is not disqualified under section 4 for driving a motor vehicle and who is not for the time being disqualified for holding or obtaining a driving licence may, subject to the provisions of section 7, apply to the licensing authority having jurisdiction in the area:

(i) in which he ordinarily resides or carries on business, or

(ii) in which the school or establishment referred to in section (12) from where he intends to receive instruction in driving a motor vehicle is situate,

for the issue to him of learner's licence.

(2) Every application under sub-section (1), shall be in such form and shall be accompanied by such documents and with such fee as may be prescribed by the Central Government.

(3) Every application under sub-section (1), shall be accompanied by a medical certificate in such form as may be prescribed by the Central Government and signed by such registered medical practitioner, as the State Government or any person authorised in this behalf by the State Government may, by notification in the Official Gazette, appoint for this purpose.

(4) If, from the application or from the medical certificate referred to in sub-section (3), in appears that the applicant is suffering from any disease or disability which is likely to cause the driving by him of a motor vehicle of the class which he would be authorised by the learner's licence applied for to drive to be a source of danger to the public or to the passengers, the licensing authorities shall refuse to issue the learner's licence:

Provided that a learner's licence limited to driving an invalid carriage may be issued to the applicant, if the licensing authority is satisfied that he is fit to drive such a carriage.

(5) No learner's licence shall be issued to any applicant unless he passes to the satisfaction of the licensing authorities such test as may be prescribed by the Central Government.

(6) When an application has been duly made to the appropriate licensing authority and the applicant has satisfied such authorities of his physical fitness under sub-section (3), and has passed to the satisfaction of the licensing authority the test referred to in sub-section (5), the licensing authority shall, subject to the provisions of section (7), issue the applicant a learner's licence unless the applicant is disqualified under section (4) for driving a motor vehicle or is for the time being disqualified for holding or obtaining a licence to drive a motor vehicle:

Provided that a licensing authority may issue a learner's licence to drive a motor cycle or a light motor vehicle notwithstanding that it is not the appropriate licensing authority, if such authority is satisfied that there is good reason for the appellant's inability to apply to the appropriate licensing authority.

(7) Where the Central Government is satisfied that it is necessary or expedient so to do, it may, by rules made in this behalf, exempt generally, either absolutely or subject to such conditions as may be specified in the rules, any class of persons from the provisions of sub-section (3), or sub-section (5), or both.

(8) Any learner's licence for directed a motor cycle in force immediately before the commencement of this Act shall, after such commencement, be deemed to be effective for driving a motor cycle with or without gear.

9. Grant of Driving Licence

(1) Any person who is not for the time being disqualified for holding or obtaining a driving licence may apply to the licensing authorities having jurisdiction in the area:

(i) in which he ordinarily resides or carries on business, or

(ii) in which the school or establishment referred to in section (12), from where

he is receiving or has received instruction in driving a motor vehicle is situated, for the issue to him of a driving licence.

(2) Every application under sub-section (7) shall be in such form and shall be accompanied by such fee and such documents as may be prescribed by the Central Government.

(3) No driving licence shall be issued to any applicant unless the passes to the satisfaction of the licensing authority such test of competence to drive as may be prescribed by the Central Government.

Provided that, where the application is for a driving licence to drive a motor cycle or a light motor vehicle, the licensing authority shall exempt the applicant from the test of competence prescribed under this sub-section, if the licensing authority is satisfied:

(*a*) (*i*) that the applicant has previously held a driving licence and that the period between the date of expiry of that licence and the date of such application does not exceed five years; or

(*ii*) that the applicant holds or has previously held a driving licence issued under section (18); or

(*iii*) that the applicant holds a driving licence issued by a competent authority of any country outside India; and

(*b*) that the applicant is not suffering from any disease or disability which is likely to cause the driving by him of a motor cycle or, as the case may be, a light motor vehicle to be a source of danger to the public; and the licensing authorities may for that purpose require the applicant to produce a medical certificate in the same form and in the same manner as is referred to in sub-section (*i*) of section (8); Provided further that where the application is for a driving licence to drive a motor vehicle (not being a transport vehicle), the licensing authority may exempt the applicant from the test of competence to drive prescribed under this sub-section, if the applicant possesses a driving certificate issued by an automobile association recognised in this behalf by the State Government.

(4) Where the application is for a licence to drive a transport vehicle, no such authorisation shall be granted to any applicant unless he possesses such minimum educational qualification as may be prescribed by the Central Government and a driving certificate issued by a school or establishment referred to in section (12).

(5) Where the applicant does not pass to the satisfaction of the licensing authority the test of competence to drive under sub-section (3), he shall not be qualified to reappear for such test:

(*a*) in the case of first three such tests, before a period of one month from the date of last test; and

(*b*) in the case of such test after the first three tests, before a period of one year from the date of last such test.

(6) The test of competence to drive shall be carried out in a vehicle of the type to which the application refers:

Provided that a person who passed a test in driving a motor cycle with gear shall be deemed also to have passed a test in driving a motor cycle without gear.

(7) When any application has been duly made to the appropriate licensing authority and the applicant has satisfied such authority of his competence to drive, the licensing authority shall issue the applicant a driving licence unless the applicant is for the time being disqualified for holding or obtaining a driving licence:

Provided that a licensing authority may issue a driving licence to drive a motor cycle or a light motor vehicle notwithstanding that it is not the appropriate licensing authority, if the licensing authorities is satisfied that there is good and sufficient reason for the applicant's inability to apply to the appropriate licensing authority:

Provided further that the licensing authority shall not issue a new driving licence to the applicant, if he had previously held a driving licence, unless it is satisfied that there is good and sufficient reason for his inability to obtain a duplicate copy of his former licence.

(8) If the licence authority is satisfied, after giving the applicant an opportunity of being heard, that he:

(a) is a habitual criminal or a habitual drunkard; or

(b) is a habitual addict to any narcotic drug or psychotropic substance within the meaning of the Narcotic Drugs and Psychotropic Substances Act, 1985 (16 of 1985.); or

(c) is a person whose licence to drive any motor vehicle has, at any time earlier, been revoked.

it may, for reasons to be recorded in writing, make an order refusing to issue a driving licence to such person and any person aggrieved by an order made by a licensing authority under this sub-section may, within thirty days of the receipt of the order, appeal to the prescribed authority.

(9) Any driving licence for driving a motor cycle in force immediately before the commencement of this Act shall, after such commencement, be deemed to be effective for driving a motor cycle with or without gear.

10. Form and Contents of Licence to Drive

(1) Every learner's licence and driving licence, except a driving licence issued under section (18), shall be in such form and shall contain such information as may be prescribed by the Central Government.

(2) A learner's licence or, as the case may be, driving licence shall also be expressed as entitling the holder to drive a motor vehicle of one or more of the following classes, namely:

(a) motor cycle without gear;

(b) motor cycle with gear;

(c) invalid carriage;

(d) light motor vehicle;

(e) medium goods vehicle;

(f) medium passenger motor vehicle;

(g) heavy goods vehicle;

(*h*) heavy passenger motor vehicle;

(*i*) road-roller;

(*j*) motor vehicle of a specified description.

11. Additions of Driving Licence

(1) Any person holding a driving licence to drive any class or description of motor vehicles, who is not for the time being disqualified for holding or obtaining a driving licence to drive any other class or description of motor vehicles, may apply to the licensing authority having jurisdiction in the area in which he resides or carries on his business in such form and accompanied by such documents and with such fees as may be prescribed by the Central Government for the addition of such other class or description of motor vehicles to the licence.

(2) Subject to such rules as may be prescribed by the Central Government, the provisions of section 9 shall apply to an application under this section as if the said application were for the grant of a licence under that section to drive the class or description of motor vehicles which the applicant desires to be added to his licence.

12. Licensing and regulations of school or establishments for imparting instruction in driving motor vehicles

(1) The Central Government may make rules for the purpose of licensing and regulating, by the State Governments, schools or establishments (by whatever name called) for imparting instruction in driving of motor vehicles and matters connected therewith.

(2) In particular, and without prejudice to the generality of the foregoing power, such rules may provide for all or any of the following matters, namely :

(*a*) licensing of such schools or establishments including grant, renewal and revocation of such licences;

(*b*) supervision of such schools or establishments;

(*c*) the form of application and the form of licence and the particulars to be contained therein;

(*d*) fee to be paid with the application for such licences;

(*e*) conditions subject to which such licences may be granted;

(*f*) appeals against the orders of refusal to grant or renew such licences and appeals against the orders revoking such licences;

(*g*) conditions subject to which a person may establish and maintain any such school or establishment for imparting instruction in driving of motor vehicles;

(*h*) nature, syllabus and duration of course or courses for efficient instruction in driving any motor vehicle;

(*i*) apparatus and equipments (including motor vehicles fitted with dual control) required for the purpose of imparting such instruction;

(*j*) suitability of the premises at which such schools or establishments may be established or maintained and facilities to be provided therein;

(*k*) qualifications, both educational and professional (including experience), which a person imparting instruction in driving a motor vehicle shall possess;

(*l*) inspection of such schools and establishments (including the services rendered by them and the apparatus, equipments and motor vehicles maintained by them for imparting such instruction);

(*m*) maintenance of records by such schools or establishments;

(*n*) financial stability of such schools or establishments;

(o) the during certificates, if any, to be issued by such schools or establishments and the form in which such driving certificates shall be issued and the requirements to be complied with for the purposes of issuing such certificates;

(*p*) such other matters as may be necessary to carry out the purposes of this section.

(3) Where the Central Government is satisfied that it is necessary or expedient so to do, it may, by rules made in this behalf, exempt generally, either absolutely or subject to such conditions as may be specified in the rules, any class of schools or establishments imparting instruction in driving of motor vehicles or matters connected therewith from the provisions of this section.

(4) A school or establishments imparting instruction in driving of motor vehicles or matters connected therewith immediately before the commencement of this Act whether under a licence or not, may continue to impart such instruction without a licence issued under this Act for a period of one month from such commencement, and if it has made in application for such licence under this Act within the said period of one month and such application is in the prescribed form, contains the prescribed particulars, and is accompanied by the prescribed fee, till the disposal of such application by the licensing authority.

13. Extent of Effectiveness of Licences, to Drive Motor Vehicles

A learner's licence or a driving licence issued under this Act shall be effective throughout India.

14. Currency of Licences to Drive Motor Vehicles

(1) A learner's licence issued under this Act shall, subject to the other provisions of this Act, be effective for a period of six months from the date of issue of the licence.

(2) A driving licence issued or renewed under this Act shall:

(*a*) in the case of a licence to drive a transport vehicle, be effective for a period of three years; and

(*b*) in the case of any other licence:

(*i*) if the person obtaining the licence, either originally or on renewal thereof, has not attained the age of forty years on the date of issue or, as the case may be, renewal thereof:

— be effective for a period of twenty years from the date of such issue or renewal; or

— until the date on which such person attains the age of forty years, whichever is earlier;

(*ii*) if the person referred to in sub-clause (*i*) has attained the age of forty years on the date of issue or, as the case may be, renewal thereof, be effective for a period of five years from the date of such issue of renewal:

Provided that every driving licence shall, notwithstanding its expiry under this sub-section, continue to be effective for a period of thirty days from such expiry.

15. Renewal of Driving Licences

(1) Any licensing authority may, on application made to renew a driving licence issued under the provisions of this Act with effect from the date of its expiry.

Provided that in any case where the application for the renewal a licence is made more than thirty days after the date of its expiry, the driving licence shall be renewed with effect from the date of its renewal:

Provided further that where the application is for the renewal of a licence to drive a transport vehicle or where in any other case the applicant has attained the age of forty years, the same shall be accompanied by a medical certificate in the same form and in the same manner as is referred to in sub-section (3) of section (8), and the provisions of sub-section (4) of section 8 shall, so far as may be, apply in relation to every such case as they apply in relation to a learner's licence.

(2) An application for the renewal of a driving licence shall be made in such form and accompanied by such documents as may be prescribed by the Central Government.

(3) Where an application for the renewal of a driving licence is made previous to, or not more than thirty days after the date of its expiry, the fee payable for such renewal shall be such as may be prescribed by the Central Government in this behalf.

(4) Where an application for the renewal of a driving licence is made more than thirty days after the date of its expiry, the fee payable for such renewal shall be such amount as may be prescribed by the Central Government:

Provided that the fee referred to in sub-section (3) may be accepted by the licensing authority in respect of an application for the renewal of a driving licence made under this sub-section if its is satisfied that the applicant was prevented by good and sufficient cause from applying within the time specified in sub-section (3):

Provided further that if the application is made more than five years after the driving licence has ceased to be effective, the licensing authority may refuse to renew the driving licence, unless the applicant undergoes and passes to its satisfaction the test of competence to drive referred to in sub-section (3) of section 9.

(5) Where the application for renewal has been rejected, the fee paid shall be refunded to such extent and in such manner as may be prescribed by the Central Government.

(6) Where the authority renewing the driving licence is not the authority which issued the driving licence it shall intimate the fact of renewal to the authority which issued the driving licence.

16. Revocation of Driving Licence on Grounds of Disease or Disability

Notwithstanding anything contained in the foregoing sections, any licensing authority may at any time revoke a driving licence or may require, as a condition of continuing to hold such driving licence, the holder thereof to produce a medical certificate in the amendment form and in the same manner as is referred to in sub-section (3) of section (8), if the licensing authority has reasonable grounds to believe that the holder of the driving licence is, by virtue of any disease or disability, unfit to drive a motor vehicle and where the authority revoking a driving licence is not the authority which issued the same, it shall intimate the fact revocation to the authority which issued that licence.

17. Orders refusing or Revoking Driving Licences and Appeals Therefrom

(1) Where a licensing authority refuses to issue any learner's licence or to issue or renew, or revokes, any driving licence, or refuses to add a class or description of motor vehicle to any driving licence, it shall do so by an order communicated to the applicator the holder, as the case may be, giving the reasons in writing for such refusal or revocation.

(2) Any person aggrieved by an order made under sub-section (1) may, within thirty days of the service on him of the order, appeal to the prescribed authority which shall decide the appeal after giving such person and the authority which made the order an opportunity of being heard and the decision of the appellate authority shall be binding on the authority which made the order.

18. Driving Licences to Drive Motor Vehicles, Belonging to the Central Government

(1) Such authority as may be prescribed by the Central Government may issue driving licence valid throughout India to persons who have completed their eighteenth year to drive motor vehicle which are the property or for the time being under the exclusive control of the Central Government and are used for the Government purposes relating to the defence of the country and unconnected with any commercial enterprise.

(2) A driving licence issued under this section shall specify the class or description of vehicle which the holder is entitled to drive and the period for which he is so entitled.

(3) A driving licence issued under this section shall not entitle the holder to drive any motor vehicle except a motor vehicle referred to in sub-section (1).

(4) The authority issuing any driving licence under this section shall, at the request of any State Government, furnish such information respecting any person to whom a driving licence is issued as that Government may at any time require.

19. Power of Licensing Authority to Disqualify from Holding a Driving Licence or Revoke Such Licence

(1) If a licensing authority is satisfied, after giving the holder of a driving licence an opportunity of being heard, that he :

(a) is a habitual criminal or a habitual drunkard; or

(b) is a habitual addict to any narcotic drug or psychotropic substance within the meaning of the Narcotic Drugs and Psychotropic Substances Act, 1985 (61 of 1985.); or

(c) is using or has used a motor vehicle in the commission of a cognizable offence; or

(*d*) has by his previous conduct as driver of a motor vehicle shown that his driving is likely to be attended with danger to the public; or

(*e*) has obtained any driving licence or a licence to drive a particular class or description of motor vehicle by fraud or misrepresentation; or

(*f*) has committed any such act which is likely to cause nuisance or danger to the public, as may be prescribed by the Central Government, having regard to the objects of this Act; or

(*g*) has failed to submit to, or has not passed, the tests referred to in the proviso to sub-section (3) of section (22); or

(*h*) being a person under the age of eighteen years who has been granted a learner's licence or a driving licence with the consent in writing of the person having the care of the holder of the licence and has ceased to be in such care,

it may, for reasons to be recorded in writing, make an order:

(*i*) disqualifying that person for a specified period for holding or obtaining any driving licence to drive all or any classes or descriptions of vehicles specified in the licence; or

(*ii*) revoke any such licence.

(2) Where an order under sub-section (1) is made, the holder of a driving licence shall forthwith surrender his driving licence to the licensing authority making the order, if the driving licence has not already been surrendered, and the licensing authority shall:

(*a*) if the driving licence is a driving licence issued under this Act, keep it until the disqualification has expired or has been removed; or

(*b*) if it is not a driving licence issued under this Act, endorse the disqualification upon it and send it to the licensing authority by which it was issued; or

(*c*) in the case of revocation of any licence, endorse the revocation upon it and if it is not the authority which issued the same, intimate the fact of revocation to the authority which issued that licence:

Provided that where the driving licence of a person authorises him to drive more than one class or description of motor vehicles and the order, made under sub-section (1), disqualifies him from driving any specified class or description of motor vehicles, the licensing authority shall endorse the disqualification upon the driving licence and return the same to the holder.

(3) Any person aggrieved by an order made by a licensing authority under sub-section (1) may, within thirty days of the receipt of the order, appeal to the prescribed authority, and such appellate authority shall give notice to the licensing authority and hear either party if so required by that party and may pass such order as it thinks fit and an order passed by any such appellate authority shall be final.

20. Power of Court Disqualify

(1) Where a person is convicted of an offence under this Act or of an offence in the commission of which a motor vehicle was used, the Court by which such person is convicted may, subject to the provisions of this Act, in addition to imposing any

other punishment authorised by law, declare the person so convicted to be disqualified, for such period as the Court may specify, from holding any driving licence to drive all classes or description of vehicles, or any particular class or description of such vehicles, as are specified in such licence:

Provided that in respect of an offence punishable under section 183 no such order shall be made for the first or second offence.

(2) Where a person is convicted of an offence under clause (c) of sub-section (7) of section 132, section 134 or section 185, the Court convicting any person of any such offence shall order the disqualification under sub-section (7), and if the offence is relatable to clause (c) of sub-section (1) of section 132 or section 134, such disqualification shall be for a period of not less than one month, and if the offence is relatable to section 185, such disqualification shall be for a period of not less than six months.

(3) A Court shall, unless for special reasons to be recorded in writing it thinks fit to order otherwise, order the disqualification of a person:

(a) who having been convicted of an offence punishable under section 184 is again convicted of an offence punishable under that section,

(b) who is convicted of an offence punishable under section 189, or

(c) who is convicted of an offence punishable under section 192:

Provided that the period of disqualification shall not exceed, in the case referred to in clause *(a)*, five years, or, in the case referred to in clause *(b)*, two years or, in the case referred to in clause *(c)*, one year.

(4) A Court ordering the disqualification of a person convicted of an offence punishable under section 184 may direct that such person shall, whether he has previously passed the test of competence to drive as referred to in sub-section (3) of section 9 or not, remain disqualified until he has subsequent to the making of the order of disqualification passed that test to the satisfaction of the licensing authority.

(5) The Court to which an appeal would ordinarily lie from any conviction of an offence of the nature specified in sub-section (1) may set aside or vary any order of disqualification made under that sub-section notwithstanding that no appeal would lie against the conviction as a result of which such order of disqualification was made.

21. Suspension of Driving Licence in Certain Cases

(1) Where, in relation to a person who had been previously convicted of an offence punishable under section 184, a case is registered by a police officer on the allegation that such person has, by such dangerous driving as is referred to in the said section 184, of any class or description of motor vehicle caused the death of, or grievous hurt to, one or more persons, the driving licence held by such person shall in relation to such class or description of motor vehicle become suspended:

(a) for a period of six months from the date on which the case is registered, or

(b) if such person is discharged or acquitted before the expiry of the period aforesaid, until such discharge or acquittal, as the case may be.

(2) Where by virtue of the provisions of sub-section (1), the driving licence held by a person becomes suspended, the police officer, by whom the case referred to in sub-

section (1) is registered, shall bring such suspension to the notice of the Court competent to take cognizance of such offence, and thereupon, such Court shall take possession of the driving licence, endorse the suspension thereon and intimate the fact of such endorsement to the licensing authority by which the licence was granted or last renewed.

(3) Where the person referred to in sub-section (1) in acquitted or discharged, the Court shall cancel the endorsement on such driving licence with regard to the suspension thereof.

(4) If a driving licence in relation to a particular class or description of motor vehicles is suspended under sub-section (1), the person holding such licence shall be debarred from holding or obtaining any licence to drive such particular class of description of motor vehicles so long as the suspension of the document remains in force.

22. Suspension or Cancellation of Driving Licence on Conviction

(1) Without prejudice to the provisions of sub-section (3) of section 20 where a person, referred to in sub-section (1) of section 21 is convicted of an offence of causing, by such dangerous driving as is referred to in section 184 of any class or description of motor vehicle the death of, or grievous hurt to, one or more persons, the Court by which such person is convicted may cancel, or suspend for such period as it may think fit, the driving licence held by such person in so far as it relates to that class or description of motor vehicle.

(2) Without prejudice to the provisions of sub-section (2) of section 20, if a person, having been previously convicted of an offence punishable under section 185 is again convicted of an offence punishable under that section, the Court, making such subsequent conviction, shall, by order, cancel the driving licence held by such person.

(3) If a driving licence is cancelled or suspended under this section, the Court shall take the driving licence in its custody, endorse the cancellation or, as the case may be, suspension, thereon and send the driving licence so endorsed to the authority by which the licence was issued or last renewed and such authority shall, on receipt of the licence, keep the licence in its safe custody, and in the case of a suspended licence, return the licence to the holder thereof after the expiry of the period of suspension on an application made by him for such return:

Provided that no such licence shall be returned unless the holder thereof has, after the expiry of the period of suspension, undergone and passed, to the satisfaction of the licensing authority by which the licence was issued or last renewed, a fresh test of competence to drive referred to in sub-section (3) of section 9 and produced a medical certificate in the same form and in the same manner as is referred to in sub-section (3) of section 8.

(4) If a licence to drive a particular class or description of motor vehicles is cancelled or suspended under this section, the person holding such a licence shall be debarred from holding, or obtaining, any licence to drive such particular class or description or motor vehicles so long as the cancellation or suspension of the driving licence remains in force.

23. Effect of Disqualification Order

(1) A person in respect of whom any disqualification order is made under section 19 or section 20 shall be debarred to the extent and for the period specified in such order from holding or obtaining a driving licence and the driving licence, if any, held by such person at the date of the order shall cease to be effective to such extend and during such period.

(2) The operation of a disqualification order made under section 20 shall not be suspended or postponed while an appeal is pending against such order or against the conviction as a result of which such order is made, unless the appellate court so directs.

(3) Any person in respect of whom any disqualification order has been made may at any time after the expiry of six months from the date of the order apply to the Court or other authority by which the order was made, to remove the disqualification; and the Court or authorities, as the case may be, may, having regard to all the circumstances, either cancel or vary the disqualification order :

Provided that where the Court or other authority refuses to cancel or vary any disqualification order under this section, a second application thereunder shall not be entertained before the expiry of a period of three months from the date of such refusal.

24. Endorsement

(1) The Court or authority making an order of disqualification shall endorse or cause to be endorsed upon the driving licence if any, held by the person disqualify, particulars of the order of disqualification and of any conviction of an offence in respect of which an order of disqualification is made; and particulars of any cancellation or variation of an order of disqualification made under sub-section (3) of section 23 shall be similarly so endorsed.

(2) A Court by which any person is convicted of an offence under this Act as may be prescribed by the Central Government, having regard to the objects of this Act, shall, whether or not a disqualification order is made in respect of such conviction, endorse or cause to be endorsed particulars of such conviction on any driving licence held by the person convicted.

(3) Any person accused of an offence prescribed under sub-section (2) shall when attending the Court bring with him his driving licence if it is in his possession.

(4) Where any person is convicted of any offence under this Act and sentenced to imprisonment for a period exceeding three months the Court awarding the sentence shall endorse the fact of such sentence upon the driving licence of the person concerned and the prosecuting authority shall intimate the fact of such endorsement to the authority by which the driving licence was granted or last renewed.

(5) When the driving licence is endorsed or caused to be endorsed by any Court, such Court shall send the particulars of the endorsement by the licensing authority by which the driving licence was granted or last renewed.

(6) Where on an appeal against any conviction or order of a Court, which has been endorsed on a driving licence, the appellate court varies or sets aside the conviction or order, the appellate court shall inform the licensing authority by which the

driving licence was granted or last renewed and such authority shall amend or cause to be amended the endorsement.

25. Transfer of Endorsement and Issue of Driving Licence Free from Endorsement

(1) An endorsement on any driving licence shall be transferred to any new or duplicate driving licence obtained by the holder thereof until the holder becomes entitled under the provisions of this section to have a driving licence issued to him free from endorsement.

(2) Where a driving licence is required to be endorsed and the driving licence is not in the possession of the Court or authority by which the endorsement is to be made, then:

(a) if the person in respect of whom the endorsement is to be made is at the time the holder of a driving licence, he shall produce the driving licence to the Court or authority within five days, or such longer time as the Court or authority may fix; or

(b) if, not being then the holder of a driving licence, he subsequently obtains a driving licence, he shall within five days after obtaining the driving licence produce it to the Court or authority, and if the driving licence is not produced within the time specified, it shall, expiration or such time, be or no effect until it is produced for the purpose of endorsement.

(3) A person whose driving licence has been endorsed shall, if during a continuous period of three years after such endorsement no further endorsement has been made against him, be entitled on surrendering his driving licence and on payment of a fee of five rupees, to receive a new driving licence free from all endorsements:

Provided that if the endorsement is only in respect of an offence contravening the speed limits referred to in section 112, such person shall be entitled to receive a new driving licence free from such endorsement on the expiration of one year of the date of the endorsement:

Provided further that in reckoning the said period of three years and one year, respectively, any period during which the said persons was disqualify for holding or obtaining a driving licence shall be excluded.

26. Maintenance of State Registers of Driving Licences

(1) Each State Government shall maintain, in such form as may be prescribed by the Central Government, a register to be known as the State Register of Driving Licences, in respect of driving licences issued and renewed by the licensing authorities of the State Government, containing the following particulars, namely:

(a) names and addresses of holders of driving licences;

(b) licence numbers;

(c) dates of issue or renewal of licences;

(d) dates of expiry of licences;

(e) classes and types of vehicles authorised to be driven; and

(f) such other particulars as the Central Government may prescribe.

(2) Each State Government shall supply to the Central Government a printed copy of the State Register of Driving Licences and shall inform the Central Government without delay of all additions to and other amendments in such register made from time to time.

(3) The State Register order Driving Licences shall be maintained in such manner as may be prescribed by the State Government.

27. Power of Central Government to Make Rules

The Central Government may make rules:

(a) regarding conditions referred to in sub-section (2) of section 3;

(b) providing for the form in which the application for learner's licence may be made, the information it shall contain and the documents to be submitted with the application referred to in sub-section (2) of section 8;

(c) providing for the form of medical certificate referred to in sub-section (3) of section 8;

(d) providing for the particulars for the test referred to in sub-section (5) of section 8;

(e) providing for the form in which the application for driving licence may be made, the information it shall contain and the documents to be submitted with the application referred to in sub-section (2) of section 9;

(f) providing for the particulars regarding test of competence to drive, referred to in sub-section (3) of section 9;

(g) specifying the minimum educational disqualifications of persons to whom licences to drive transport vehicles may be issued under this Act and the time within which such disqualifications are to be acquired by such persons;

(h) providing for the form and contents of the licences referred to in sub-section (1) of section 10;

(i) providing for the form and contents of the application referred to in sub-section (1) of section 11 and documents to be submitted with the application and the fee to be charged;

(j) providing for the form and contents subject to which section 9 shall apply to an application made under section 11;

(k) providing for the form and contents of the application referred to in sub-section (1) of section 15 and the documents to accompany such application under sub-section (2) of section 15;

(l) providing for the authority to grant licences under sub-section (1) of section 18;

(m) specifying the fees payable under sub-section (2) of section 3, sub-section (2) of section 9 and sub-sections (3) and (4) of section 15 for the grant of learner's licences, and for the grant the renewal of driving licences and licences for the purpose of regulating the schools or establishments for imparting instructions in driving motor vehicles;

(n) specifying the acts for the purposes of clause *(f)* of sub-section (1) of section 19;

(o) specifying the offences under this Act for the purposes of sub-section (2) of section 24;

(*p*) to provide for all or any of the matters referred to in sub-section (1) of section 26;

(*q*) any other matter which is, or has to be, prescribed by the Central Government.

28. Power of State Government to Make Rules

(1) A State Government may make rules for the purposes of carrying into effect the provisions of this Chapter other than the matters specified in section 27.

(2) Without prejudice to the generality of the foregoing power, such rules may provide for:

(*a*) the appointment, jurisdiction, control and functions of licensing authorities and other prescribed authorities;

(*b*) the conduct and hearing of appeals that may be preferred under this Chapter, the fees to be paid in respect of such appeals and the refund of such fees; provided that no fee so fixed shall exceed twenty-five rupees;

(*c*) the issue of duplicate licences to replace licences lost, destroyed or mutilated, the replacement of photographs which have become obsolete and the fees to be charged therefor;

(*d*) the badges and uniform to be worn by drives of transport vehicles and the fees to be paid in respect of badges;

(*e*) the fee payable for the issue of a medical certificate under sub-section (3) of section 8;

(*f*) the exemption of prescribed persons, or prescribed classes of persons, from payment of all or any portion of the fees payable under this Chapter;

(*g*) the communication of particulars of licences granted by one licensing authority to other licensing authorities;

(*h*) the duties, functions and conduct of such persons to whom licences to drive transport vehicles are issued;

(*i*) the exemption of drivers of road-rollers from all or any of the provisions of this Chapter or of the rules made thereunder;

(*j*) the manner in which the State Register of Driving Licences shall be maintained under section 26;

(*k*) any other matter which is to be, or may be, prescribed.

CHAPTER III

LICENSING OF CONDUCTORS OF STAGE CARRIAGES

29. Necessity for Conductors Licence

(1) No person shall act as a conductor of a stage carriage unless be holds an effective conductor's licence issued to him authorising him to act as such conductor; and no person shall employ or permit any person who is not so licensed to act as a conductor of a stage carriage.

(2) A State Government may prescribe the conditions subject to which sub-section (1) shall not apply to a driver of a stage carriage performing the functions of a conductor

or to a person employed to act as a conductor for a period not exceeding one month.

30. Grant to Conductors Licence

(1) Any person who possesses such minimum educational disqualification as may be prescribed by the State Government and is not disqualified under sub-section (1) of section 31 and who is not for the time being disqualified for holding or obtaining a conductor's licence may apply to the licensing authority having jurisdiction in the area in which he ordinarily resides or carries on business for the issue to him of a conductor's licence.

(2) Every application under sub-section (1) shall be in such form and shall contain such information as may be prescribed.

(3) Every application for conductor's licence shall be accompanied by a medical certificate in such form as may be prescribed, signed by a registered medical practitioner and shall also be accompanied by two clear copies of a recent photograph of the applicant.

(4) A conductor's licence issued under this Chapter shall be in such form and contain such particulars as may be prescribed and shall be effective throughout the State in which it is issued.

(5) The fee for a conductor's licence and for each renewal thereof shall be one-half of that for a driving licence.

31. Disqualifications for the Grant of Conductors Licence

(1) No person under the age of eighteen years shall hold, or be granted, a conductor's licence.

(2) The licensing authority may refuse to issue a conductor's licence:

(a) if the applicant does not posses the minimum educational disqualification:

(b) if the medical certificate produced by the applicant discloses that he is physically unfit to act as a conductor; and

(c) if any previous conductor's licence held by the applicant was revoked.

32. Revocation of a Conductors Licence on Grounds of Disease or Disability

A conductor's licence may at any time be revoked by any licensing authority if that authorities has reasonable grounds to believe that the holder of the licence is suffering from any disease or disability which is likely to render him permanently unfit to hold such a licence and where the authority revoking a conductor's licence is not the authority which issued the same, it shall intimate the fact of such revocation to the authority which issued that licence:

Provided that before revoking any licence, the licensing authority shall give the person holding such licence a reasonable opportunity of being heard.

33. Orders Refusing etc., Conductors Licences and Appeals Therefrom

(1) Where a licensing authority refuses to issue or renew, or revokes any conductor's licence, it shall do so by an order communicated to the applicant or the holder, as the case may be, giving the reasons in writing for such refusal or revocation.

(2) Any person aggrieved by an order made under sub-section (1) may, within thirty days of the service on him of the order, appeal to the prescribed authority which shall decide the appeal after giving such person and the authority which made the order an opportunity of being heard and the decision of the appellate authorities shall be binding on the authority which made the order.

34. Power of Licensing Authority to Disqualify

(1) If any licensing authority is of opinion that it is necessary to disqualify the holder of a conductor's licence for holding or obtaining such a licence on account of his previous conduct as a conductor, it may, for reasons to be recorded, make an order disqualifying that person for a specified period, not exceeding one year, for holding or obtaining a conductor's licence:

Provided that before disqualifying the holder of a licence, the licensing authority shall give the person holding such licence a reasonable opportunity of being heard.

(2) Upon the issue of any such order, the holder of the conductor's licence shall forthwith surrender the licence to the authority making the order, if the licence has not already been surrendered, and the authority shall keep the licence until the disqualification has expired or has been removed.

(3) Where the authority disqualifying the holder of a conductor's licence under this section is not the authority which issued the licence, it shall intimate the fact of such disqualification to the authority which issued the same.

(4) Any person aggrieved by an order made under sub-section (1) may, within thirty days of the service on him of the order, appeal to the prescribed authority which shall decide the appeal after giving such person and the authority which made the order an opportunity of being heard and the decision of the appellate authority shall be binding on the authority which made the order.

35. Power of Court to Disqualify

(1) Where any person holding a conductor's licence is convicted of an offence under this Act, the Court by which such person is convicted may, in addition to imposing any other punishment authorised by law, declare the person so convicted to be disqualified for such period as the Court may specify for holding a conductor's licence.

(2) The Court to which an appeal lie from any conviction of an offence under this Act may set aside or vary any order of disqualification made by the Court below, and the Court to which appeals ordinarily lie from such Court, may set aside or vary any order of disqualification made by that Court, notwithstanding that no appeal lies against the conviction in connection with which such order was made.

36. Certain Provisions of Chapter II to Apply to Conductors Licence

The provisions of sub-section (2) of section 6, section 14, 15 and 23, sub-section (1) of section 24 and sections 25 shall, so far as may be, apply in relation to a conductor's licence, as they apply in relation to a driving licence.

37. Savings

If any licence to act as a conductor of a stage carriage (by whatever name called) has been issued in any State and is effective immediately before the commencement of this Act, it shall continue to be effective, notwithstanding such commencement, for the period for which it would have been effective, if this Act had not been passed, and every such licence shall be deemed to be a licence issued under this Chapter as if this Chapter had been in force on the date on which that licence was granted.

38. Power of State Government to Make Rules

(1) A State Government may make rules for the purposes of carrying into effect the provisions of this Chapter.

(2) Without prejudice to the generality of the foregoing power, such rules may provide for:

(a) the appointment, jurisdiction, control and functions of licensing authorities and other prescribed authorities under this Chapter;

(b) the conditions subject to which drivers of stage carriages performing the functions of a conductor and persons temporarily employed to act as conductors may be exempted from the provisions of sub-section (1) of section 20;

(c) the minimum educational disqualifications of conductors; their duties and functions and the conduct of persons to whom conductor's licences are issued;

(d) the form of application for conductor's licences or for renewal of such licences and the particulars it may contain;

(e) the form in which conductor's licences may be issued or renewed and the particulars it may contain;

(f) the issue of duplicate licences to replace licences lost, destroyed or mutilated, the replacement of photographs which have become obsolete and the fees to be charged therefor;

(g) the conduct and hearing of appeals that may be preferred under this Chapter, the fees to be paid in respect of such appeals and the refund of such fees :

Provided that no fee so fixed shall exceed twenty-five rupees;

(h) the badges and uniform to be worn by conductors of stage carriages and the fees to be paid in respect of such badges;

(i) the grant of the certificates referred to in sub-section (3) of section 30 by registered medical practitioners and the form of such certificates;

(j) the conditions subject to which, and the extend to which, a conductor's licence issued in another State shall be effective in the Stage;

(k) the communication of particulars of conductor's licences from one authority to other authorities; and

(l) any other matter which is to be, or may be, prescribed.

CHAPTER IV

REGISTRATION OF MOTOR VEHICLES

39. Necessity for Registration

No person shall drive any motor vehicle and no owner of a motor vehicle shall cause or permit the vehicle to be driven in any public place or in any other place unless the vehicle is registered in accordance with this Chapter and the certificate of registration of the vehicle has not been suspended or cancelled and the vehicle carries a registration mark displayed in the prescribed manner:

Provided that nothing in this section shall apply to a motor vehicle in possession of a dealer subject to such conditions as may be prescribed by the Central Government.

40. Registration, Where to be Made

Subject to the provisions of section 42, section 43 and section 60, every owner of a motor vehicle shall cause the vehicle to be registered by a registering authority in whose jurisdiction he has the residence or place of business where the vehicle is normally kept.

41. Registration, How to be Made

(1) An application by or on behalf of the owner of a motor vehicle for registration shall be in such form and shall be accompanied by such documents, particulars and information and shall be made within such period as may be prescribed by the Central Government :

Provided that where a motor vehicle is jointly owned by more persons that one, the application shall be made by one of them on behalf of all the owners and such applicant shall be deemed to be the owner of the motor vehicle for the purposes of this Act.

(2) An application referred to in sub-section (1) shall be accompanied by such fee as may be prescribed by the Central Government.

(3) The registering authority shall issue to the owner of a motor vehicle registered by it a certificate of registration in such form and containing such particulars and information and in such manner as may be prescribed by the Central Government.

(4) In addition to the other particulars required to be included in the certificate of registration, it shall also specify the type of the motor vehicle, being a type as the Central Government may, having regard to the design, construction and use of the motor vehicle, by notification in the Official Gazette, specify.

(5) The registering authority shall enter the particulars of the certificate referred to in sub-section (3) in a register to be maintained in such form and manner as may be prescribed by the Central Government.

(6) The registering authority shall assign to the vehicle, for display thereon, a distinguishing mark (in this Act referred to as the registration by mark) consisting of one of the groups of such of those letters and followed by such letters and figures as are allotted to the State by the Central Government from time to time by notification in the Official Gazette, and displayed and shown on the motor vehicle in such form and in such manner as may be prescribed by the Central Government.

(7) A certificate of registration issued under sub-section (3), whether before or after the commencement of this Act, in respect of a motor vehicle, other than a transport vehicle, shall, subject to the provisions contained in this Act, be valid only for a period of fifteen years from the date of issue of such certificate and shall be renewable.

(8) An application by or on behalf of the owner of a motor vehicle, other than a transport vehicle, for the renewal of a certificate of registration shall be made

within such period and in such form, containing such particulars and information as may be prescribed by the Central Government.

(9) An application referred to in sub-section (8) shall be accompanied by such fee as may be prescribed by the Central Government.

(10) Subject to the provisions of section 56, the registering authority may, on receipt of an application under sub-section (8), renew the certificate of registration for a period of five years and intimate the fact to the original registering authority, if it is not the original registering authority.

(11) If the owner fails to make an application under sub-section (1), or, as the case may be, under sub-section (8) within the period prescribed, the registering authority may, having regard to the circumstances of the case, required the owner of pay, in lie of any action that may be taken against him under section 177, such amount not exceeding one hundred rupees as may be prescribed under sub-section (73) :
Provided that action under section 177 shall be taken against the owner where the owner fails to pay the said amount.

(12) Where the owner has paid the amount under sub-section (11), no action shall be taken against him under section 177.

(13) For the purposes of sub-section (11), State Government may prescribe different amounts having regard to the period of delay on the part of the owner in making an application under sub-section (1) or sub-section (8).

(14) An application for the issue of a duplicate certificate of registration shall be made to the original registering authority in such form, containing such particulars and information along with such fee as may be prescribed by the Central Government.

42. Special Provisions for Registration of Motor Vehicles of Diplomatic Officers, etc.

(1) Where an application for registration of a motor vehicle is made under sub-section (1) of section 41 by or on behalf of any diplomatic officer or consular officer, then, notwithstanding anything contained in sub-section (3) or sub-section (6) of that section, the registering authority shall register the vehicle in such manner and in accordance with such procedure as may be provided by rules made in this behalf by the Central Government under sub-section (3) and shall assign to the vehicle for display thereon a special registration mark in accordance with the provisions contained in those rules and shall issue a certificate (hereafter in this section referred to as the certificate of registration) that the vehicle has been registered under this section; and any vehicle so registered shall not, so long as it remains the property of any diplomatic officer or consular officer, require to be registered otherwise under this Act.

(2) If any vehicle registered under this section ceases to be the property of any diplomatic officer or consular officer, the certificate of registration issued under this section shall also cease to be effective, and the provisions of sections 39 and 40 shall thereupon apply.

(3) The Central Government may make rules for the registration of motor vehicles belonging to diplomatic officers and consular officers regarding the procedure to be followed by the registering authority for registering such vehicles, the form in

which the certificates of registration of such vehicles are to be issued, the manner in which such certificates of registration are to be sent to the owners of the vehicles and the special registration marks to be assigned to such vehicles.

(4) For the purposes of this section, "diplomatic officer" or "consular officer" means any person who is recognised as such by the Central Government and if any question arises as to whether a person is or is not such an officer, the decision of the Central Government thereon shall be final.

43. Temporary Registration

(1) Notwithstanding anything contained in section 40 the owner of a motor vehicle may apply to any registering authority or other prescribed authority to have the vehicle temporarily registered in the prescribed manner and for the issue in the prescribed manner of a temporary certificate of registration and a temporary registration mark.

(2) A registration made under this section shall be valid only for a period not exceed one month, and shall not be renewable :

Provided that where a motor vehicle so registered is a chassis to which a body has not been attached and the same is detained in a workshop beyond the said period of one month for being fitted with a body, the period may, on payment of such fees, if any, as may be prescribed, be extended by such further period or periods as the registering authority or other prescribed authority, as the case may be, may allow.

44. Production of Vehicle at the Time of Registration

The registering authority shall before proceeding to register a motor vehicle or renew the certificate of registration in respect of a motor vehicle, other than a transport vehicle, require the person applying for registration of the vehicle or, as the case may be, for renewing the certificate of registration to produce the vehicle either before itself or such authority as the State Government may by order appoint in order that the registering authority may satisfy itself that the particulars contained in the application are true and that the vehicle complies with the requirements of this Act and of the rules make thereunder.

45. Refusal of Registration or Renewal of the Certificate of Registration

The registering authority may, by order, refuse to register any motor vehicle, or renew the certificate of registration in respect of a motor vehicle (other than a transport vehicle), if in either case, the registering authority has reason to believe that it is a stolen motor vehicle or, the vehicle is mechanically defective or fails to comply with the requirements of this Act or of the rules made thereunder, or if the applicant fails to furnish particulars of any previous registration of the vehicle or furnishes inaccurate particulars in the application for registration of the vehicle or, as the case may be, for renewal of the certificate or registration thereof and the registering authority shall furnish the applicant whose vehicle is refused registration, or whose application for renewal of the certificate of registration is refused, a copy of such order, together with the reason for such refusal.

46. Effectiveness in India of Registration

Subject to the provisions of section 47, a motor vehicle registered in accordance with this Chapter in any State shall not required to be registered elsewhere in India and a certificate of registration issued or in force under this Act in respect of such vehicle shall be effective throughout India.

47. Assignment of New Registration Mark on Removal to Another State

(1) When a motor vehicle registered in one State has been kept in another State, for a period exceeding twelve months, the owner of the vehicle shall, within such period and in such form containing such particulars as may be prescribed by the Central Government, apply to the registering authority, within whose jurisdiction the vehicle then is, for the assignment of a new registration mark and shall present the certificate of registration to that registering authority:

Provided that an application under this sub-section shall be accompanied:

(i) by the no such certificate obtained under section 48, or

(ii) in a case where no such certificate has been obtained, by:

(a) the receipt obtained under sub-section (2) of section 48; or

(b) the postal acknowledgment received by the owner of the vehicle if he has sent an application in this behalf by registered post acknowledgment due to the registering authority referred to in section 48, together with a declaration that he has not received any communication from such authority refusing to grant such certificate or requiring him to comply with any direction subject to which certificate may be granted:

Provided further that, in a case where a motor vehicle is held under a hire-purchase, lease or hypothecation agreement, an application under this sub-section shall be accompanied by a no such certificate from the person with whom such agreement has been entered into, and the provisions of section 51, so far as may be, regarding obtaining of such certificate from the person with whom such agreement has been entered into, shall apply.

(2) The registering authority, to which application is made under sub-section (1), shall after making such verification, as it thinks fit, of the returns, if any, received under section 62, assign the vehicle a registration mark as specified in sub-section (6) of section 41 to be displayed and shown thereafter on the vehicle and shall enter the mark upon the certificate of registration before returning it to the applicant and shall, in communication with the registering authority by whom the vehicle was previously registered, arrange for the transfer of the registration of the vehicle from the records of that registering authority to its own records.

(3) Where a motor vehicle is held under a hire-purchase or lease or hypothecation agreement, the registering authority shall, after assigning the vehicle a registration mark under sub-section (2), inform the person whose name has been specified in the certificate of registration as the person with whom the registered owner has entered into the hire-purchase or lease or hypothecation agreement (by sending to such person a notice by registered post acknowledgment due at the address of such person entered in the certificate of registration the fact of assignment of the said registration mark).

(4) A State Government may make rules under section 65 requiring the owner of a motor vehicle not registered within the State, which is brought into or is for the time being in the State, to furnish to the prescribed authority in the State such

information with respect to the motor vehicle and its registration as may be prescribed.

(5) If the owner fails to make an application under sub-section (1) within the period prescribed, the registering authority may, having regard to the circumstances of the case, require the owners to pay, in lieu of any action that may be taken against him under section 177, such amount not exceeding one hundred rupees as may be prescribed under sub-section (7):

Provided that action under section 177 shall be taken against the owner where the owner fails to pay the said amount.

(6) Where the owner has paid the amount under sub-section (5), no action shall be taken against him under section 177.

(7) For the purposes of sub-section (5), the State Government may prescribe different amounts having regard to the period of delay on the part of the owner in making an application under sub-section (1).

48. No Objection Certificate

(1) The owner of a motor vehicle when applying for the assignment of a new registration mark under sub-section (1) of section 47, or where the transfer of a motor vehicle is to be effected in a State other than the State of its registration, the transferor of such vehicle when reporting the transfer under sub-section (1) of section 50, shall make an application in such form and in such manner as may be prescribed by the Central Government to the registering authority by which the vehicle was registered for the issue of a certificate (hereafter in this section referred to as the no such certificate), to the effect that the registering authority has no objection for assigning a new registration mark to the vehicle or, as the case may be, for entering the particulars of the transfer of ownership in the cover of registration.

(2) The registering authority shall, on receipt of an application under sub-section (1), issue a receipt in such form as may be prescribed by the Central Government.

(3) On receipt of an application under sub-section (1), the registering authority may, after making such inquiry and requiring the applicant to comply with such directions as it deems fit and within thirty days of the receipt thereof, by order in writing, communicate to the applicant that it has granted or refused to grant the no objection certificate:

Provided that a registering authority shall not refuse to grant the no objection certificate unless it has recorded in writing the reasons for doing so and a copy of the same has been communicated to the applicant.

(4) Where while a period of thirty days referred to in sub-section (3), the registering authority does not refuse to grant that no objection certificate or does not communicate the refusal to the applicant, the registering authority shall be deemed to have granted the no objection certificate.

(5) Before granting or refusing to grant the no objection certificate, the registering authority shall obtain a report in writing from the police that no case relating to the theft of the motor vehicle concerned has been reported or is pending, verify whether all the amounts due to Government including road tax in respect of that

motor vehicle have been paid and take into account such other factors as may be prescribed by the Central Government.

(6) If the owner of a motor vehicle fails to intimate his new address to the concerned registering authority within the period specified in sub-section (1), the registering authority may, having regard to the circumstances of the case, require the owner of pay, in lieu of any action that may be taken against him under section 177, such amount not exceeding one hundred rupees as may be prescribed under sub-section (4):

Provided that action under section 177 shall be taken against the owner where he fails to pay the said amount.

(7) Where a person has paid the amount under sub-section (2), no action shall be taken against him under sub-section 177.

(8) For the purposes of sub-section (2), a State Government may prescribe different amounts having regard to the period of delay in intimating his new address.

(9) On receipt of intimation under sub-section (1), the registering authority may, after making such verification as it may think fit, cause the new address to be entered in the certificate of registration.

(10) A registering authority other than the original registering authority making any such entry shall communicate the altered address to the original registering authority.

(11) Nothing in sub-section (1) shall apply where the change of the address recorded in the certificate of registration is due to a temporary absence not intended to exceed six months in duration or where the motor vehicle is neither used nor removed from the address recorded in the certificate of registration.

49. Transfer of Ownership

(1) Where the ownership of any motor vehicle registered under this Chapter is transferred:

(a) the transferor shall:

(i) in the case of a vehicle registered within the same State, within fourteen days of the transfer, report the fact of transfer, in such form with such documents and in such manner, as may be prescribed by the Central Government to the registering authority within where jurisdiction the transfer is to be effected and shall simultaneously send a copy of the said report of the transferee; and

(ii) in the case of a vehicle registered outside the State, within forty-five days of the transfer, forward to the registering authorities referred to in sub-clause *(i)*:

(A) the no objection certificate obtained under section 48; or

(B) in a case where no such certificate has been obtained:

(i) the receipt obtained under sub-section (2) of section 48; or

(ii) the postal acknowledgment received by the transferred if he has sent an application in this behalf by registered post acknowledgment due to the registering authority referred to in section 48, together with a declaration that he has not received any communication from such authority refusing to grant

such certificate or requiring him to comply with any direction subject to which such certificate may be granted;

(*b*) the transferee shall, within thirty days of the transfer, report the transfer to the registering authority within whose jurisdiction he has the residence or place of business where the vehicle is normally kept, as the case may be, and shall forward the certificate of registration to that registering authority together with the prescribed fee and a copy of the report received by him from the transferor in order that particulars of the transfer of ownership may be entered in the certificate of registration.

(2) Where:

(*a*) the person in whose name a motor vehicle stands registered dies, or

(*b*) a motor vehicle has been purchased or acquired at a public auction conducted by, or an behalf of, Government, the person succeeding to the possession of the vehicle or, as the case may be, who has purchased or acquired the motor vehicle, shall make an application for the purpose of transferring the ownership of the vehicle in his name, in the registering authority in whose jurisdiction he has the residence or place of business where the vehicle is normally kept, as the case may be, in such manner, accompanied with such fee, and within such period as may be prescribed by the Central Government.

(3) If the transferor or the transferee fails to report to the registering authority the fact of transfer within the period specified in clause (*a*) or clause (*b*) of sub-section (1), as the case may be, or if the person who is required to make an application under sub-section (2) (hereafter in this section referred to as the other person) fails to make such application within the period prescribed, the registering authority may, having regard to the circumstances of the case, require the transferor or the transferee, or the other person, as the case may be, to pay, in lieu of any action that may be taken against him under sub-section 177 such amount not exceeding one hundred rupees as may be prescribed under sub-section (5):

Provided that action under section 177 shall be taken against the transferor or the transferee or the other person, as the case may be, where he fails to pay the said amount.

(4) Where a person has paid the amount under sub-section (3), no action shall be taken against him under section 177.

(5) For the purposes of sub-section (3), a State Government may prescribe different amounts having regard to the period of delay on the part of the transferor or the transferee in reporting the fact of transfer of ownership of the motor vehicle or of the other person in making the application under sub-section (2).

(6) On receipt of a report under sub-section (1), or an application under sub-section (2), the registering authority may cause the transfer of ownership to be entered in the certificate of registration.

(7) A registering authority making any such entry shall communicate the transfer of ownership to the transferor and to the original registering authority, if it is not the original registering authority.

50. Special Provisions Regarding Motor Vehicle Subject to Hire-Purchase Agreement, etc.

(1) Where an application for registration of a motor vehicle which is held under a hire-purchase, lease or hypothecated.

MOTOR VEHICLES LEGISLATION

Transport wing of the Department of Road Transport and Highways is responsible for following legislations:

1. **Motor Vehicles Act, 1988.**
2. **Central Motor Vehicles Rules, 1989.**
3. **The Carriers Act, 1865.**

The Motor Vehicles Act, 1988 has so far been amended three times in the years 1994, 2000 and 2001.

Amendment in the year 1994. This included

- Rationalization of the definition of the various categories of motor vehicles;
- Mandating of a minimum one year experience of driving a light motor vehicle before a person can be granted a licence for transport vehicle tightening of norms for drivers transporting dangerous or hazardous goods;
- Encouraging use of battery, CNG and solar energy as an auto fuel by exempting vehicles using such fuel from the requirement of permit or fixation of fare by the State Government;
- Empowering Central Government to make rule for standardizing components in Motor Vehicles;
- Increasing the amount of compensation in the event of death from Rs. 25,000/- to Rs. 50,000/- in respect of no fault liability etc.

Amendment in the year 2000. This included

- Authorized use of LPG as an auto fuel.
- Buses used by educational institutions brought under the purview of permit regime.
- Alterations made in transport vehicle without prior approval of the Registering Authority were barred.

Amendment in the year 2001 was necessitated by

- Need to bring the buses plying on CNG within the purview of State Transport Authority in respect of fixation of fares and route permits.

The Central Motor Vehicles Rules, 1989 have been amended from time to time to meet the emerging requirements :

Amendment in the year 2004. Notified amendments include

- Safety norms for various components of agricultural tractors such as power steering, lamps, light, parking light, etc.;

- Extension of Bharat Stage-ll emission norms for four wheeled vehicles in Solapur and Lucknow from 1-6-2004;
- Specifications of smart card and related peripherals to be used for smart card based Driving Licence(DL) and Registration Certificate (RC); and
- Introduction of Bharat Stage-Ill emission norms in 11 mega cities from 1-4-2005.

Amendment in the year 2005. Notified amendments include

- Updation of the list of dangerous and hazardous goods in the Central Motor Vehicles Rules. The terms "Battery Operated Vehicle" and "Power Tiller" have been defined.
- The emission norms, overall dimension and other related norms for "Power Tiller" have been prescribed.
- Time limits for various functions discharged by Licencing and Registering authorities and Appellate authorities under the Motor Vehicles Acts/Rules have been specified.
- It has been made mandatory for the manufacturer to supply a protective headgear conforming to BIS Standards at the time of sale of the two wheelers, subject to the exceptions under Section 129 of Motor Vehicles Act, 1988.
- Type approval Rules for CNG/LPG vehicles have been rationalized.
- States have been empowered to prescribe special provisions such as fog lamp, power steering, defogging and demisting systems in transport vehicles plying in hill areas.

The Carriers Act, 1865 was enacted on 14-2-1865 regulating the liability of Carriers. The Act enabled common carriers to limit the liability for the loss of or damage to property delivered to them to be carried but also to declare their liability for loss of or damage to such property occasioned by the negligence or criminal acts of themselves, their servants or agents. Since 1865 many changes have taken place and the Road Transport scenario has totally changed.

- A review of the Act has been done.
- A Bill to repeal the Carriers Act, 1865 and to enact the Carriage by Road Act, 2005 has been introduced in the Rajya Sabha on 7-12-2005.
- This legislation would help to make the transport system transparent and modernise the systems and procedures of the transportation trade through registration of common carrier and equitable apportionment of liability between the common carrier and the consignor.
- The Bill was referred to Department Related Parliamentary Standing Committee on Transport, Tourism and Culture.
- The Committee has submitted its report to Rajya Sabha on 21-3-2006.
- The recommendations of the Committee are under examination in the Department.

INDEX

D

U

V

W